Lecture Notes in Computer Science 6687

Commenced Publication in 1973
Founding and Former Series Editors:
Gerhard Goos, Juris Hartmanis, and Jan

T0074366

José Manuel Ferrández
José Ramón Álvarez Sánchez
Félix de la Paz
F. Javier Toledo (Eds.)

New Challenges on Bioinspired Applications

4th International Work-Conference on the Interplay
Between Natural and Artificial Computation, IWINAC 2011
La Palma, Canary Islands, Spain, May 30 - June 3, 2011
Proceedings, Part II

 Springer

Volume Editors

José Manuel Ferrández
F. Javier Toledo
Universidad Politécnica de Cartagena
Departamento de Electrónica
Tecnología de Computadoras y Proyectos
Pl. Hospital, 1
30201 Cartagena, Spain
E-mail: info@iwinac.org

José Ramón Álvarez Sánchez
Félix de la Paz
Universidad Nacional de Educación a Distancia
E.T.S. de Ingeniería Informática
Departamento de Inteligencia Artificial
Juan del Rosal, 16, 28040 Madrid, Spain
E-mail: info@iwinac.org

ISSN 0302-9743   e-ISSN 1611-3349
ISBN 978-3-642-21325-0   e-ISBN 978-3-642-21326-7
DOI 10.1007/978-3-642-21326-7
Springer Heidelberg Dordrecht London New York

Library of Congress Control Number: Applied for

CR Subject Classification (1998): F.1, F.2, I.2, G.2, I.4, I.5, J.3-4, J.1

LNCS Sublibrary: SL 1 – Theoretical Computer Science and General Issues

Typesetting: Camera-ready by author, data conversion by Scientific Publishing Services, Chennai, India

Printed on acid-free paper

Springer is part of Springer Science+Business Media (www.springer.com)

Preface

Searching for the Interplay between Natural and Artificial Computation

The general aim of these volumes, continuing with ideas from Professor José Mira and with neurocybernetic concepts from Wiener and W.S. McCulloch, is to present a wider and more comprehensive view of the computational paradigm (CP), proposed by Alan Turing, than usual in computer science and artificial intelligence (AI) and to propose a way of using that which makes it possible: (1) to help neuroscience and cognitive science, by explaining the latter as a result of the former, (2) to establish an interaction framework between natural system computation by posing a series of appropriate questions in both directions of the interaction, from artificial systems to natural systems (in computational neuroscience), and from natural systems to artificial systems (in bioinspired computation). This is the main motivation of the International Work-conference on the Interplay between Natural and Artificial Computation, trying to contribute to both directions of the interplay:

I: From Artificial to Natural Computation. What can computation, artificial intelligence (AI) and knowledge engineering (KE) contribute to the understanding of the nervous system, cognitive processes and social behavior? This is the scope of computational neuroscience and cognition, which uses the computational paradigm to model and improve our understanding of natural science.

II: From Natural Sciences to Computation, AI and KE. How can computation, AI and KE find inspiration in the behavior and internal functioning of physical, biological and social systems to conceive, develop and build up new concepts, materials, mechanisms and algorithms of potential value in real-world applications? This is the scope of the new bionics, known as bioinspired engineering and computation, as well as of natural computing.

To address the two questions exposed in the scope of IWINAC 2011, we will make use of a wide and comprehensive view of the computational paradigm that first considers three levels of description for each calculus (physical mechanisms, symbols and knowledge) and then distinguishes between two domains of description (the level "own" domain and the domain of the external observer).

This wider view of the computational paradigm allows us more elbow room to accommodate the results of the interplay between nature and computation. The IWINAC forum thus becomes a methodological approximation (set of intentions, questions, experiments, models, algorithms, mechanisms, explanation procedures, and engineering and computational methods) to the natural and artificial perspectives of the mind embodiments problem, both in humans and in artifacts. This is the philosophy of the IWINAC meetings, the "interplay" movement between the natural and artificial, facing this same problem every two years. We want to know how to model biological processes that are associated with measurable physical magnitudes and, consequently, we also want to design and build robots that imitate the corresponding behaviors based on that knowledge. This synergistic approach will permit us not only to build new computational systems based on the natural measurable phenomena, but also to understand many of the observable behaviors inherent to natural systems.

The difficulty of building bridges over natural and artificial computation was one of the main motivations for the organization of IWINAC 2011. In this edition, the conference was simultaneously coorganized with the Joint Workshop and Summer School: Astrostatistics and Data Mining in Large Astronomical Databases 2011, that aims to apply AI techniques to astronomical data. The IWINAC 2011 proceedings volumes include the 108 works selected by the Scientific Committee after a refereeing process. The first volume, entitled *Foundations on Natural and Artificial Computation*, includes all the contributions mainly related to the methodological, conceptual, formal, and experimental developments in the fields of neurophysiology and cognitive science. The second volume entitled *New Challenges on Bioinspired Applications* contains the papers related to bioinspired programming strategies and all the contributions related to the computational solutions to engineering problems in different application domains, especially health applications, including the CYTED "Artificial and Natural Computation for Health" (CANS) research network papers.

An event like IWINAC 2011 cannot be organized without the collaboration of a group of institutions and people, whom we would like to thank now, starting with *UNED* and *Universidad Politécnica de Cartagena*. The collaboration of the *UNED associated center* was crucial, as was the efficient work of the Local Committee, Francisco Javier Neris Paz and Juan Antonio González Arnaez, with the close collaboration of the *Instituto de Astrofísica de Canarias*, and the essential support of Rafael Rebolo and Juan Carlos Pérez. In addition to our universities, we received financial support from the Spanish *Ministerio de Ciencia e Innovación, CYTED, Red Nacional en Computación Natural y Artificial* and *APLIQUEM S.L.*

We want to express our gratefulness to our invited speakers, Changjiu Zhou from Singapore Polytechnic, Paul Cull, Oregon State University, Rüdiger Dillmann from Karlsruhe Institute of Technology (KIT) and Jon Hall, Open University, for accepting our invitation and for their magnificent plenary talks.

We would also like to thank the authors for their interest in our call and the effort in preparing the papers, condition *sine qua non* for these proceedings, and to all the Scientific and Organizing Committees, in particular, the members of these committees that have acted as effective and efficient referees and as promoters and managers of pre-organized sessions on autonomous and relevant topics under the IWINAC global scope.

Our sincere gratitude also goes to Springer and to Alfred Hofmann and his collaborators, Anna Kramer and Leonie Kunz, for the continuous receptivity, help, and collaboration in all our joint editorial ventures on the interplay between neuroscience and computation.

Finally, we want to express our special thanks to *ESOC S.L.*, our technical secretariat, and to Victoria Ramos, for making this meeting possible, arranging all the details that comprise the organization of this kind of event.

All the authors of papers in this issue, as well as the IWINAC Program and Organizing Committees, would like to pay tribute to the memory of Professor Mira, both as a great scientist and as a good friend. We still greatly miss him.

June 2011

José Manuel Ferrández Vicente
José Ramón Álvarez Sánchez
Félix de la Paz López
Fco. Javier Toledo Moreo

Organization

General Chairman

José Manuel Ferrández Vicente

Organizing Committee

José Ramón Álvarez Sánchez
Félix de la Paz López
Fco. Javier Toledo Moreo

Local Organizing Committee

Francisco Javier Neris Paz
Juan Antonio González Arnaez

Invited Speakers

Paul Cull, USA
Rüdiger Dillmann, Germany
Jon Hall, UK
Changjiu Zhou, Singapore

Field Editors

Diego Andina, Spain
José M. Azorín, Spain
Mª Consuelo Bastida Jumilla, Spain
Francisco Bellas, Spain
Carlos Cotta Porras, Spain
Verónica Dahl, Canada
Richard Duro, Spain
Eduardo Fernández, Spain
Antonio Fernández Caballero, Spain
Antonio J. Fernández Leiva, Spain
José Manuel Ferrández, Spain
Vicente Garcerán Hernández, Spain
Pedro Gómez Vilda, Spain
Juan Manuel Górriz Sáez, Spain
M. Dolores Jiménez, Spain
Javier de Lope, Spain

Dario Maravall, Spain
Rafael Martínez Tomás, Spain
Félix de la Paz, Spain
Mariano Rincón Zamorano, Spain
Javier Ropero Peláez, Brazil
Daniel Ruiz Fernández, Spain
Andreas Schierwagen, Germany
Antonio Soriano, Spain
M. Jesús Taboada, Spain

Scientific Committee (Referees)

Andy Adamatzky, UK
Michael Affenzeller, Austria
Abraham Ajith, Norway
Igor Aleksander, UK
Amparo Alonso Betanzos, Spain
Jose Ramon Alvarez Sanchez, Spain
Shun ichi Amari, Japan
Diego Andina, Spain
Davide Anguita, Italy
Margarita Bachiller Mayoral, Spain
Antonio Bahamonde, Spain
Dana Ballard, USA
Emilia I. Barakova, The Netherlands
Alvaro Barreiro, Spain
Senen Barro Ameneiro, Spain
Francisco Bellas, Spain
Guido Bologna, Switzerland
Juan Botia, Spain
François Bremond, France
Giorgio Cannata, Italy
Enrique J. Carmona Suarez, Spain
Joaquin Cerda Boluda, Spain
Enric Cervera Mateu, Spain
Antonio Chella, Italy
Santi Chillemi, Italy
Eris Chinellato, Spain
Emilio S. Corchado, Spain
Carlos Cotta, Spain
Erzsebet Csuhaj Varju, Hungary
Jose Manuel Cuadra Troncoso, Spain
Veronica Dahl, Canada
Felix de la Paz Lopez, Spain
Javier de Lope, Spain
Erik De Schutter, Belgium

Angel P. del Pobil, Spain
Ana E. Delgado Garcia, Spain
Gines Domenech, Spain
Jose Dorronsoro, Spain
Gerard Dreyfus, France
Richard Duro, Spain
Reinhard Eckhorn, Germany
Patrizia Fattori, Italy
Juan Pedro Febles Rodriguez, Cuba
Paulo Felix Lamas, Spain
Eduardo Fernandez, Spain
Antonio Fernandez Caballero, Spain
Manuel Fernandez Delgado, Spain
Miguel A. Fernandez Graciani, Spain
Antonio J. Fernandez Leiva, Spain
Abel Fernandez Laborda, Spain
Jose Manuel Ferrandez, Spain
Kunihiko Fukushima, Japan
Cristina Gamallo Solorzano, Spain
Jose A. Gamez, Spain
Vicente Garceran Hernandez, Spain
Jesus Garcia Herrero, Spain
Juan Antonio Garcia Madruga, Spain
Francisco J. Garrigos Guerrero, Spain
Tamas D. Gedeon, Australia
Charlotte Gerritsen, The Netherlands
Marian Gheorghe, UK
Pedro Gomez Vilda, Spain
Juan M Gorriz, Spain
Manuel Graña Romay, Spain
Francisco Guil Reyes, Spain
John Hallam, Denmark
Juan Carlos Herrero, Spain
Cesar Hervas Martinez, Spain
Tom Heskes, The Netherlands
Roberto Iglesias, Spain
Fernando Jimenez Barrionuevo, Spain
M. Dolores Jimenez Lopez, Spain
Kok Joost N., The Netherlands
Jose M. Juarez, Spain
Kostadin Koroutchev, Spain
Elka Korutcheva, Spain
Yasuo Kuniyoshi, Japan
Petr Lansky, Czech Republic
Markus Lappe, Germany

Maria Longobardi, Italy
Maria Teresa Lopez Bonal, Spain
Ramon Lopez de Mantaras, Spain
Pablo Lopez Mozas, Spain
Tino Lourens, The Netherlands
Max Lungarella, Japan
Manuel Luque Gallego, Spain
Francisco Macia Perez, Spain
George Maistros, UK
Saturnino Maldonado, Spain
Vincenzo Manca, Italy
Daniel Mange, Switzerland
Riccardo Manzotti, Italy
Dario Maravall, Spain
Roque Marin, Spain
Jose Javier Martinez Alvarez, Spain
Rafael Martinez Tomas, Spain
Jesus Medina Moreno, Spain
Jose del R. Millan, Switzerland
Victor Mitrana, Spain
Jose Manuel Molina Lopez, Spain
Javier Monserrat Puchades, Spain
Juan Morales Sanchez, Spain
Federico Moran, Spain
Roberto Moreno Diaz, Spain
Arminda Moreno Diaz, Spain
Ana Belen Moreno Diaz, Spain
Isabel Navarrete Sanchez, Spain
Nadia Nedjah, Brazil
Taishin Y. Nishida, Japan
Richard A. Normann, USA
Manuel Ojeda Aciego, Spain
Lucas Paletta, Austria
Jose T. Palma Mendez, Spain
Juan Pantrigo, Spain
Alvaro Pascual Leone, USA
Miguel Angel Patricio Guisado, Spain
Gheorghe Paun, Spain
Juan Pazos Sierra, Spain
Mario J. Perez Jimenez, Spain
Jose Manuel Perez Lorenzo, Spain
Franz Pichler, Austria
Jose M. Puerta, Spain
Carlos Puntonet, Spain
Alexis Quesada Arencibia, Spain

Andonie Razvan, USA
Luigi M. Ricciardi, Italy
Mariano Rincon Zamorano, Spain
Victoria Rodellar, Spain
Jesus Rodriguez Presedo, Spain
Jose Carlos Rodriguez Rodriguez, Spain
Camino Rodriguez Vela, Spain
Javier Ropero Pelaez, Brazil
Daniel Ruiz, Spain
Ramon Ruiz Merino, Spain
Pedro Salcedo Lagos, Chile
Juan Vicente Sanchez Andres, Spain
Angel Sanchez Calle, Spain
Eduardo Sanchez Vila, Spain
Jose Luis Sancho Gomez, Spain
Gabriella Sanniti di Baja, Italy
Jose Santos Reyes, Spain
Ricardo Sanz, Spain
Shunsuke Sato, Japan
Andreas Schierwagen, Germany
Guido Sciavicco, Spain
Radu Serban, The Netherlands
Igor A. Shevelev, Russia
Juan A. Sigüenza, Spain
Jordi Solé i Casals, Spain
Antonio Soriano Paya, Spain
Maria Jesus Taboada, Spain
Settimo Termini, Italy
Javier Toledo Moreo, Spain
Rafael Toledo Moreo, Spain
Jan Treur, The Netherlands
Enric Trillas Ruiz, Spain
Ramiro Varela Arias, Spain
Marley Vellasco, Brazil
Lipo Wang, Singapore
Stefan Wermter, UK
Hujun Yin, UK
Changjiu Zhou, Singapore

Table of Contents – Part II

Table of Contents – Part I

Neuromorphic Detection of Vowel Representation Spaces

Pedro Gómez-Vilda[1], José Manuel Ferrández-Vicente[2],
Victoria Rodellar-Biarge[1], Agustín Álvarez-Marquina[1],
Luis Miguel Mazaira-Fernández[1], Rafael Martínez-Olalla[1],
and Cristina Muñoz-Mulas[1]

[1] Grupo de Informática Aplicada al Tratamiento de Señal e Imagen,
Facultad de Informática, Universidad Politécnica de Madrid,
Campus de Montegancedo, s/n, 28660 Madrid
`pedro@pino.datsi.fi.upm.es`
[2] Dpto. Electrónica, Tecnología de Computadoras,
Univ. Politécnica de Cartagena,
30202, Cartagena

Abstract. In this paper a layered architecture to spot and characterize vowel segments in running speech is presented. The detection process is based on neuromorphic principles, as is the use of Hebbian units in layers to implement lateral inhibition, band probability estimation and mutual exclusion. Results are presented showing how the association between the acoustic set of patterns and the phonologic set of symbols may be created. Possible applications of this methodology are to be found in speech event spotting, in the study of pathological voice and in speaker biometric characterization, among others.

1 Introduction

Speech processing is evolving from classical paradigms more or less statistically oriented to psycho- and physiologic paradigms more inspired in speech perception facts [1]. Especially important within speech perception are vowel representation spaces. These may be formally defined as applications between the space of acoustic representations at the cortical level to the set of perceptual symbols defined as vowels at the phonologic or linguistic level [12]. These relations can be expressed using graphs and Self Organizing Maps [10]. In the present work the aim is placed in mimicking some of the most plausible physiological mechanisms used in the Auditory Pathways and Centres of Human Perception for vowel spotting and characterization [11]. The detection and characterization of vowel spaces is of most importance in many applications, as in pathological characterization or forensic speaker recognition, therefore the present work will concentrate in specific vowel representation space detection and characterization by neuromorphic methods. The paper is organized as follows: A brief description of vowel nature based in formant characteristics and dynamics is given in section 2. In section 3 the layers of a Neuromorphic Speech Processing Architecture

J.M. Ferrández et al. (Eds.): IWINAC 2011, Part II, LNCS 6687, pp. 1–11, 2011.

based on Hebbian Units [7] implementing the detection paradigms is presented. In section 4 some results are given from simulations, accompanied by a brief discussion. Conclusions are presented in section 5.

2   Nature and Structure of Vowels

Speech may be described as a time-running acoustic succession of events (or phonetic sequence, see Fig.2.top) [7]. Each event is associated with an oversimplified phonation paradigm composed of vowels, and non-vowels. The acoustic-phonetic nature of these beads is based on the association of the two first resonances of the Vocal Tract, which are referred to as 'formants', and described as F_1 and F_2. F_1 in the range of 200-800 Hz is the lowest,. F_2 sweeps a wider range, from 500 to 3000 Hz. Under this point of view the nature of vowels may be described by formant stability during a time interval larger than 30 ms, and relative position in the F_2 vs F_1 space, in which is often called the 'Vowel Triangle' (see Fig.1).

Non-vowel sounds are characterized by unstable formants (dynamic), by not having a representation inside the vowel triangle, or by lacking a neat F_2 vs F_1 pattern. Sounds as $[\omega, j, b, d, J, g, p, t, c, k, \beta, \delta, \zeta, \gamma, r, r]$ are included in the first class. The second class comprises vowel-like sounds by their stability as $[l, \lambda, \Upsilon, v, z, m, n, n, \eta]$ but with representation spaces out of the area delimited by the triangle [i, a, u]. The third group includes unvoiced sounds as $[f, s, \varphi, \theta, \int, \chi,]$ which are articulated without phonation (vocal fold vibration) and produce smeared pseudo-formants in the spectrum resulting from turbulent air flow in the vocal tract. The International Phonetic Alphabet (IPA) [2] has been used,

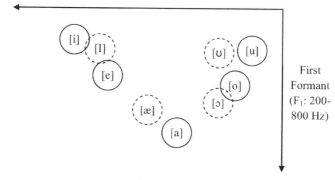

Fig. 1. Subset of the Reference Vowel Triangle for the case under study. The plot of F_2 (ordinate) vs F_2 (abscissa) is the one classically used in Linguistics. The vowel set i, e, a, o, u is sometimes referred as the *cardinal set*. The number of vowels differentiated by a listener (full line) depends on the phonologic coding of each language. Other acoustic realizations (dash line) are commonly assigned to nearby phonologic representations. For instance, in the case of study the acoustic realization [æ] in Spanish could be perceptually assigned by a listener to /a/.

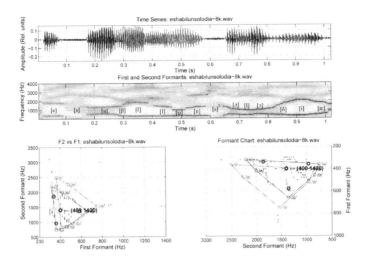

Fig. 2. Top: time series of the utterance *-es hábil un solo día-* ([*esaβIωυnsolodiæ*]) uttered by a male speaker. Middle: Adaptive Linear Prediction Spectrogram (grey background) and first two formants (superimposed in color). The color dots mark the positions of each pair (F_1,F_2) from green (the oldest) to red (the most recent). An approximate phonetic labeling is given as a reference. Bottom Left: Formant plot of F_2 vs F_1. Bottom Right: Same plot as a Formant Chart commonly used in Linguistics. The black circles give the centroids of the vowel triangle extremes and its center of gravity. The blue triangle and circles give the limit positions of the five cardinal vowels /i/, /e/, /a/, /o/, /u/ (male speaker in blue, female in magenta). These plots show the formant trajectories of the utterance. There is color correspondence between the bottom and middle templates to track formant trajectories on the time axis.

with symbols between square brackets [a] and bars /a/ are phonemes (acoustic representations) and phonologic representations, respectively. A target sentence is used as an example in Fig.2 which reproduces a spectrogram with both static and dynamic formant patterns. The sentence *-es hábil un solo día-* represents the full vowel triangle in Spanish, although acoustically some of the vowels are not extreme. Formants are characterized in this spectrogram (middle template) by darker energy envelope peaks. What can be observed in the figure is that the vowels and vowel-like sounds correspond to stable positions of the formants.

3   Neuromorphic Computing for Speech Processing

The term 'neuromorphic' is used for emulating information processing by neurologic systems. As far as speech is concerned, it has to see with neuronal units and circuits found in the Auditory Pathways and Centres. The functionality of these structures is becoming better understood as neurophysiology is deepening in functionality [3][13][15]. Preliminary work has been carried out on the characterization of speech dynamics by the Auditory Cortex for consonant description [4][5], where a Neuromorphic Speech Processing Architecture (NSPA) based in Hebbian Units [7] was proposed and widely discussed. The present paper is focussed on the sections of the NSPA specifically devoted to vowel characterization. A general description is given in Fig.3.

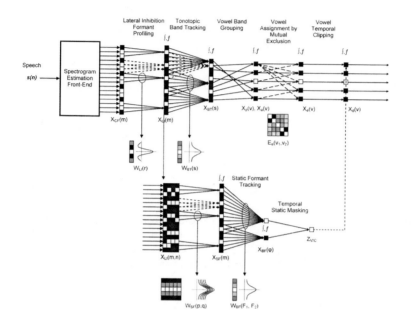

Fig. 3. Vowel processing and representation sections of the Neuromorphic Speech Processing Architecture described in [4][5]. Upper data-flow pipeline: Spectrogram Estimation Front-End, Lateral Inhibition Formant Profiling, Tonotopic Band Tracking, Vowel Band Grouping, Vowel Assignment by Mutual Exclusion and Vowel Temporal Clipping. Lower data-flow pipeline: Static Formant Tracking and Temporal Static Masking (see text for a detailed description).

Spectrogram Estimation Front-End. This section provides a spectral description of speech s(n) evolving in the time domain (spectrogram as the one in Fig.2.middle). A matrix $X_{CF}(m,n)$ is produced describing frequency activity in time (where n is the time index) as a result of a linear layer of characteristic frequency (CF) units. These units may be seen as roughly related to nerve fibres in the Auditory Periphery, each one reacting to a specific channel in frequency (where m is the frequency index). In the present case Linear Predictive Coding have been used to build the spectrogram:

$$X_{CF}(m, n) = 20 \cdot log_{10} \left| 1 - \sum_{k=1}^{K} a_{k,n} e^{-jmk\Omega\tau} \right|^{-1} \tag{1}$$

where $a_{k,n}, 1 \leq k \leq K$ is the set of coefficients of the equivalent K-order Inverse Filter, Ω the frequency resolution (separation between channels) and τ the sampling interval.

Lateral Inhibition Formant Profiling. The activity of neighbour fibres is reduced to represent formant descriptions at the lowest cost by lateral inhibition [6] as:

$$X_{LI}(m) = u \left(\sum_{i=-r}^{r} w_{LI}(i) X_{CF}(m+i) - \vartheta_{LI}(m) \right) \qquad (2)$$

where w_i are the weights in the lateral inhibition connections. Typically, for a set of five weights (r=2) these may be set up to configurations such as $-1/6, -1/3, 1, -1/3, -1/6$, reproducing the classical Mexican Hat. The function implicit in (2) may be seen as a Hebbian Unit modelling membrane integration and threshold $(\int,)$ by weighted average and nonlinear conforming. Therefore $u(.)$ is a nonlinear activation function (step or sigmoid) firing if membrane activity overcomes a specific threshold $\vartheta_{LI}(m)$.

Tonotopic Band Tracking. Vowel detection is based on the combination of activity by band tracking units (BTU's) from neighbour CF fibers by Hebbian Units as:

$$X_{BT}(s) = u \left(\sum_{i=-\beta_s}^{\beta} w_{BT}(i,s) X_{LI}(\gamma_s + i) - \vartheta_{LI}(s) \right) \qquad (3)$$

where s is the band index, γ_s and β_s are the indices to the center frequency and half the bandwidth respectively. In this case, the weights of the summation w_{BT} are selected to reproduce the output probability of the band according to a marginal probability density function (gaussian, with μ_s and σ_s the band mean and standard deviation):

$$X_{BT}(i,s)) = \Gamma(\xi_i \mid \mu_s, \sigma_s) = \frac{1}{\sigma_s \sqrt{2\pi}} e^{-\frac{(\xi_i - \mu_s)^2}{2\sigma_s^2}}$$

$$-\beta_s \Omega \leq \xi_i \leq -\beta_s \Omega; \xi_i = i\Omega; \mu_s = \gamma_s \Omega; \sigma_s = \beta_s \Omega \qquad (4)$$

Vowel Band Grouping. Once a sufficient number of BTU's have tuned their respective frequency spaces, they must be somehow combined among themselves to represent vowel activity as ordered pairs $X_{BT}(i), X_{BT}(j)$. This combination strategy is very much language-dependent, based on a previous agreement among the speakers of the language. As a matter of fact each language has developed its own encoding table, which finds its counterpart in the representation spaces to be found in the Auditory Centers. As an example, the encoding table for the five cardinal vowels [a, e, i, o, u] for standard Spanish is shown in Table 1. Other languages are known to have a larger symbol system, in which case the phonological vowel set would be correspondingly larger.

Table 1. Phonol. Formant Association Table for Spanish

BTU's F_2/F_1 (Hz)	220-440	300-600	550-950
550-850	/u/	void	void
700-1100	aliased	/o/	void
900-1500	aliased	aliased	/a/
1400-2400	aliased	/e/	void
1700-2900	/i/	void	void

This configuration is the result of averaging estimations from 8 male speakers, a similar table for female speakers could be produced. The positions marked as 'void' correspond to non-vowel sounds (second class), whereas the positions marked as 'aliased' may be ascribed to nearby valid vowel representation spaces showing a larger probability function with respect to the acoustic model.

Vowel Assignment by Mutual Exclusion. The vowel representation spaces must be unambiguously coded to bear plausible meaning to the listener. Therefore a strong exclusion mechanism is proposed, which would be activated each time enough activity is detected simultaneously by several units in a specific acoustic space, thus the vowel showing the largest activity or detection probability reacts as a 'winner-takes-all' silencing other possible vowel candidates. A neural circuit combines each two band activities by pairs according to the following paradigm:

$$X_p(\nu) = w_{p1}(X_{BT}(s_1) \times w_{p2}(X_{BT}(s_2)$$
$$X_a(\nu))) = u(X_p(\nu) - \vartheta_a(\nu)) \tag{5}$$

where $X_p(\nu)$ may be seen as the activation probability for vowel v given the input template $X_{CF}(m)$, and ν is the index to the set of vowels in the phonological system:

$$X_p(\nu) = p(\nu \mid X_{CF}(m)); \nu \in \{u, o, a, i, e\} \tag{6}$$

On its turn, weights w_{p1} and w_{p2} encode the relative probabilities of the respective formants in the detection of the vowel. The symbol ($\times$) represents the logical operator and, and may be implemented also by a Hebbian Unit. The mutual exclusion among representation spaces is governed by the following combination paradigm:

$$X_a(\nu))) = u(E(\nu_1, \nu_2)X_a(\nu) - \vartheta_e(\nu)) \tag{7}$$

where $E(v_1, v_2)$ is the mutual exclusion matrix, pre-wired as in the present case:

$$E(v_1, v_2) = \begin{pmatrix} +1.0 & -1.0 & +0.0 & 0.0 & -0.2 \\ -1.0 & +1.0 & -0.2 & -0.2 & +0.0 \\ +0.0 & -1.0 & +1.0 & -1.0 & +0.0 \\ +0.0 & -0.2 & -0.2 & +1.0 & -1.0 \\ -0.2 & +0.0 & +0.0 & -1.0 & +1.0 \end{pmatrix} \tag{8}$$

The elements in the main diagonal are set to +1.0, each vowel probability exciting the next unit (solid arrow in Fig.3) whereas it acts as a strong, weak or neuter inhibitory input (-1.0, -0.2, +0.0) to other vowels (dash arrows). Equation (7) is a discriminant function [8] based on Bayesian Decision Theory using log likelihood ratios:

$$L_e(\nu) = log \left\{ \frac{p(x_{CF} \mid \nu)}{p(x_{CF} \mid \bar{\nu})} \right\} ; X_e(\nu) = \begin{cases} 1; L_e(\nu) > \xi_e(\nu) \\ 0; L_e(\nu) \le \xi_e(\nu) \end{cases} \tag{9}$$

Vowel Temporal Clipping. This step adds the stability property demanded for vowel sounds. A control signal as $Z_{V/C}(n)$ marking the temporal segments or intervals where formants are stable within some limits is used to inhibit or enable the expression of each vowel by logical and functions ($\times$) as defined in (5):

$$X_d(\nu) = u \left(Z_{V/C} \times X_e(\nu) - \vartheta_d(\nu) \right) \tag{10}$$

Static Formant Tracking. The temporal clipping signal is estimated by tracking the segments where the first two formants remain relatively stable. This activity is captured using mask-based neuromorphic units as already explained in [4][5] which process the spectrogram as a true auditory image [8]:

$$X_{SF}(m, n) = u \left[\sum_{p=-P}^{P} \sum_{q=0}^{Q} w_{SF}(p, q) X(m + p, n - q)) - \vartheta_{SF}(m) \right] \tag{11}$$

The weight matrix $w_{SF}(p, q)$ is a bell-shaped histogram displaced in the time index (q). Practical values for P and Q are 4 and 8, respectively, resulting in a 9x9 mask.

Temporal Static Masking. Stability has to be detected separately on the two first formants and further combined. Two independent units, φ_1 and φ_1 will be tuned to two frequency bands centred at (γ_1, γ_2) with half bandwidths (β_1, β_2) similarly to (3):

$$X_{BF}(\varphi) = u \left(\sum_{i=-\beta_s}^{\beta} w_{BF}(i, \varphi) X_{SF}(\gamma_\varphi + i) - \vartheta_{BF}(\varphi) \right) \tag{12}$$

The weights of the integration function are fixed as gaussian distributions following (4). The fusion of formant masking units is carried out by a classical *and* operator:

$$Z_{V/C} = u \left(X_{BF}(\varphi_1) \times X_{BF}(\varphi_1) - \vartheta_{V/C} \right) \tag{13}$$

This signal is used in (10) to validate the intervals of formant stable activity which can be associated to vowel representation spaces.

4   Results and Discussion

In what follows some results from processing the model sentence in Fig.2 with the proposed structure will be shown. The details of the architecture are the following: $1 \leq m \leq M = 512$ CF fibre units are used, defining a resolution in frequency of 16 Hz for a sampling frequency of 8000 Hz. A spectrum frame is produced each 2 ms to define a stream of approximately 500 frames per second. The dimensions of the BTU's are defined as in Table 1. An example of the operation of BTU's $X_{BT}(220 - 440)$ and $X_{BT}(1800 - 3000)$ and the formant fusion unit $X_a(/i/)$ is shown in Fig.4.

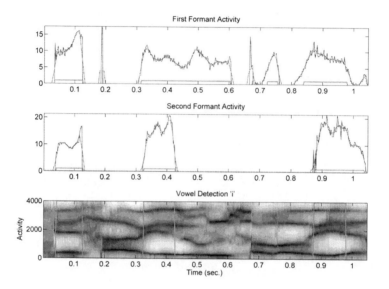

Fig. 4. Top: Activity of BTU $X_{BT}(220 - 440)$. Input activity at the unit membrane before (blue) and after integration (red), and firing after threshold (green). Middle: Idem for $X_{BT}(1800 - 3000)$. Bottom: Fusion of both BTU's in unit $X_a(/i/)$ (in green). The spectrogram is given as a reference.

This unit selects vowel segments corresponding to [I] or [i], and to [e] (first segment between 0.04 and 0.13 s). This is compliant with the ability of any BTU to capture activity from acoustic spaces overlapping in part with neighbour units as explained before. When the respective activities of both $X_a(/e/)$ and $X_a(/i/)$ are subject to mutual exclusion the first segment will be assigned to /e/ (cyan) and the two last ones will be captured by /i/ (blue) as seen in Fig.5. Vowel detection is evident after this operation.

The use of the temporal static masking signal $Z_{V/C}$ helps in removing certain ambiguities in vowel-consonant assignments as it may be seen in Fig.6. The vowel intervals have been delimited to the most stable segments of the utterance. Table 2 gives a detailed description of the detection process.

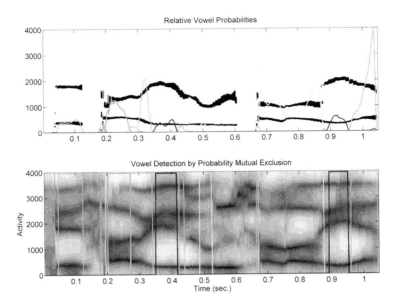

Fig. 5. Top: Probability estimates for the five vowels at layer $X_a(\nu)$. The first two formants are superimposed for reference as by layer $X_{LI}(m)$. Bottom: Activity of layer $X_e(\nu)$. Vowel color reference: /i/-blue, /e/-cyan, /a/-green, /o/-yellow, /u/-red.

Table 2. Vowel detection results

Interval (ms)	Observations
0.04-0.13	[e] is detected
0.13-0.21	void (sibilant [s])
0.21-0.27	[a] is detected
0.27-0.30	[æ] is detected as /e/
0.31-0.35	void (approximant [β])
0.35-0.41	[i] is detected
0.41-0.50	void (lateral [l])
0.50-0.53	[υ] is detected as /o/
0.53-0.69	void (nasal [n] and a sibilant [s])
0.69-0.76	[o] is detected as /o/
0.76-0.77	void (lateral [l])
0.77-0.86	[o] is detected as /o/
0.86-0.89	void (approximant [δ])
0.89-0.96	[i] is detected
0.96-0.1.05	unstable [$i \to e \to$ æ] is fragmentarily detected as /e/

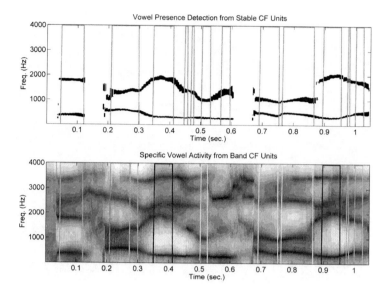

Fig. 6. Top: Output activity of the temporal masking unit $Z_{V/C}$. Bottom: Activity of layer $X_d(\nu)$

5   Conclusions

Through the present work it has been shown that vowel characterization can be carried out based on the criteria of formant stability and relative position inside the vowel triangle of the speaker using neuromorphic (Hebbian) processing units (neurons). It has also been shown that band categorization is carried out using gaussians as marginal distributions. Under this point of view the membrane activity of band categorization neurons (after integration) may receive the consideration of conditional probabilities. Output firing rates are to be seen as results of decision-making algorithms when mutual exclusion is used on competing conditional probabilities. The process relies strongly on the use of lateral inhibition to profile formants and to establish vowel representation spaces in a "winner-takes-all" strategy. This implies a decision problem which may produce unexpected results, as in the interval 0.50-0.53, where a rather obscure vowel [ʋ] is mistaken as /o/. This fact demands a small explanation: although the resulting vowel space is not fully represented by /o/ the acoustic-phonetic space controlled by this symbol is very ubiquitous, as to be able of seizing the surrounding space, which is not very much questioned by any of the other vowel representations except /u/, -see the mutual exclusion matrix in (8). This result is left deliberately 'as-is' to put into evidence eager seizing (aliasing or usurpation) of unclaimed representation spaces by strongly implanted vowels under the phonological point of view. This behaviour may explain difficulties in speakers of reduced vowel representation spaces to recognize much richer vowel systems from foreign origin. The utility of these results is to be found in automatic phonetic labeling of the speech trace for speech spotting, as well as in the detection of the speaker's identity [14], where stable characteristic vowel segments are sought for contrastive similarity tests.

Acknowledgements

This work is being funded by grants TEC2009-14123-C04-03 from Plan Nacional de I+D+i, Ministry of Science and Technology of Spain and CCG06-UPM/TIC-0028 from CAM/UPM.

References

1. Acero, A.: New Machine Learning Approaches to Speech Recognition. In: FALA 2010, Vigo, Spain, November 10-12 (2010); ISBN: 978-84-8158-510-0
2. http://www.arts.gla.ac.uk/IPA/ipachart.html
3. Barbour, D.L., Wang, X.: Temporal Coherence Sensitivity in Auditory Cortex. J. Neurophysiol. 88, 2684–2699 (2002)
4. Gómez, P., Ferrández, J.M., Rodellar, V., Fernández, R.: Time-frequency Representations in Speech Perception. Neurocomputing 72, 820–830 (2009)
5. Gómez, P., Ferrández, J.M., Rodellar, V., Alvarez, A., Mazaira, L.M., Olalla, R., Muñoz, C.: Neuromorphic detection of speech dynamics. Neurocomputing 74(8), 1191–1202 (2011)
6. Greenberg, S., Ainsworth, W.H.: Speech processing in the auditory system: an overview. In: Greenberg, W.A.S. (ed.) Speech Processing in the Auditory System, pp. 1–62. Springer, New York (2004)
7. Hebb, D.O.: The Organization of Behavior. Wiley, New York (1949)
8. Huang, X., Acero, A., Hon, H.W.: Spoken Language Processing. Prentice-Hall, Upper Saddle River (2001)
9. Jahne, B.: Digital Image Processing. Springer, Berlin (2005)
10. Kohonen, T.: Self-Organizing Maps. Springer, Heidelberg (1997)
11. Munkong, R., Juang, B.H.: Auditory Perception and Cognition. IEEE Signal Proc. Magazine, 98–117 (May 2008)
12. O'Shaughnessy, D.: Speech Communication. Human and Machine. Addison-Wesley, Reading (2000)
13. Palmer, A., Shamma, S.: Physiological Representation of Speech. In: Greenberg, S., Ainsworth, W., Popper, A. (eds.), pp. 163–230. Springer, New York (2004)
14. Rose, P., Kinoshita, Y., Alderman, T.: Realistic Extrinsic Forensic Speaker Discrimination with the Diphthong /aI/. In: Proc. 11th Austr. Int. Conf. on Speech Sci. and Tech., pp. 329–334 (December 2006)
15. Shamma, S.: Physiological foundations of temporal integration in the perception of speech. J. Phonetics 31, 495–501 (2003)

Speaker Recognition Based on a Bio-inspired Auditory Model: Influence of Its Components, Sound Pressure and Noise Level

Ernesto A. Martínez–Rams[1] and Vicente Garcerán–Hernández[2]

[1] Universidad de Oriente, Avenida de la América s/n, Santiago de Cuba, Cuba
eamr@fie.uo.edu.cu
[2] Universidad Politécnica de Cartagena, Antiguo Cuartel de Antiguones
(Campus de la Muralla), Cartagena 30202, Murcia, España
vicente.garceran@upct.es

Abstract. In the present work an assessmet of the influence of the different components that form a bioinspired auditory model in the speaker recognition performance by means of neuronal networks, at different sound pressure levels and Gaussian white noise of the voice signal, was made. The speaker voice is processed through three variants of an auditory model. From its output, a set of psychophysical parameters is extracted, with which neuronal networks for speaker recognition will be trained. Furthermore, the aim is to compare three standardization methods of parameters. As a conclusion, we can observed how psycophysical parameters characterize the speaker with acceptable rates of recognition; the typology of auditory model has influence on speaker recognition.

1 Introduction

The present work carries out an evaluation of the influence of the different components that are part of a bioinspired auditory model on the speaker recognition performance. It also studies and assesses how the sound pressure level influences on speaker recognition, and the combination of both, the pressure level of sound and the noise level of speech signal. Traditionally, automatic speaker recognition (ASR) is carried out by means of pattern recognition techniques, Fig. 1. Among the oldest techniques of ASR there are: direct comparison of long-term average speech spectrum (LTASS), dynamic time warping (DTW) and vector quantification (VQ), all them based on the distance measurement/distortion in the parametric field. Other recognition techniques are based on Hidden Markov Model (HMM) [1,2,3], in these the speech utterance is fit into a probabilistic framework, modeled as phonetic classes; the Gaussian Mixture Models (GMM) [4,5,6,7]. This is a technique based on maximum likelihood estimation; the discriminative term allows for the use of a binary classifier, such as a support vector machine (SVM) [4,8,9,10]. ASR, both identification and verification, are based on the analysis and parameters extraction of a speech utterance from an unknown speaker and compared with speech models of known speakers. In identification, the unknown speaker is identified as the speaker whose model best matches the

J.M. Ferrández et al. (Eds.): IWINAC 2011, Part II, LNCS 6687, pp. 12–24, 2011.
© Springer-Verlag Berlin Heidelberg 2011

input utterance. In verification, if the match is good enough, that is, above a threshold, the identity claim is accepted. It is necessary, then, to have a system of classification and decision.

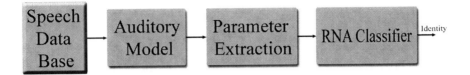

Fig. 1. Block diagram of a speaker recognition system using an auditory model

Figure 2 shows the auditory model's main block, it is composed of three components: middle and outer ear model (OEM) [11,12], basilar membrane model (BM) [11,12,13], and inner hair cell model (IHC) [11,12,13,14]. Three configurations like variants of this master model we will use, Table 1.

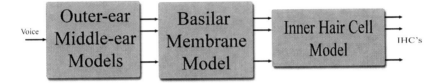

Fig. 2. The auditory model used in the present work

Table 1. Auditory model variants

Variants	OEM	BM	IHC
1	Without filter	Linear Filter Bank	Rectifier
2	Pass Band Filter	Linear Filter Bank	Rectifier
3	Pass Band Filter	Linear Filter Bank	Biophysical

In this work we used the same analysis and extraction of voice parameters performed in the article [11]. These parameters are related to psychophysical sensations that evoke the fluctuations of loudness, roughness and virtual tone. In the same way as in the previous article [11], only amplitude parameters were considered (formerly defined as LA, RA, VTA), since they provide greater speaker recognition rates.

2   Methods

2.1   Speaker Database and Locutions

The vocabulary used in the experiments comes from the 'Ahumada' database [15], which was designed and collected for speaker recognition tasks in Spanish.

In order to compute our feature sets, 656 utterances of the digit "uno" (which means "one" in Spanish) spoken by eleven speakers were used as input signals. All of them with a duration greater than 200 ms. Finally, we add Gaussian white noise with different values of signal to noise ratio, obtaining five set of noisy signals with 5, 10, 20 and 30 dB SNR and one without noise.

2.2   Auditory Model and Variants Description

OEM pass band filter, Fig. 3 a), is synthesized from the combined action of two finite impulse response filter (FIR), which characterize the amplitude response of outer and middle ear [11]. The linear filter bank, which represents the basilar membrane of the cochlear duct, 3 b), it is synthesized by means of 30 filters [16] (4th-order Butterworth band pass). The half wave rectifier that models the IHC function only represents the non linear characteristic of this cochlea region. The biophysical model used for IHC calculates both, the displacement of inner hair cell sterocilia for a given basilar membrane velocity and the intracellular IHC potential as a response to any given stereocilia displacement [14], showing the characteristics of neuronal adaptation, rectification and saturation effects of the IHC.

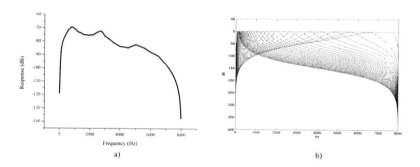

Fig. 3. Frequency response: a) OEM band pass filter; b) MB filter bank

For all three variant we are selected: work frequency range between 100 and 7,000 Hz with logarithmic separation from central frequency; the input signal is divided into 30 channels; sound pressure level of 30, 60 and 80 dB SPL according to the corresponding tests performed in [11].

2.3   Feature Extractions and Neuronal Network Architecture

The psychophysical parameters are extracted for each auditory model variant, using three standardization methods showed in [11], named A, B and C. Also, the first 20 ms of the envelopes are not considered for the computation of parameters. We have used two neural network architectures for speaker recognition. On the first one, the parameters LA (Loudness Amplitude), RA (Roughness Amplitude) and VTA (Virtual Tone Amplitude) are processed in an independent way; the second architecture is used for process the combination of these three parameters (LRVTA)[19].

3   Results

This section shows the results for each variant of auditory model and for each of the standardization methods previously mentioned. The goal of these tests is to analyze what is the influence on the success rate of speaker recognition which presents each of the components that make up the auditory model. To this aim, we proposed a battery of three tests:

- Influence of the sound pressure on speaker identification. Neural training and simulation are performed with voices without noise.
- Influence of noise level in speakers identification. Both training and simulation are performed with voices without and with noise.
- Evolution of recognition performance according to the noise level. The neural training is done with voices without noise, and simulation with voices without and with noise.

3.1   Influence of the Sound Pressure on the Speakers Identification

The results of this analysis are shown in Figures 4, 5 and 6. From Figure 4 we can conclude that, for the first variant of the auditory model, the highest levels of recognition (with the training vectors)were obtained using the LA and LRVTA parameters with methods A and B. These levels imply a 100.00% of success. also, in the generalization test (with the validation & test vectors) the highest success levels were obtained using LA and LRVTA parameters with methods A and B.

In Figure 5 we can conclude that, for the second variant of the auditory model the highest levels of recognition (with the training vectors) were obtained using the LA and LRVTA parameters with methods A and B. These levels imply a 100.00% of success. also, in the generalization test (with the validation & test vectors) the highest success levels were obtained using LA and LRVTA parameters with methods A and B.

In Figure 6 we can conclude that, for the third variant of the auditory model, the highest levels of recognition (with the training vectors) were obtained using the LA and LRVTA parameters with methods A and B. These levels imply a 100.00% of success. also, in the generalization test (with the validation & test vectors) the highest success levels were obtained using LA and LRVTA parameters with methods A and B.

Comparing these results we can conclude that the inclusion of OEM bandpass filter on the second variant did not significantly influence on the rate of recognition in relation to the first variant. The reason for this is that their function is only capture the incident sound waves and provide an initial filtering

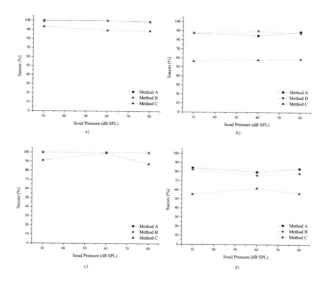

Fig. 4. Measure of success (%) in the training and validation + test, using the first variant of the auditory model, methods A, B and C. Parameters: a) LA, training; b) LA, validation + test; c) LRVTA, training; LRVTA, validation + test

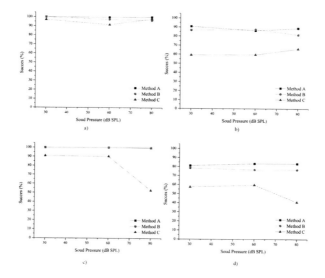

Fig. 5. Measure of success (%) in the training and validation + test, using the second variant of the auditory model, methods A, B and C. Parameters: a) LA, training; b) LA, validation + test; c) LRVTA, training; LRVTA, validation + test

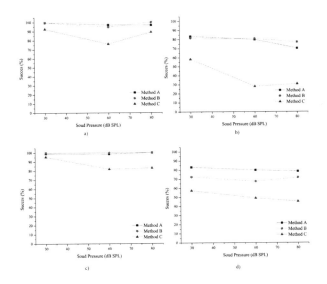

Fig. 6. Measure of success (%) in the training and validation + test, using the third variant of the auditory model, methods A, B and C. Parameters: a) LA, training; b) LA, validation + test; c) LRVTA, training; LRVTA, validation + test

of signal and help sound localization [12,13,18]. The best results in the success rate obtained with the first and the second variant with respect to the third one, is possibly due to the biophysical model used [14]. As regards the type of parameter and standardization method used, we can express that the best hit rates were achieved with LA and LRVTA parameters and normalization methods A and B, with rates close to 100% of success in the neural training. However in the generalization test, with validation & test vectors, the decrease is more noticeable with the LA parameter than with the LRVTA parameter. In contrast, with RA and VTA parameters, and with normalization method C the best results are not achieved.

3.2   Influence of Noise on the Speaker Identification

Figures 7, 8 and 9 show the rates of correct answers on tests. In the case of the first variant of auditory model, the best results were obtained with the parameters LA and LRVTA, methods A and B. Training with vectors from the peak noise level, 5 dB SNR, the minimum of success rate is over 90% and 98.85%, LA and LRVTA parameters, respectively. In the generalization test with validation & test vectors, higher levels of success were achieved with LA and LRTVA parameters, methods A and B, Fig. 7.

In the case of the second variant of the auditory model, the best results were obtained with the parameters LA and LRVTA, and methods A and B. Training with vectors from the peak noise level, 5 dB SNR, the minimum of success rate is over 90% and 99.14%, LA and LRVTA parameters, respectively. In the

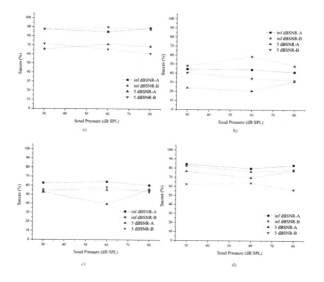

Fig. 7. Measure of success (%) with validation + test and with noise, using the first variant of the auditory model, both methods A and B. Parameters: a) LA; b) RA; c) VTA, d) LRVTA

generalization test with validation & test vectors, higher levels of success were achieved with LA and LRTVA parameters, methods A and B, Fig. 8.

In the case of the third variant of the auditory model, the best results were obtained with the parameters LA and LRVTA, and methods A and B. Training with vectors from the peak noise level, 5 dB SNR, the minimum of success rate is over 44.54% and 49.43%, LA and LRVTA parameters, respectively. In the generalization test with validation & test vectors, higher levels of success were achieved with LA and LRTVA parameters, methods A and B, Fig. 9.

With the first and second variants of the auditory model, both standardization methods A and B, and both parameters LA and LRVTA, good results were obtained in the training tests with a success rate above 90%, and even high levels of noise are above 5 dB SNR. In contrast, with the third variant of the auditory model, it experienced a decrease in success rate. However, in the generalization test with the three variants of the auditory model, that shows a hit rate decrease with increase in the noise level. These changes are manifested in different ways depending on the model variant, parameter, and method of normalization, Figures 7, 8 and 9. When making a comparison between the auditory model variants with the parameters LA, RA, VTA and LRVTA, it was showed that the best hit rates are achieved with variants 1 and 2, and with methods of standardization A and B. Furthermore, rate hits behave similarly in both variants 1 and 2 of the auditory model. The inclusion or not of the outer and middle ear function does not significantly influence on speaker recognition rate, whereas the IHC model type used does. The IHC model, used in the third variant of the auditory model [14], decreases the hit rate as opposed to that used in both the

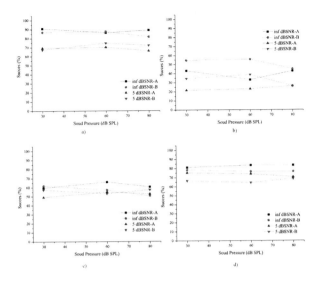

Fig. 8. Measure of success (%) with validation + test and with noise, using the second variant of the auditory model, both methods A and B. Parameters: a) LA; b) RA; c) VTA, d) LRVTA.

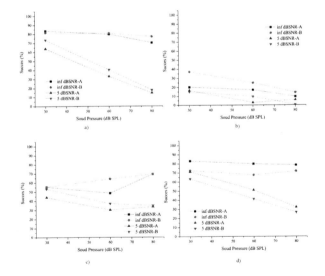

Fig. 9. Measure of success (%) with validation + test and with noise, using the third variant of the auditory model, both methods A and B. Parameters: a) LA; b) RA; c) VTA, d) LRVTA.

first and the second variant of the auditory model. We can also observe that the parameters that most affect the success rate are the LA and LRVTA. The LRVTA combines the three psychophysical parameters of speaker, and that is why it is more immune to noise and to variations in sound pressure. It can also be seen that the standardization methods A and B are those that show a better performance on the speaker recognition.

3.3   Analysis of the Evolution of Recognition Performance as a Function of Noise

In this analysis, we conduct a study of the evolution of recognition performance as a function of Gaussian white noise levels. The network trained with locutions without noise is simulated with locutions at different levels of noise and sound pressure. The results obtained for each parameter, standardization methods and variants of models, are summarized in Figures 10 to 13.

Then we make a comparison, using the rates of speaker recognition, of the three variants of the auditory model, using the same parameters and standardization methods at a time. In Fig. 10 we can see that, for LA parameter, standardization methods A and B and the three variants of the auditory model, the speaker recognition rates with locutions without noise is above 80%. Compared to 30 dB SNR (low noise) there is a sudden decrease in success rate, and hit rate is low at higher levels of noise. As can be seen that at 30 dB SNR, that is, the first and second variants of the auditory model, the hit rate is higher than the third variant.

In the case of the RA parameter, Fig. 11, if we compare the results, we seen that the first and second variant of the auditory model would be more immune from noise to 30 dB SNR, but not at higher intensities of noise. AS can also be see, the first and second variants of the model show a smaller decrease in performance in relation to the third one.

In the case of VTA parameter, Fig. 12, comparing the results of the three variants of the auditory model, it seems that options 1 and 2 showed greater immunity to noise up to 30 dB SNR, and not to higher levels of noise. As can be seen, that up to 30 dB SNR, the first and second variants of the model, its hit rate is higher in relation to third one.

In the case of LRVTA parameter, Fig. 13, comparing the results of the three variants of the auditory model, it seems that the first variant shows a high immunity to noise up to 30 dB SNR for all pressure levels, but not to higher levels of noise. The third variant of model showed a high immunity to noise, with signals loudness of 30 dB SPL and noise levels of up to 10 dB SNR.

It can be concluded that the first and second variants of the model showed a lower degradation of recognition with noise and with the psychophysical parameters LA, RA, VTA and LRVTA. However, the third variant of the model and the LRVTA parameter showed a high immunity to noise to 10 dB SNR.

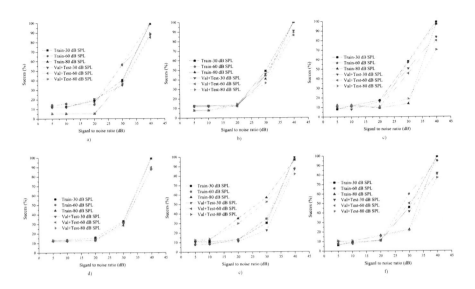

Fig. 10. Measure of success (%) of the LA parameter, training and validation + test at different levels of noise, using: a) first variant of model, method A; b) second variant of model, method A; c) third variant of model 3, method A; d) first variant of model, method B; e) second variant of model, method B; f) third variant of model, method B

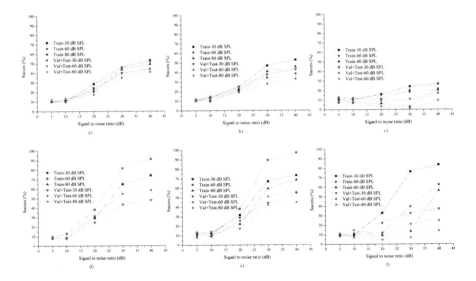

Fig. 11. Measure of success (%) of the RA parameter, training and validation + test at different levels of noise, using: a) first variant of model, method A; b) second variant of model, method A; c) third variant of model 3, method A; d) first variant of model, method B; e) second variant of model, method B; f) third variant of model, method B

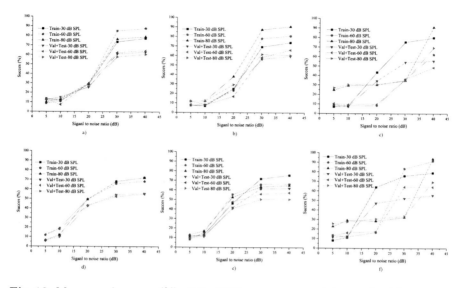

Fig. 12. Measure of success (%) of the VTA parameter, training and validation + test at different levels of noise, using: a) first variant of model, method A; b) second variant of model, method A; c) third variant of model 3, method A; d) first variant of model, method B; e) second variant of model, method B; f) third variant of model, method B

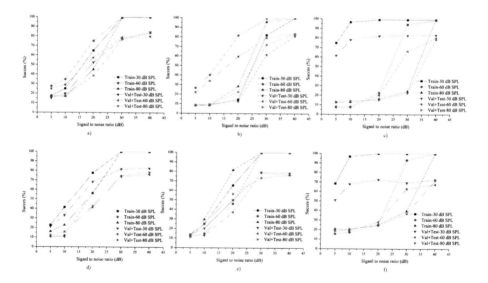

Fig. 13. Measure of success (%) of the LRVTA parameter, training and validation + test at different levels of noise, using: a) first variant of model, method A; b) second variant of model, method A; c) third variant of model 3, method A; d) first variant of model, method B; e) second variant of model, method B; f) third variant of model, method B

4   Conclusions an Future Works

The results on speaker recognition with a connectionist approach using artificial neural networks and an auditory model of perception are very dependent on both parameters used as the model variants. While the inclusion or otherwise of the outer and middle ear filters shows no significant influence on the speaker recognition, the IHC model does have an influence. Although only three parameters are used and their combination, significant results are achieved in speaker recognition. It is possible to achieve similar or better success rate than using a different auditory model for IHC. It can also be concluded that there is no significant degradation of recognition according to the noise level up to 30 dB SNR, from which point there is a fall monotonically decreasing the level of recognition.

Acknowledgements

This project is funded by the Spanish Ministry of Education and Science (Profit ref. CIT-3900-2005-4), and the Thematic Network "Computación Natural y Artificial para Salud"(CANS) of CYTED, Spain.

References

1. Resch, B.: Automatic Speech Recognition with HTK (A Tutorial for the Course Computational Intelligence), pp. 1–6 (2004)
2. Richardson, M., Bilmes, J., Diorio, C.: Hidden-Articulator Markov Models for Speech Recognition. Speech Communication 41, 511–529 (2003)
3. Reynolds, D.A.: Automatic Speaker Recognition, pp. 1–42. MIT Lincoln Laboratory (2002)
4. Wan, V.: Speaker Verification using Support Vector Machines (2003)
5. Godino–Llorente, J.I., Aguilera–Navarro, S., Gómez–Vilda, P.: Detección automática de patología por abuso vocal mediante modelos estadísticos de mezclas Gaussianas. In: URSI, pp. 1–2 (2001)
6. Reynolds, D.A., Rose, R.C.: Robust text-independent speaker identification using Gaussian Mixture Speaker Models. IEEE Transactions on Speech and Audio Processing 3(1), 72–83 (1995)
7. Reynolds, D.A.: Speaker identification and verification using Gaussian mixture speaker models. Speech Communication, 91–108 (1995)
8. Ganapathiraju, A.: Support vector machines for speech recognition (2002)
9. Clarkson, P., Moreno, P.J.: On the use of support vector machines for phonetic classification, 1–4 (2005)
10. Chen, P., Lin, C., Schölkopf, B.: A Tutorial on v-Support Vector Machines, 1–29 (2003)
11. Martínez–Rams, E.A., Garcerán–Hernández, V.: Assessment of a speaker recognition system based on an auditory model and neural nets. In: Mira, J., Ferrández, J.M., Álvarez, J.R., de la Paz, F., Toledo, F.J. (eds.) IWINAC 2009. LNCS, vol. 5602, pp. 488–498. Springer, Heidelberg (2009)
12. Martínez–Rams, E., Cano–Ortiz, S.D., Garcerán–Hernández, V.: Implantes Cocleares: Desarrollo y Perspectivas. Revista Mexicana de Ingeniería Biomédica 27(1), 45–54 (2006)

13. Martínez–Rams, E., Garcerán–Hernández, V., Ferrández–Vicente, J.M.: Low Rate Stochastic Strategy for Cochlear Implants. Neurocomputing Letters 72(4-6), 936–943 (2009)
14. Lopez Poveda, E.A., Eustaquio-Martín, A.: A biophysical model of the Inner Hair Cell: The contribution of potassium currents to peripherical auditory compression. Journal of the Association for Research in Otolaryngology JARO 7, 218–235 (2006)
15. Ortega-Garcia, J., González-Rodriguez, J., Marrero-Aguiar, V.: Ahumada: A large speech corpus in Spanish for speaker identification and verification. Speech Communication 31(2-3), 255–264 (2004)
16. Martínez–Rams, E., Cano–Ortiz, S.D., Garcerán–Hernández, V.: Diseño de banco de filtros para modelar la membrana basilar en una prótesis coclear. In: Conferencia Internacional FIE, pp. 1–6 (2006)
17. Lopez-Poveda, E.A., Meddis, R.: A human nonlinear cochlear filterbank. J. Acoust. Soc. Am. 110(6), 3107–3118 (2001)
18. Martínez–Rams, E., Garcerán–Hernández, V.: ANF Stochastic Low Rate Stimulation. In: Mira, J., Álvarez, J.R. (eds.) IWINAC 2007. LNCS, vol. 4527, pp. 103–112. Springer, Heidelberg (2007)
19. Martínez–Rams, E., Garcerán–Hernández, V.: A Speaker Recognition System based on an Auditory Model and Neural Nets: performance at different levels of Sound Pressure and of Gaussian White Noise. In: Ferrández, J.M., et al. (eds.) IWINAC 2011. LNCS, vol. 6687, pp. 157–166. Springer, Heidelberg (2011)

Inner-Hair Cells Parameterized-Hardware Implementation for Personalized Auditory Nerve Stimulation

Miguel A. Sacristán-Martínez[1], José M. Ferrández-Vicente[1],
Vicente Garcerán-Hernández[1], Victoria Rodellar-Biarge[2],
and Pedro Gómez-Vilda[2]

[1] Universidad Politécnica de Cartagena, Dpto. Electrónica,
Tecnología de Computadoras y Proyectos,
Cartagena, 30202, Spain
miguel.sacristan@upct.es, jm.ferrandez@upct.es,
vicente.garceran@upct.es
[2] Universidad Politécnica de Madrid,
Dpto. Arquitectura y Tecnología de Sistemas Informáticos,
Boadilla, 28660, Spain
victoria@pino.datsi.fi.upm.es,
pedro@datsi.fi.upm.es

Abstract. In this paper the hardware implementation of an inner hair cell model is presented. Main features of the design are the use of Meddis' transduction structure and the methodology for Design with Reusability. Which allows future migration to new hardware and design refinements for speech processing and custom-made hearing aids.

1 Introduction

Many people suffering deafness would find relief to their disabilities by receiving cochlear implants. They deliver electric stimuli proportional to the sound or speech captured from the environment to the auditory nerve, [1]. Artificial bio-inspired systems behave adequately in general, but demand very high computational costs, which render them inadequate for many real-time applications.

In addition, Speech processing has advanced tremendously in recent years, however, still fall short for performance and noise tolerance level when compared with biological recognition systems [2]. For this reason, biologically inspired algorithms are not only important in the design field of hearing aids, also to improve the interfaces between humans and machines.

The work herein described is substantially focussed to the implementation of a first prototype in a FPGA showing the behaviour of the inner auditory system, with the function of converting the movements of the basilar membrane in trains of pulses to be carried to the brain for its processing. The proposed system may serve as a front-end processor in bio-inspired speech processing and recognition applications. The utilization of Design with Reusability methodology

J.M. Ferrández et al. (Eds.): IWINAC 2011, Part II, LNCS 6687, pp. 25–32, 2011.

allows not only accelerating the cycle of prototyping, but also adapting the design to the specific needs of a particular problem, for example, varying the amount and frequency of transduction channels as a function of the patient's needs or recognition system.

1.1   Auditory System Description

Auditory periphery can be described as the concatenation of three stages: outer, middle and inner ear, as shown in Fig. 1a. Sound is the result of the vibrations of the air surrounding the ear. They reach the outer ear, consisting of the pinea and the ear canal, which adapts the sound pressure wave and is partly responsible for the location of the source. Vibrations reach the middle ear through the auditory canal and cause displacements in the tympanic membrane where the middle ear begins. These movements are transmitted through a chain of three small bones (malleus, incus and stapes) and reach to the cochlea which is the main part of the inner ear. Cochlea can be considered as divided into two scales by the basilar membrane, the tympanic and the vestibular scales, fig. 1b. The organ of Corti, located in the basilar membrane, transduces the mechanical vibrations into electrical pulses transmitted to the brain by the auditive nerve [3]. High frequency waves stimulate hair cells near the organ of Corti base and low frequency ones act close to the apex.

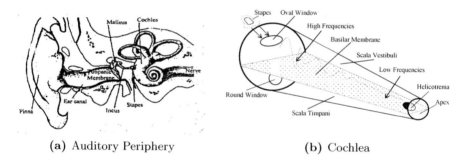

(a) Auditory Periphery   (b) Cochlea

Fig. 1. Structure of the Auditory Periphery, (a) extracted from [4]

1.2   Bio-inspired Models

Figure 2 shows the typical block structure based on the auditory periphery. Outer and Middle ears capture the sound waves and perform an initial filtering of the signal in order to increase the pressure of the sound in the region from 2 to 7 kHz. The cochlea is typically modelled by two stages. The first one models the mechanical selectivity taken place in the cochlear structure. The vibration velocity of several points along the basilar membrane may be simulated by a set of filters like the dual resonance nonlinear filter (DRNL) proposed in [5,6]. The second one simulates the behaviour of the inner hair cell transducing the vibrations into electrical signals.

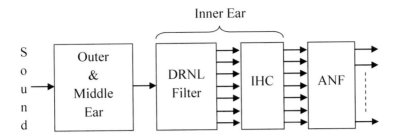

Fig. 2. Bio-inspired block diagram of the processor

2 Inner Hair Cells (IHC)

The IHC perceives basilar membrane movements and stimulates the afferent neural fibbers (ANF) by neurotransmitters libration in the synaptic cleft. Each IHC rectifies and compresses signals coming from the basilar membrane reducing heavily the frequencies above 1 kHz and the occurrence of the non-instantaneous compressions in the synaptic cleft.The behaviour of the IHC can be modelled as a function of the neurotransmitters flux through three reservoirs, fig. 3.

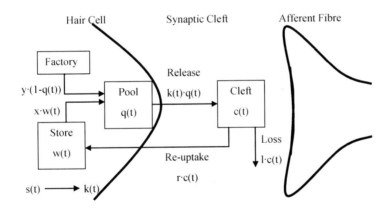

Fig. 3. IHC function characterized by the Meddis model B

The first one $q(t)$ represents the amount of neurotransmitters ready to be released as a function of the membrane displacement $s(t)$. The second reservoir represents the amount of neurotransmitters released in the synaptic cleft $c(t)$, which determines the impulse rate in the postsynaptic afferent fibbers. The third reservoir $w(t)$, represents the amount of neurotransmitters recovered from the synaptic cleft contributing to the net amount of free neurotransmitters.

The first step is to calculate the membrane permeability $k(t)$ as function of the acoustic stimulus $s(t)$. The following relations hold:

$$k(t) = \begin{cases} [g(s(t)dt + A)]/[s(t) + A + B]; & s(t) + A > 0 \\ 0; & s(t) + A \le 0 \end{cases} \tag{1}$$

where A and B are positive constants and $B >> A$. It must be noticed that when $s(t) = 0$ some spontaneous activity is allowed.

The total amount of free neurotransmitters will depend on the presence of the generated, released and re-processed neurotransmitters within a given interval dt. The product between the free transmitters available $q(t)$ and the membrane permeability $k(t)$, gives the amount of transmitters released into the cleft. On the other hand, the cell has the capability for generating transmitters at the rate of $y \, dt[M - q(t)]$, y represents the replenishing rate and M, the upper limit of free transmitters, for $M > q(t)$; otherwise this value will be zero. The recovered transmitters amount is the input to a re-processing store $w(t)$ and it will contribute in a proportion x to the free transmitters total amount $q(t)$.

$$q(t + 1) = q(t) + y[(M - q(t)]dt - q(t)k(t)dt + xw(t)dt \tag{2}$$

The re-processing reservoir stores the difference between the returned and re-processed transmitters within dt:

$$w(t + 1) = w(t) + rc(t)dt - xw(t)dt \tag{3}$$

The total amount of transmitters presents into the cleft is denoted by $c(t)$. Part of these transmitters will return into the hair cell in a proportion of $r \, c(t)$ according to a rate r and part of it will be lost: $l \, r(t)$, l representing the loss rate. And finally the amount of neurotransmitters in the cleft will depend on the released, returned and lost amounts:

$$c(t + 1) = c(t) + q(t)k(t)dt - lc(t)dt - rc(t)dt \tag{4}$$

The parameters, q, A, B, r, l, x, y and M are constants.

3 Implementation

Dataflow shown in figure 4a is the transformation of the previous equations into an implementable structure that improves the solution proposed in [7]. The main difference corresponds with the first left branch where the adders have been reordered to provide the operands to the divider as soon as possible. This improvement minimizes the delay to start the division operation.

Parallel hardware implementation, shown in fig. 4b, includes several buffers, represented by the black boxes. These delaying elements have been included in order to adjust the latency of all paths. So, the structure has been transformed into a pipeline.

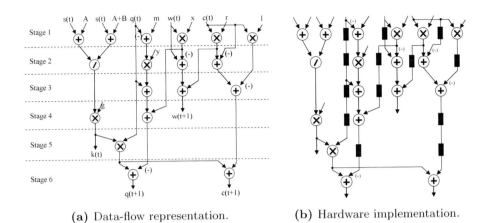

(a) Data-flow representation.   **(b)** Hardware implementation.

Fig. 4. Parallel implementation

Also a serial implementation has been explored to evaluate its performance in terms of area, which could be critical in the design of low-power nerve stimulators with a big number of channels. This implementation consists in a single arithmetical block for each operation involved in the algorithm and the controller for sequencing the operations execution, fig. 5.

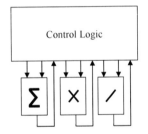

Fig. 5. Serial implementation

As the arithmetic blocks are internally pipelined, the controller sequentially programs same type operations that haven't dependencies among them. Operations scheduling for the serial implementation is shown in table 1. The complete algorithm is computed in six stages. By comparing this information with the parallel implementation shown in fig. 4b we can notice that the number of stages is the same in both cases, but the parallel implementation is fully pipelined meanwhile the serial one needs to completely cover one stage before starting the computation of the following one. It means that the results of the first stage needs 2 more clock tics in the serial implementation. For stage 2 and 3 the results availability is the same in both implementations because the division operation is the bottleneck. Since there is only one addition and one multiplication in stages 4 and 5, the delay is the same. And finally stage 6 needs one clock cycle more in the serial implementation.

Table 1. Operations scheduling for the serial hardware

	ADDER		MULT.	DIVIDER
Stage 1	**[1]** $s(t) + A$ **[2]** $s(t) + (A + B)$ **[3]** $q(t) + M$		**[4]** $x \cdot w(t)$ **[5]** $r \cdot c(t)$ **[6]** $l \cdot c(t)$	
Stage 2	**[7]** $w(t) - [4]$ **[8]** $c(t) - [5]$		**[9]** $y \cdot [3]$	**[10]** $[1]/[2]$
Stage 3	**[11]** $[9] + q(t)$ **[12]** $[7] + [5]$ **[13]** $[8] - [6]$			
Stage 4	**[14]** $[11] + [4]$		**[15]** $g \cdot [10]$	
Stage 5			**[16]** $[15] \cdot q(t)$	
Stage 6	**[17]** $[16] + [14]$ **[18]** $[16] + [13]$			

4 Results

The design was implemented using IEEE 754 double precision arithmetical blocks available at the CoreGen tool of $Xilinx^{TM}$. These blocks can be configured to perform addition, multiplication and division operations. Subtraction will be calculated by inverting the sign bit of the subtrahend. Overflow control is accomplished inside the blocks according to the mentioned IEEE standard.

Each block generated by the CoreGen tool is encapsulated within a VHDL *entity*, which will contain only the input and output ports and control signals. In this way, just changing the description of the components we can easily evaluate the performance of different particular structures and implementations of the basic components.

The structures were implemented over VIRTEX-5 devices using ISE 10.1 SP3 synthesis software and simulated for the 17 channels as proposed in [8]. The design was tested under the same conditions than [8], the simulation results were fully coincident which validates the functional behaviour of the structures.

Related with the hardware implementation, all the arithmetical blocks were generated by the CoreGen utility in two different ways: **Full logic**, using only LUTs; and **DSP48E**, exploding the internal arithmetic structures inside the

Virtex5 family. As DSP48E slices are mainly intended for multiplication and inner product, multipliers benefits from them doubling their working frequency and reducing to one fourth the amount of needed LUTs. Adders have 10% of improvement in speed and divider doesn't benefit from these slices.

The two structures have almost the same working frequencies regardless if they use full logic or DSP48E slices as can be seen in table 2. The use of DSP48E slices don't present a significative impact on the working frequency due to the critical path is inside the divider. The differences in the frequencies are due to the routing of the signals, bigger circuits let less free routes. DSP48E slices increase the latency of the arithmetic blocks because the control logic for normalizing the result or checking overflows can not be integrated within the arithmetical logic. The second line in table 2 shows that the working frequency for calculating 17 IHCs is in the range 1.5 to 2 MHz. Both structures are able to work in real time considering a data sampling frequency of 16 kHz for the input sound.

Table 2. Working frequency

	Parallel		Serial	
	Full logic	DSP48E	Full logic	DSP48E
Clock cycle (MHz)	203	211	250	252
17 IHC channels (MHz)	1.46	1.45	2	1.92

Table 3 shows the synthesis results for both structures using either LUTs or DSP48E blocks. The parallel structure needs the same resources to calculate just one IHC or the 17 used in [8] due to its pipeline operation.

Table 3. Hardware allocation

	Parallel		Serial	
	Full logic	DSP48E	Full logic	DSP48E
1 IHC channel	23870 LUTs	$14803 LUTs$ $81 DSP48E$	7037 LUTs	$4973 LUTs$ $16 DSP48E$
17 IHC channels	23870 LUTs	$14803 LUTs$ $81 DSP48E$	119629 LUTs	$84541 LUTs$ $272 DSP48E$

If only LUTs are used, the design fits in the second smallest Virtex5 device, XC5VLX50. But if DSP48E blocks are used, the high number of them require jumping to the second largest Virtex5 although only a 10% of the LUTs will be used. The number of LUTs in the full logic implementation of the parallel structure is smaller than expected because the constant operands cause that the inner structure of the arithmetical can be simplified.

Serial structure demonstrates being the smaller of the two for just one IHC. But with only three IHCs serial structure equals parallel results and it's impracticable for 17 IHCs.

5   Conclusions

The parallel structure can be synthesized on the smallest devices of the Virtex5 family. And its hardware implementation cost is independent of the number of stimulation channels. The results of the serial implementation explodes in physical resources demand as the number of channels increases. But serial implementation has the facility of easily customizing the constants of each ICH involved in the complete model.

DSP48E blocks utilization has no advantage in any case because they can not be harnessed to optimize the critical path. Also the use of DSP48E slices in the parallel structure avoids its internal simplification although there are many arithmetical blocks with constant operands.

The design methodology allows the design modification in order to introduce more or less transduction channels in different basilar membrane sections. Thus the design can be tailored for a specific nerve stimulation problem.

Acknowledgment

This work has been funded by grant TEC2009-14123-C04-03 from Plan Nacional de I+D+i, Ministry of Science and Technology, by grant CCG06-UPM/TIC-0028 from CAM/UPM, and by project HESPERIA (http.//www.proyectohesperia .org) from the Programme CENIT, Centro para el Desarrollo Tecnológico Industrial, Ministry of Industry, Spain.

References

1. Bruce, I.C., White, M.W., Irlicht, L.S., O'Leary, S.J., Dynes, S., Javel, E., Clark, G.M.: A stochastic model of the electrically stimulated auditory nerve: single-pulse response. IEEE Transactions on Biomedical Engineering 46(6), 617–629 (1999)
2. Martínez–Rams, E.A., Garcerán–Hernández, V.: ANF stochastic low rate stimulation. In: Mira, J., Álvarez, J.R. (eds.) IWINAC 2007. LNCS, vol. 4527, pp. 103–112. Springer, Heidelberg (2007)
3. Zigmond, M.J., Bloom, F.E., Landis, S.C., Roberts, J.L., Squire, L.R.: Fundamental Neuroscience. Academic Press, London (1999)
4. Rodellar, V., Gómez, P., Sacristán, M.A., Ferrández, J.M.: An inner ear hair cell parametrizable implementation. In: 42th Midwest Symposium on Circuits and Systems (Las Cruces), vol. 2, pp. 652–655 (1999)
5. Lopez-Poveda, E.A., Meddis, R.: A human nonlinear cochlear filter bank. The Journal of the Acoustical Society of America 110(6), 3107–3118 (2001)
6. Meddis, R., O'Mard, L.P., Lopez-Poveda, E.A.: A computational algorithm for computing nonlinear auditory frequency selectivity. The Journal of the Acoustical Society of America 109(6), 2852–2861 (2001)
7. Ferrández-Vicente, J.M., Sacristán-Martínez, M.A., Rodellar-Biarge, V., Gómez-Vilda, P.: A High Level Synthesis of an Auditory Mechanical to Neural Transduction Circuit. In: Mira, J., Álvarez, J.R. (eds.) IWANN 2003. LNCS, vol. 2686, pp. 678–685. Springer, Heidelberg (2003)
8. Martinez-Rams, E., Garceran-Hernandez, V., Ferrandez-Vicente, J.M.: Low rate stochastic strategy for cochlear implants. Neurocomputing, 936–943 (2009)

Semiautomatic Segmentation of the Medial Temporal Lobe Anatomical Structures

M. Rincón[1], E. Díaz-López[1], F. Alfaro[1], A. Díez-Peña[2], T. García-Saiz[1],
M. Bachiller[1], A. Insausti[4], and R. Insausti[3]

[1] Dept. Inteligencia Artificial. E.T.S.I. Informática,
Universidad Nacional de Educación a Distancia, Madrid, Spain
mrincon@dia.uned.es
[2] DEIMOS Space S.L.U., Ronda de Poniente, 19, Edificio Fiteni VI,
2-2ª 28760 Tres Cantos, Madrid, Spain
[3] Human Neuroanatomy Laboratory, School of Medicine,
University of Castilla-La Mancha, Albacete, Spain
[4] Departamento de Ciencias de la Salud, Campus de Tudela,
Universidad Pública de Navarra, Pamplona, Spain

Abstract. Medial temporal lobe (MTL) is a region of the brain related with processing and declarative memory consolidation. Structural changes in this region are directly related with Alzheimer's disease and other dementias. Manual delimitation of these structures is very time consuming and error prone. Automatic methods are needed in order to solve these problems and make it available in the clinical practice. Unfortunately, automatic methods are not robust enough yet. The use of semiautomatic methods provides an intermediate solution with the advantages of automatic methods under the supervision of the expert. This paper propose two semiautomatic methods oriented to make the delineation of the MTL structures easy, robust and fast.

Keywords: semiautomatic segmentation; medial temporal lobe.

1 Introduction

It is known that human memory systems cover many brain areas. However, neuroanatomical studies of the medial temporal lobe (MTL) have shown its relationship with cognitive functions, in particular with the processing and declarative memory consolidation[14,15]. Disorders in the MTL are related with diseases such as Alzheimer and other dementias (schizophrenia, depression, psychotic disorders, epilepsy, etc). The study of this region allows the characterization of different groups, the classification of a particular case with respect to those groups or the early detection of changes which are a sign indicating the possibility of disease development.

Currently, the most precise procedures for the delimitation of the MTL structures in magnetic resonance imaging (MRI) are manual [13,4]. As always, when working with medical images, manual analysis of MRI by the expert to delineate

J.M. Ferrández et al. (Eds.): IWINAC 2011, Part II, LNCS 6687, pp. 33–40, 2011.

the structures of interest is a heavy and rather subjective task given the blurring of the contours and imprecise definition of the structures of interest. Therefore, for clinical use, some automation of the analysis is needed, so that the results can be reproduced, more accurate and less economically and temporarily expensive. Although there are automatic segmentation of the whole brain attempts (SPM, freesurfer, FSL, CARET, etc.), the results are not entirely satisfactory in terms of accuracy and reproducibility. Success does not exceed 90% [8] in subcortical structures and is even lower when focusing on MTL cortical structures [13,6].

This paper describes two methods for semiautomatic delineation of MTL structures in T1-weighted structural MRI (T1-w MRI). In both cases, Freesurfer [2,1] is used to obtain an initial automatic segmentation of many brain structures which is subsequently refined. The main advantage of semiautomatic methods is that they have the speed and reliability of automatic segmentation methods but under the supervision of the expert.

2   Methods

Two alternative methods are proposed for tracing MTL structures. The first method is intended to help the manual tracing of the structures of interest by the expert, while the second method automates the procedure described by Insausti et al. [4] and requires minimal expert intervention, just to confirm some key points in which the algorithm is based on.

2.1   Method A: Manual Tracing Procedure

One of the most common problems after the automatic segmentation obtained by Freesurfer is the presence of mislabeled areas. For example, as shown in Figure 1, the white matter should grow in the area surrounded by yellow, and gray matter must also grow in the area surrounded by blue.

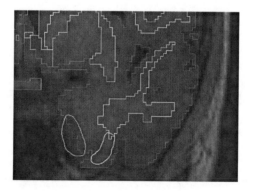

Fig. 1. Segmentations problems after Freesurfer

In order to refine this segmentation and obtain more robust results, a protocol for MTL manual tracing has been defined. This protocol consists of four stages: 1) Initialization, 2) Image adaptation to the expert, 3) semiautomatic assistance and 4) postprocessing.

1. Initialization: our method is initialized with a preliminary segmentation made with Freesurfer. Starting with a segmentation with less structures facilitates manual tracing. Therefore, this method starts from *aseg.mgz* segmentation, which identifies in MTL the following structures: WM, GM, Amygdala, Hippocampus and lateral ventricle.

2. Image adaptation to the expert: Now, the initial segmentation (*aseg.mgz*) and the original anatomical image T1-w MRI are adapted to the expert. First, a resample of the volumes according to the sagittal, coronal and transverse proyections is carried out and the resolution is set to $0.5x0.5x0.5mm$. This resample is justified because 1) expert's declarative knowledge is expressed along the coronal axis and 2) a big difference in segmentation results was detected between $0.5mm^2$ and $1mm^2$ resolutions. Second, in order to facilitate the annotation and allow correspondence between experts, only 1 out of every 4 coronal annotated slices is left because experts annotate images every $2mm$ in their manual protocol.

3. Semiautomatic assistance: the expert improves the segmentation using the *editor* module of the open source platform *3DSlicer* [12]. New semiautomatic operators have been implemented to repair some of the most common errors made by Freesurfer. This operators complement the tools already available in the 3DSlicer Editor module. We have detected that Freesurfer uses a conservative approach in its segmentation. The most common errors are failure to consider some narrow WM regions, which are classified as GM, and some GM regions, which are classified as background. Fast marching and Level set methods are very popular for interactive image segmentation. As an example, Fig. 2 shows the results after applying the *Extend-WM* operator, which is based on the fast marching algorithm implemented by ITK [3]. This method uses some seeds (initialization and growth) to control the segmentation of a certain part of the WM, which is added to the previous WM region. Fig. 3 shows the results after applying the *Extend-GM* operator, also based on the fast marching algorithm. Once the segmentation is corrected, following the protocol described in [4,5], the expert identifies the boundaries of the MTL structures of interest: temporopolar cortex (TP), entorhinal cortex (EC) and the perirhinal cortex (PR).

4. Postprocessing: Finally, the annotations (1 out of 4 slices) are extended in order to generate a complete 3D volume and 3D surface models are built with 3DSlicer Model Maker module. Figure 4 shows the final results after postprocessing.

2.2   Method B: Semiautomatic Parcellation

This method automates the protocol described in [4,5] for the automatic parcelation of the cortical structures of the medial temporal lobe. This method consists

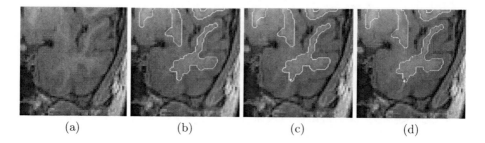

(a)   (b)   (c)   (d)

Fig. 2. Example using *Extend-WM* operator. a) T1MRI; b) initial segmentation; c) operator configuration; d) region to add.

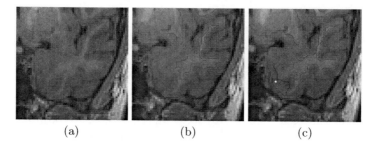

(a)   (b)   (c)

Fig. 3. Example using *Extend-GM* operator. a) T1MRI; b) initial segmentation; c) region to add 1 seed.

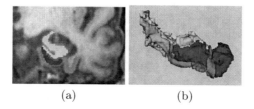

(a)   (b)

Fig. 4. Manual segmentation: a) Left MTL slice segmentation ; b) MTL 3D surface model

of the following stages: 1) Initialization; 2) Sulci and gyri 2D localization; 3) Sulci and gyri 3D localization; 4) Sulci labeling and 5) MTL parcelation.

1. Initialization: First, the structures that will form the basis of the algorithm are selected starting from Freesurfer segmentation *aparc+aseg2009.mgz*. 3D images for MTL region, amygdala, hippocampus, sulci (superior and inferior temporal sulci, collateral sulcus and hippocampal fissure), background and WM are created.

2. Sulci and gyri 2D localization: On each slice in the coronal plane the temporal lobe contour is obtained from its mask and sulci and gyri are calculated using the degree of curvature at each point. The median curvature for the point

set is calculated and used to distinguish between sulci and gyri. Those points whose curvature exceeds this threshold are considered as gyrus and the rest as sulcus. Subsequently, in each segment (gyri and sulci), the bottom of the sulcus or the farthest point of the gyrus is obtained as the largest curvature point. Two kind of segments are deleted from the generated list: very short segments (shorter than two voxels) and non-exterior segments (segments that do not have any voxel in contact with the background).

3. Sulci and gyri 3D localization: To obtain a 3D representation of sulci and gyri, 2D sulci detected in step 2 in each slice are associated with those from neighboring slices considering continuity. The continuity is defined in terms of coincident points between two sulci and/or minimum distance from the points of maximum curvature between two sulci.

4. Sulci labeling: Then, to label the sulci that have been located previously, a set of heuristic rules that describe the sulci and gyri of interest are used in this phase. These rules are evaluated on all the slices in order to get a list of sulci/gyri with more probability to correspond to each label. This stage is likely to contain errors due to shape and position variability of sulci and gyri in every brain, so the results are presented to the user in order to be validated. Figure 5 shows an example in which the collateral sulcus is highlighted as suggestion of the program. Thanks to the graphical user interface, the user can change the decision of the program and choose any of the suggested sulci. This change can be made over all the available slices and allows choosing more than one label.

5. MTL parcelation: Finally, once the gyri and sulci have been identified, as well as the most important structures, the protocol described in [4,5] is applied to identify the boundaries of the MTL structures of interest: temporopolar cortex (TPC), entorhinal cortex (EC) and the perirhinal cortex (PRC). This protocol was defined empirically by the experts while tracing manually a large number of cases. Basically, the method consists of a set of topographic rules in which the brain is divided in sequences of slices in the coronal plane. In each sequence of slices, the borders of TP, PR and EC are defined according to the topographic characteristics of the sulci and gyri that are present in those slices.

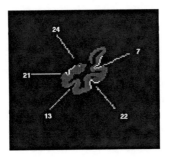

Fig. 5. 3D sulci labelling

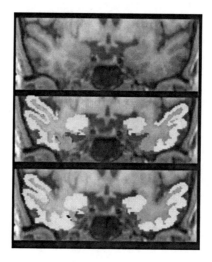

Fig. 6. MTL location example: (up) Original brain slice. (midle) Freesurfer segmentation for EC and PRC. (down) Segmentation for EC and PRC with our algorithm.

3   Experimental Results

Interobserver and intraobserver volumetric differences were evaluated on 3 cases. MRI scans were obtained in a GE Healthcare scan at 3T and medial temporal lobe structures were manually delineated by two experts using the protocol described in method A. This protocol saves enough time in the LTM segmentation of a case (5h per case). The editor module extensions are very helpful because they simplify the task. However, the time required is still significant. Although the number of cases is small to be statistically significative, interobserver and intraobserver differences shown in Table 1 are high, which makes necessary some refinements of the manual protocol.

The aim of the automatic method (method B) is to obtain similar results to method A in less time. Currently, there are some problems in the 2D sulcus localization and therefore in 3D mapping. The average similarity between volumes segmented automatically and by an expert is 73'9%.

4   Conclusions

Unfortunately, automatic methods for MTL segmentation and parcellation are not robust enough yet. This article describes two protocols designed to assist the expert in the segmentation of the structures of interest automating the process as much as possible.

A protocol for manual tracing assisted by semiautomatic tools has been defined. This protocol allows to decrease the time required by the expert and uncertainties in the protocol followed by the experts have been detected. Semiautomatic tools, adapted to the particular problem by modeling declarative

Table 1. Volumetric interobserver and intraobserver variability in MTL delineation (%)

		Intraobserver	Interobserver
Left	Temporopolar cortex	4.3	8.3
	Perirhinal cortex	5.4	11.0
	Entorhinal cortex	5.6	29.4
	Hippocampus	13.1	4.2
	Amygdala	18.1	12.8
	Parahippocampal cortex	6.3	5.5
Right	Temporopolar cortex	6.8	15.7
	Perirhinal cortex	16.4	14.5
	Entorhinal cortex	20.9	18.5
	Hippocampus	7.8	8.2
	Amygdala	12.5	14.4
	Parahippocampal cortex	22.3	15.0

and procedural knowledge used by the expert, present the advantages of automatic methods in terms of speed and precision, but they are only applicable under expert's supervision. The automatic parcelation of the cortical structures of the medial temporal lobe shows promising results, in spite of the fact that the method is in an early development stage.

Acknowledgments

The authors would like to thank to Ministry of Industry, Tourism and Trade for financial support via project TSI-020110-2009-362 (PLAN AVANZA 2009-2010).

References

1. Dale, A.M., Fischl, B., Sereno, M.I.: Cortical surface-based analysis I: Segmentation and surface reconstruction. Neuroimage 9, 179–194 (1999)
2. Fischl, B., Kouwe, A., Destrieux, C., Halgren, E., Segonne, F., Salat, D.H., Busa, E., Seidman, L.J., Goldstein, J., Kennedy, D., Caviness, V., Makris, N., Rosen, B., Dale, A.M.: Automatically Parcellating the Human Cerebral Cortex. Cerebral Cortex 14, 11–22 (2004)
3. Ibáñez, L., Schroeder, W., Ng, L., Cates, J., Consortium, T.I.S., Hamming, R.: The ITK Software Guide. Kitware, Inc. (January 2003)
4. Insausti, R., Juottonen, K., Soininen, H., Insausti, A., Partanen, K., Vainio, P., Laakso, M.P., Pitkanen, A. M.: VolumetricAnalysis of the Human Entorhinal, Perirhinal, and Temporopolar Cortices. AJNR Am. J. Neuroradiol. 19, 656–671 (1998)
5. Insausti, R., Insausti, A.M., Mansilla, F., Abizanda, P., Artacho-Pérula, E., Arroyo-Jimenez, M.M., Martinez-Marcos, A., Marcos, P.: The human parahippocampal gyrus. In: 3th Annual Meeting of Society for Neuroscience Anatomical and MRI correlates, New Orleans, USA (November 2003)

6. Insausti, R., Rincón, M., González-Moreno, C., Artacho-Pérula, E., Díez-Peña, A., García-Saiz, T.: Neurobiological significance of automatic segmentation: Application to the early diagnosis of alzheimer's disease. In: Mira, J., Ferrández, J.M., Álvarez, J.R., de la Paz, F., Toledo, F.J. (eds.) IWINAC 2009. LNCS, vol. 5602, pp. 134–141. Springer, Heidelberg (2009)
7. Juottonen, K., Laakso, M.P., Insausti, R., Lehtovirta, M., Pitkanen, A., Partanen, K., Soininen, H.: Volumes of the Entorhinal and Perirhinal Cortices in Alzheimer'sDisease. Neurobiology of Aging 19 (1998)
8. Klauschen, F., Goldman, A., Barra, V., Meyer-Lindenberg, A., Lundervold, A.: Evaluation of Automated Brain MR Image Segmentation and Volumetry Methods. Human Brain Mapping 30, 1310–1327 (2009)
9. Hu, Y.J., Grossberg, M.D., Mageras, G.S.: Semiautomatic medical image segmentation with adaptive local statistics in conditional random fields framework. In: 30th Annual International Conference of the IEEE Engineering in Medicine and Biology Society, pp. 3099–3102 (2008)
10. Nowinski, et al.: A New Presentation and Exploration of Human Cerebral Vasculature Correlated With Surface and Sectional. Neuroanatomy Anat. Sci. Ed. 2, 24–33 (2009)
11. Pichon, E., Tannenbaum, A., Kikinis, R.: A statistically based flow for image segmentation. Med. Image Anal. 8(3), 267–274 (2004)
12. Pieper, S., Lorensen, B., Schroeder, W., Kikinis, R.: The NA-MIC Kit: ITK, VTK, Pipelines, Grids and 3D Slicer as an Open Platform for the Medical Image Computing Community. In: Proceedings of the 3rd IEEE International Symposium on Biomedical Imaging: From Nano to Macro 2006, vol. 1, pp. 698–701 (2006)
13. Sánchez-Benavidesab, G., Gómez-Ansónc, B., Sainzd, A., Vivesd, Y., Delfinod, M., Peña-Casanova, J.: Manual validation of FreeSurfer's automated hippocampal segmentation in normal aging, mild cognitive impairment. Alzheimer Disease subjects 181(3), 219–225 (2010)
14. Scoville, W.B., Milner, B.: Loss of recent memory after bilateral hippocampal lesions. J. Neurol. Neurosurg. Psychiatry 20, 11–21 (1957)
15. Squire, L.R., Stark, C., Clark, R.E.: The medial temporal lobe. Annu. Rev. Neurosci. 27, 279–306 (2004)
16. Smith, S.M., Jenkinson, M., Woolrich, M.W., Beckmann, C.F., Behrens, T.E.J., Johansen- Berg, H., Bannister, P.R., De Luca, M., Drobnjak, I., Flitney, D.E., Niazy, R., Saunders, J., Vickers, J., Zhang, Y., De Stefano, N., Brady, J.M., Matthews, P.M.: Advances in functional and structural MR image analysis and implementation as FSL. Neuroimage 23, 208–219 (2004)

Analysis of Spect Brain Images Using Wilcoxon and Relative Entropy Criteria and Quadratic Multivariate Classifiers for the Diagnosis of Alzheimer's Disease

F.J. Martínez[1], D. Salas-González[1], J.M. Górriz[1], J. Ramírez[1],
C.G. Puntonet[2], and M. Gómez-Río[3]

[1] Dept. of Signal Theory, Networking and Communications,
University of Granada, Spain
[2] Dept. of Computers Architecture and Technology
[3] Virgen Nieves Hospital, Deptartment Nuclear Medicine, Granada, Spain

Abstract. This paper presents a computer aided diagnosis technique for improving the accuracy of the early diagnosis of the Alzheimer's disease. 97 SPECT brain images from the "Virgen de las Nieves" Hospital in Granada are studied. The proposed method is based on two different classifiers that use two different separability criteria and a posterior reduction of the feature dimension using factor analysis. Factor loadings are used as features of two multivariate classifiers with quadratic discriminant functions. The result of these two different classifiers is used to figure out the final decision. An accuracy rate up to 92.78% when NC and AD are considered is obtained using the proposed methodology.

1 Introduction

Alzheimer's disease (AD) is one of the most common cause of dementia in the elderly people, and affects approximately 30 millions individuals worldwide. With a growing older population in developed nations, AD affected people is expected to quadruple over the next 50 years [1].

AD early diagnosis remains being a difficult task. Furthermore, in this stage, there are more opportunities to treat the disease [11]. Single Photon Emission Computed Tomography (SPECT) is a widely used technique to study the functional properties of the brain. Clinicians usually evaluate these images via visual inspection, therefore an automatic computer aided diagnosis (CAD) tool is desirable [6].

Many efforts have been done lately in this direction [18,5,2,16,17], using different analysis and clustering methods. In this work, we choose a different approach, based on two different classifiers with two different feature selection criteria, and involving Factor Analysis and Multivariate Normal Classifiers.

The work is organised as follows: in Section 2 the image dataset, feature selection, feature extraction and the quadratic discriminant function used for classification purposes are presented. In Section 3, we summarize and discuss the

J.M. Ferrández et al. (Eds.): IWINAC 2011, Part II, LNCS 6687, pp. 41–48, 2011.

classification performance obtained applying the proposed methodology. Lastly, the conclusions are drawn in Section 4.

2   Material and Methods

To study and test the classification of SPECT brain images, we use a set of 97 images (41 Normal Controls (NC) and 56 Alzheimer's Disease (AD)) from the "Virgen de las Nieves" Hospital in Granada. Demographic information about the dataset can be found at Table 1.

Table 1. Demographic details of the dataset. NC = Normal Controls, AD 1 = possible AD, AD 2 = probable AD, AD 3 = certain AD. μ and σ stands for population mean and standard deviation respectively.

	#samples	Sex(M/F)(%)	μ[range/σ]
NC	43	32.95/12.19	71.51[46-85/7.99]
AD1	30	10.97/18.29	65.29[23-81/13.36]
AD2	20	13.41/9.76	65.73[46-86/8.25]
AD3	4	0/2.43	76[69-83/9.90]

In SPECT imaging, the dimension of the feature space (number of voxels) is very large compared to the number of available training samples (usually ~ 100 images). This scenario leads to the so-called small sample size problem [3], as the number of available samples is greater than the number of images. Therefore, a reduction in the dimension of the feature vector is desirable before to perform classification.

Each SPECT image has $67,200$ voxels ($48 \times 40 \times 35$) with intensity values ranging from 0 to 255. Some of them corresponds to positions outside the brain. We initially discard those voxels which present an intensity value lower than 70 ($< 27.5\%$ of maximum intensity value). Basically, positions outside the brain and some very low-intensity regions inside the brain are discarded.

Voxels are ranked using the absolute value obtained from Mann-Whitney-Wilcoxon test and Relative Entropy. Fig. 1 shows the brain image I^t with the absolute value of the Mann-Whitney statistic and Relative entropy in each voxel. In this example, Normals and AD images were considered in the calculation of the image I^t. Mann-Whitney-Wilcoxon test and Relative Entropy gives us information about voxel class separability. Note that the ranking criteria are different from Mann-Whitney to Relative Entropy. This aspect would help making a decision when building the classifiers.

Subsequently, we select a fixed number N of voxels from the ordered data. This N value is obtained by sweeping a wide range and analyzing the results in terms of accuracy, specificity and sensitivity. In this work, those selected voxels will be modeled using factor analysis.

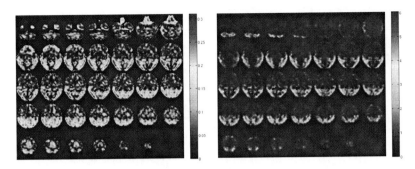

Fig. 1. *Left*: Mann-Whitney-Wilcoxon U values for each voxel. *Right*: Relative Entropy for each voxel.

After this selection, we have N selected voxels, but N is still too high (thousands of voxels) compared with the number of samples. So, we need another feature reduction algorithm that allow us prevent the small-size problem.

To do this, a number of key features (K) are extracted using Factor Analysis technique [9]. Factor analysis is a statistical method used to describe variability among observed variables in terms of fewer unobserved variables called factors. The K observed variables are modeled as linear combinations of the factors, plus error. We use factor analysis to reduce the feature dimension. Factor analysis estimates how much of the variability in the data is due to common factors.

Once the K factor loadings have been estimated, we rotate them using a varimax approach which is a change of coordinates that maximizes the sum of the variance of the squared loadings. This method attempts to maximize the variance on the new axes. Thus, we obtain a pattern of loadings on each factor that is as diverse as possible. These factor loadings will be used as features for classification purposes.

Finally, we try to classify the factor loadings into AD or NC. In this work, we use the multivariate normal classifier [12], which was selected among others options because of its good performance when number of features is small.

Suppose that $\mathbf{v}$ denotes a p-component random vector of observations made on any individual; $\mathbf{v}_0$ denotes a particular observed value of $\mathbf{v}$, and π_1, π_2 denote the two populations involved in the problem. The basic assumption is that $\mathbf{v}$ has different probability distributions in π_1 and π_2. Let the probability density of $\mathbf{v}$ be $f_1(\mathbf{v})$ in π_1, and $f_2(\mathbf{v})$ in π_2. The simplest intuitive argument, termed the likelihood ratio rule, classifies $\mathbf{v}_0$ as π_1 whenever it has greater probability of coming from π_1 than from π_2. This classification rule can be written as:

$$\mathbf{v}_0 \in \pi_1 \text{ if } f_1(\mathbf{v}_0)/f_2(\mathbf{v}_0) > 1 \tag{1}$$

$$\mathbf{v}_0 \in \pi_2 \text{ if } f_1(\mathbf{v}_0)/f_2(\mathbf{v}_0) \leq 1. \tag{2}$$

The most general form of the model is to assume that π_i is a multivariate normal population with mean μ_i and dispersion matrix Σ_i for $i = 1, 2$. Thus $f_i(\mathbf{v}) = (2\pi)^{-p/2} |\Sigma_i|^{-1/2} \exp\{\frac{1}{2}(\mathbf{v} - \mu_i)' \Sigma_i^{-1} (\mathbf{v} - \mu_i)\}$, so that we obtain

$$
\begin{aligned}
\frac{f_1(\mathbf{v})}{f_2(\mathbf{v})} &= |\Sigma_2|^{1/2} |\Sigma_1|^{-1/2} \exp[-\frac{1}{2}\{\mathbf{v}'(\Sigma_1^{-1} - \Sigma_2^{-1})\mathbf{v} \\
&\quad - 2\mathbf{v}'(\Sigma_1^{-1}\mu_1 - \Sigma_2^{-1}\mu_2) \\
&\quad + \mu_1'\Sigma_1^{-1}\mu_1 - \mu_2'\Sigma_2^{-1}\mu_2\}]
\end{aligned}
\tag{3}
$$

Hence, on taking logarithms in (1), we find that the classification rule for this model is: allocate $\mathbf{v}_0$ to π_1 if $Q(\mathbf{v}_0) > 0$, and otherwise to π_2, where $Q(\mathbf{v})$ is the discriminant function

$$
\begin{aligned}
Q(\mathbf{v}) &= \frac{1}{2}\log\{|\Sigma_2|/|\Sigma_1|\} - \frac{1}{2}\{\mathbf{v}'(\Sigma_1^{-1} - \Sigma_2^{-1})\mathbf{v} \\
&\quad - 2\mathbf{v}'(\Sigma_1^{-1}\mu_1 - \Sigma_2^{-1}\mu_2) + \mu_1'\Sigma_1^{-1}\mu_1 \\
&\quad - \mu_2'\Sigma_2^{-1}\mu_2\}.
\end{aligned}
\tag{4}
$$

Since the terms in $Q(\mathbf{v})$ include the quadratic form $\mathbf{v}'(\Sigma_1^{-1} - \Sigma_2^{-1})\mathbf{v}$, which will be a function of the squares of elements of $\mathbf{v}$ and cross-products between pairs of them, this discriminant function is known as the quadratic discriminant.

In any practical application, the parameters μ_1, μ_2, Σ_1 and Σ_2 are not known. Given two training sets, $\mathbf{v}_1^{(1)}, ..., \mathbf{v}_{n_1}^{(1)}$ from π_1, and $\mathbf{v}_1^{(2)}, ..., \mathbf{v}_{n_2}^{(2)}$ from π_2 we can estimate these parameters using the sample mean and sample standard deviation.

The performance of the classification is tested using the leave-one-out method: the classifier is trained with all but one images of the database and we categorize the remaining one, which is not used to define the classifier. In that way, all SPECT images are classified and accuracy, sensibility and specificity parameters are calculated. These measures are obtained from Eq. 5, 6 and 7.

$$
\text{Sensibility} = \frac{TP}{TP + FN}
\tag{5}
$$

$$
\text{Specificity} = \frac{TN}{TN + FP}
\tag{6}
$$

$$
\text{Accuracy} = \frac{TP + TN}{TP + FP + FN + TN}
\tag{7}
$$

where TN and TP means true negative and true positive respectively (number of correctly classified negative and positive samples), and FN and FP means false negative and false positive respectively (number of wrong classified negative and positive samples).

This cross-validation strategy has been used to assess the discriminative accuracy of different multivariate analysis methods applied to the early diagnosis of Alzheimer's disease [15,13,14,7], discrimination of frontotemporal dementia from AD [10] and in classifying atrophy patterns based on magnetic resonance imaging images [4].

3   Results and Discussion

Firstly, we study the behavior of the two different voxel selection criteria in function of the number of voxels selected, N. Fig. 2 shows that the two methods used have almost the same behavior in function of N, although Wilcoxon performs better when number of selected voxels is less than 13000 voxels and Relative Entropy is the best when using more than 15000 selected voxels. In this case, the best values are 6000 and 13000 for Wilcoxon (Classifier A) and $15000-17000$ for Relative Entropy (Classifier B).

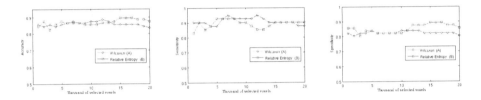

Fig. 2. Values of accuracy (left), sensibility (center) and specificity (right) versus N for each classifier

Fig. 2 depicts the sensibility and specificity versus the number of selected voxels. In this case, we see that in overall, the best values are similar to the values obtained for accuracy. Best performance results are obtained when $N = 6000$ voxels for classifier A and $N = 16000$ for classifier B.

When we look at the behavior in function of the number of factor loadings extracted in the Factor Analysis step, we found the pattern on Fig. 3, where we can see that the best values for accuracy are $K = 6$ for Classifier B and $K = 8$ for Classifier A.

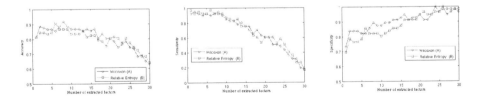

Fig. 3. Values of accuracy (left), sensibility (center) and specificity (right) versus K for each classifier

Fig. 3 shows that the best sensibility values are found at these values of K too. Furthermore, this figure shows that the specificity increases as the number of factors increase, but that it has local maximums in the same values of K that sensibility and accuracy, so we choose these values for the final model.

The final chosen model is composed of two different classifiers, with these variables:

- **Classifier A**: Mann-Whitney-Wilcoxon criterion for ranking and selecting the best 6000 voxels. Factor analysis to extract 8 factor loadings, used as input to a multivariate normal classifier .
- **Classifier B**: Relative Entropy for ranking and selecting the best 16000 voxels. Factor analysis to extract 6 factor loadings, used as input to a multivariate normal classifier.

The mean of these two results is taken as the final result, and with this method, we find the validation results given in Table 2. This table also shows the performance of some other methods used in Alzheimer's Disease Computer Aided Diagnosis [8,18,16], in order to see that the proposed method based on factor analysis and quadratic discriminative functions outperforms previous methods like Voxels-as-Features (VAF) [18], Principal Component Analysis (PCA) [16] or Gaussian Mixture Models (GMM) [8] in accuracy, sensitivity and specificity. This is due to the two different voxel selection criteria, which selects different areas in the brain that can be related to different patterns in AD patients.

Table 2. Results of the proposed quadratic classifier, obtained by leave-one-out method, and compared to some other methods proposed in bibliography: Voxel-as-Features (VAF) with linear SVM classifier, Principal Component Analysis (PCA) with linear SVM classifier and Gaussian Mixture Models (GMM) with Radial Basis Function SVM classifier.

	Quadratic	VAF-LSVM	PCA-LSVM	GMM-SVM
Accuracy	92.78%	83.51%	89.69%	89.28%
Sensibility	92.68%	83.93%	89.28%	89.29%
Specificity	92.86%	82.93%	90.24%	90.24%

4   Conclusions

In this work, an automatic procedure to classify SPECT brain images for the diagnosis of Alzheimers Disease is presented. The proposed methodology takes one normalized three-dimensional SPECT brain image input. Then, most discriminative voxels are selected by two different criteria. As different patients may show different patterns of AD, the use of these two different criteria let us to find different areas in the brain that otherwise would not be selected. Then, after a feature extraction step, the result of two different multivariate quadratic classifiers is combined to obtain the final result. The proposed methodology achieves better results than recent published articles for SPECT images, with sensibility, specificity and accuracy values greater than 90%.

Acknowledgement

This work was developed under a "Initiation Research Grant" funded by the University of Granada (Spain). It also was partly supported by the MICINN under the PETRI DENCLASES (PET2006-0253), TEC2008-02113, NAPOLEON (TEC2007-68030-C02-01) and HD2008-0029 projects and the Consejería de Innovación, Ciencia y Empresa (Junta de Andalucía, Spain) under the Excellence Projects P07-TIC-02566 and P09-TIC-4530.

References

1. Brookmeyer, R., Johnson, E., Ziegler-Graham, K., Arrighi, M.: Forecasting the global burden of alzheimer's disease. Alzheimer's & dementia 3, 186–191 (2007)
2. Chaves, R., Ramírez, J., Górriz, J.M., López, M., Salas-Gonzalez, D., Álvarez, I., Segovia, F.: Svm-based computer-aided diagnosis of the alzheimer's disease using t-test nmse feature selection with feature correlation weighting. Neuroscience Letters 461(3), 293–297 (2009)
3. Duin, R.P.W.: Classifiers in almost empty spaces. In: Proceedings 15th International Conference on Pattern Recognition, vol. 2, pp. 1–7. IEEE, Los Alamitos (2000)
4. Fan, Y., Batmanghelich, C., Clark, C., Davatzikos, C.: Spatial patterns of brain atrophy in mci patients, identified via high-dimensional pattern classification, predict subsequent cognitive decline. NeuroImage (2008)
5. Fung, G., Stoeckel, J.: Svm feature selection for classification of spect images of alzheimer's disease using spatial information. Knowledge and Information Systems 11(2), 243–258 (2007)
6. Goethals, I., van de Wiele, C., Slosman, D., Dierckx, R.: Brain spet perfusion in early alzheimer disease: where to look? European Journal of Nuclear Medicine 29(8), 975–978 (2002)
7. Górriz, J.M., Ramírez, J., Lassl, A., Salas-Gonzalez, D., Lang, E.W., Puntonet, C.G., Álvarez, I., López, M., Gómez-Río, M.: Automatic computer aided diagnosis tool using component-based svm. In: IEEE Nuclear Science Symposium Conference Record, Medical Imaging Conference, Dresden, Germany, pp. 4392–4395. IEEE-NSS (2008)
8. Górriz, J., Segovia, F., Ramírez, J., Lassl, A., Salas-Gonzalez, D.: Gmm based spect image classification for the diagnosis of alzheimer's disease. Applied Soft Computing (2010),
   http://www.sciencedirect.com/science/article/B6W86-0RVNF8-3/2/
   8783119887bbadbbb4e630523d075336 doi:10.1016/j.asoc.2010.08.012
9. Harman, H.H.: Modern Factor Analysis. University of Chicago Press, Chicago (1976)
10. Higdon, R., Foster, N.L., Koeppe, R.A., DeCarli, C.S., Jagust, W.J., Clark, C.M., Barbas, N.R., Arnold, S.E., Turner, R.S., Heidebrink, J.L., Minoshima, S.: A comparison of classification methods for differentiating fronto-temporal dementia from Alzheimer's disease using FDG-PET imaging. Statistics in Medicine 23, 315–326 (2004)
11. Johnson, E., Brookmeyer, R., Ziegler-Graham, K.: Modeling the effect of alzheimer's disease on mortality. The International Journal of Biostatistics 3(1), Article 13 (2007)

12. Krzanowski, W.J. (ed.): Principles of multivariate analysis: a user's perspective. Oxford University Press, New York (1988)
13. López, M., Ramírez, J., Górriz, J., Álvarez, I., Salas-González, D., Segovia, F., Chaves, R.: Svm-based cad system for early detection of the alzheimer's disease using kernel pca and lda. Neuroscience Letters 464, 233–238 (2009)
14. Ramírez, J., Górriz, J., Chaves, R., López, M., Salas-González, D., Álvarez, I., Segovia, F.: Spect image classification using random forests. Electronics Letters 45, 604–605 (2009)
15. Ramírez, J., Górriz, J., Salas-González, D., Romero, A., López, M., Ávarez, I., Gómez-Río, M.: Computer-aided diagnosis of alzheimer's type dementia combining support vector machines and discriminant set of features. Information Sciences (2009) (in Press) Corrected Proof,
    http://www.sciencedirect.com/science/article/B6V0C-4WDGCSR-1/2/
    13d69ec63ef1f8b72b26a4f55efeb55c
16. Ramírez, J., Górriz, J., Segovia, F., Chaves, R., Salas-González, D., López, M., Álvarez, I., Padilla, P.: Computer aided diagnosis system for the alzheimer's disease based on partial least squares and random forest spect image classification. Neuroscience Letters 472, 99–103 (2010)
17. Salas-Gonzalez, D., Górriz, J.M., Ramírez, J., López, M., Alvarez, I., Segovia, F., Chaves, R., Puntonet, C.G.: Computer aided diagnosis of alzheimer's disease using support vector machines and classification trees. Physics in Medicine and Biology 55(10), 2807–2817 (2010)
18. Stoeckel, J., Ayache, N., Malandain, G., Malick Koulibaly, P., Ebmeier, K.P., Darcourt, J.: Automatic classification of SPECT images of alzheimer's disease patients and control subjects. In: Barillot, C., Haynor, D.R., Hellier, P. (eds.) MICCAI 2004. LNCS, vol. 3217, pp. 654–662. Springer, Heidelberg (2004),
    http://www.springerlink.com/content/vrb0kua41ktjtl35/

MRI Brain Image Segmentation with Supervised SOM and Probability-Based Clustering Method

Andres Ortiz[1], Juan M. Gorriz[2], Javier Ramirez[2], and Diego Salas-Gonzalez[2]

[1] Communications Engineering Department
University of Malaga. 29004 Malaga, Spain
[2] Department of Signal Theory, Communications and Networking
University of Granada. 18060 Granada, Spain

Abstract. Nowadays, the improvements in Magnetic Resonance Imaging systems (MRI) provide new and aditional ways to diagnose some brain disorders such as schizophrenia or the Alzheimer's disease. One way to figure out these disorders from a MRI is through image segmentation. Image segmentation consist in partitioning an image into different regions. These regions determine diferent tissues present on the image. This results in a very interesting tool for neuroanatomical analyses. Thus, the diagnosis of some brain disorders can be figured out by analyzing the segmented image. In this paper we present a segmentation method based on a supervised version of the Self-Organizing Maps (SOM). Moreover, a probability-based clustering method is presented in order to improve the resolution of the segmented image. On the other hand, the comparisons with other methods carried out using the IBSR database, show that our method ourperforms other algorithms.

1 Introduction

Nowadays, Magnetic Resonance Imaging systems (MRI) provide an excellent spatial resolution as well as a high tissue contrast. Nevertheless, since actual MRI systems can obtain 16-bit depth images corresponding to 65535 gray levels, the human eye is not able to distinguish more than several tens of gray levels. On the other hand, MRI systems provide images as slices which compose the 3D volume. Thus, computer aided tools are necessary to exploit all the information contained in a MRI. These are becoming a very valuable tool for diagnosing some brain disorders such as the Alzheimer's disease [1]. Moreover, modern computers which contain a large amount of memory and several processing cores, have enough process capabilities for analyzing the MRI in reasonable time. Image segmentation consist in partitioning an image into different regions. In MRI, segmentation consist of partitioning the image into different neuroanatomical structures which corresponds to different tissues. Hence, analyzing the neuroanatomical structures and the distribution of the tissues on the image, brain disorders or anomalies can be figured out. Hence, the importance of having effective tools for grouping and recognizing different anatomical tissues, structures and fluids is growing with the improvement of the medical

J.M. Ferrández et al. (Eds.): IWINAC 2011, Part II, LNCS 6687, pp. 49–58, 2011.
© Springer-Verlag Berlin Heidelberg 2011

imaging systems. These tools are usually trained to recognize the three basic tissue classes found on a on a healthy brain MR image: white matter (WM), gray matter (GM) and cerebrospinal fluid (CSF). All of the non-recognized tissues or fluids are classified as suspect to be pathological. The segmentation process it can be performed in two ways. The first, consist of manual delimitation of the structures present within an image by an expert. The second consist of using an automatic segmentation technique. As commented before, computer image processing techniques allow exploiting all the information contained in a MRI. There are several automatic segmentation techniques. Some of them use the information contained in the image histogram [2, 3, 11]. This way, since different contrast areas should correspond with different tissues, the image histogram can be used for partitioning the image. In the ideal case, three different image intensities should be found in the histogram corresponding to Gray Matter (GM), White Matter (WM) and CerebroSpinal Fluid (CSF). Moreover, the resolution should be high enough that each voxel is composed by a single tissue type. Nevertheless, variations on the contrast of the same tissue are found in an image due to RF noise or shading effects due to magnetic field variations. These effects which affects to the tissue homogeneity on the image are a source of errors for automatic segmentation methods. Other methods use statistical classifiers based on the expectation-maximization algorithms (EM) [1, 5, 6], maximum likelihood (ML) estimation [7] or Markov random fields [8, 9]. Nevertheless, artificial intelligence based techniques, have been proved to be noise-tolerant [10–16] and provide promising results. Some of these methods are based on the Kohonen's Self-Organizing Maps (SOM) [12–15]. In this paper, we present a segmentation method based on a supervised version of SOM, which is trained by using one of the volumes present on the IBSR [14] repository. Moreover, multiobjective optimization is used for feature selection and a probability-based clustering method over the SOM aids to improve the segmentation process resolution. After this introduction, Section 2 describes the feature extraction and selection process, Section 3 introduces the supervised version of SOM, and Section 4 describes the use of supervised SOM and the probability-based clustering method. Finally Section 5 presents the results obtained by using the IBSR database [14] and Section 6 concludes this paper.

2   Supervised Self-Organizing Maps (SOM)

SOM [17] is one of the most used artificial neural network models for unsupervised learning. The main purpose of SOM is to group the similar data instances close in into a two or three dimensional lattice (output map). On the other hand, different data instances will be apart in the output map. SOM consist of a number or neurons also called units which are arranged following a previously determined lattice. During the training phase, the distance between an input vector and the weights associated to the units on the output map are calculated. Usually, the Euclidean distance is used as shown in Equation 1. Then, the unit closer to the input vector is referred as winning unit and the associated weight is

updated. Moreover, the weights of the units in the neighbor of the winning unit
are also updated as in Equation 2. The neighbor function defines the shape of the
neighborhood and usually, a Gaussian function which shrinks in each iteration
is used as shown in Equation 3. This deals with a competitive process in which
the winning neuron each iteration is called Best Matching Unit (BMU).

$$U_\omega(t) = argmin_i \|x(t) - \omega_i(t)\| \tag{1}$$

$$\omega_i(t+1) = \omega_i(t) + \alpha_i(t)h_{U_i}(t)\Big(x(t) - \omega_i(t)\Big) \tag{2}$$

$$h_{U_i}(t) = e^{-\frac{\|r_U - r_i\|}{2\sigma(t)^2}} \tag{3}$$

In Equation 3, r_i represents the position on the output space (2D or 3D) and
$\|r_U - r_i\|$ is the distance between the winning unit and the i-neuron on the
output space. On the other hand, $\sigma(t)$ controls the reduction of the Gaussian
neighborhood on each iteration. $\sigma(t)$ Usually takes the form of exponential decay
function as in Equation 4.

$$\sigma(t) = \sigma_0 e^{\left(\frac{-t}{\tau_1}\right)} \tag{4}$$

In the same way, the learning factor $\alpha(t)$ in Equation 2, also diminishes in time.
However, α may decay in a linear or exponential fashion. Unsupervised SOM are
frequently used for classification. Nevertheless, it does not use class information
in the training process. As a result, the performance with high-dimensional input
data highly depends on the specific features and the calculation of the clusters
borders may be not optimally defined. Therefore, we used a supervised version of
the SOM, by adding an output layer composed by four neurons (one per class).
This architecture is shown in Fig. 2. In this structure, each weight vector ω_{ij} on
the SOM is connected to each neuron on the output layer y_k.

After all the input vectors have been presented to the SOM layer, some of the
units remain unlabeled. At his point, a probability-based relabeling method is

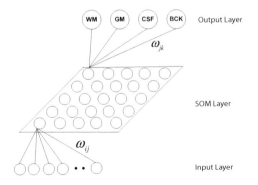

Fig. 1. Architecture of the supervised SOM

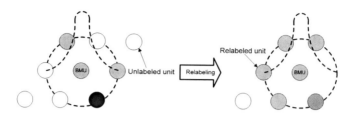

Fig. 2. SOM layer relabeling method

applied by using a 2D Gaussian kernel centered in each BMU. Thus, a majority-voting scheme with the units inside the Gaussian kernel is used to relabel the unlabeled units.

In Equation 5, the Gaussian kernel used to estimate the label for unlabeled units is shown. In this equation, σ determines the width of the Gaussian kernel. In other words, it is the neighborhood taken into account for the relabeling process. On the other hand, (x, y) is the position of the BMU in the SOM grid.

$$L(x, y, \sigma) = \frac{1}{2\pi\sigma^2} e^{-\frac{(x^2+y^2)}{2\sigma^2}} \tag{5}$$

Regarding the calculation of the quality of the output map, there exist two measures. The first is the quantization error, which is a measure of the resolution of the map. This can be calculated by computing the average distance between all the BMUs and the input data vectors. There is another measure of the goodness of the SOM. This measure is the topographic error, which measures how the SOM preserves the topology. This error can be computed with the Equation 6, where N is the total number of input vectors and $\bar{u}(x_i)$ is 1 if first and second BMU for the input vector $\bar{x}(i)$ are adjacent units (0 otherwise) [5].

$$t_e = \frac{1}{N} \sum_{i=1}^{N} u(\bar{x}_i) \tag{6}$$

Then, the lower q_e and q_t, the better the SOM is adapted to the input patterns.

3   Feature Extraction and Selection for MRI Segmentation

The segmentation method we have implemented consist of four stages as shown in Fig. 3.

Fig. 3. Feature extraction and selection for training the SOM classifier

Since we deal with a supervised version of SOM classifier, we use one of the volumes on the IBSR repository [14]. The segmented volume on the IBSR repository is considered as ground truth and therefore, a reference for SOM training. The first for training the SOM which will be used for segmenting further images consist of splitting the image by using overlapping windows. There are works which deals with the influence of the window size with image processing [21]. However, after several trials, we chose square windows of 7x7 pixels. As we only use square windows, the size is referred to the dimension of one side or $w = 7$. As a result of this preprocessing stage, we obtain a matrix where the number of rows is the number of the center pixels on each window and another matrix which stores the coordinates of the central pixel of each window. This way, first and second order features as well as moment invariants are computed from the window data. First order features are the intensity of the central pixel, and mean and variance of the intensity on each window. On the other hand, second order features are computed by using the Gray Level Co-Occurrence Matrix (GCLM) method for calculating the 14 textural features proposed by Haralick et al. in [18]. Having into account the 17 computed features, we compose the input vectors for the SOM classifier. As a result we have a feature vector which dimension is equal to the number of extracted features. These features will be the input to the classifier which will work with the feature space ($\mathbb{R}^{17}$). Hence they have to describe the image and if possible, not to content redundant information. Thus, the feature vector has to be enough different from one segment to another for the classifier. Thus, the features have to be properly selected in order to keep only the more discriminant. However, selecting a set of features is not a trivial task since it varies from one image to another and the feature extraction process plays a decisive role in the segmentation performance. In order to select the more discriminant features we use multiobjective optimization. This way, we keep only the features which maximize the SOM performance. As commented in Section 1, the quality of the SOM can be evaluated by two measurements. The first is the quantization error, q_e, and the second is the topographical error t_e. Hence, the lower the quantization and topographical errors, the better the quality of the SOM. Thus, the features can be selected in order to provide the lower qe and te quantities. This can be accomplished by using evolutive computing multiobjective optimization. In order to minimizing both, qe and te, the multiobjective optimization problem can be reduced to a single objective problem by minimizing the function shown in Equation 7.

$$F_{qt} = \left(\frac{q_t}{2} + \frac{q_e}{2} \right) \tag{7}$$

As a result of the selection process, we obtain a feature set which depends on the plane the segmentation is carried out.

4   Segmentation Process

Once the SOM has been trained with one of the volumes on the IBSR database and using the features selected by the algorithm described in Section 2, we

Fig. 4. Feature Selection and SOM training

proceed to segment new images. These new images are also taken from the IBSR database in order to be able to compare with the ground truth segmentation. For segmenting a new image, the first step is to extract windows of size w as commented in Section 2. Then, the optimal set of features calculated in the feature selection stage are extracted from each pixel, depending on the image plane (coronal, sagital or axial). After presenting the computed feature vectors to the architecture in Fig. 2, each pixel is determined to belong to a tissue class (White Matter, Gray Matter, Cerebrospinal Fluid or Background).

5   Experimental Results

In this Section, we present the segmentation results obtained with several images from the IBSR database. Moreover, the Jaccard/Tanimoto [14] coefficient has been calculated in order to compare the overlap average metric with other segmentation protocols.

These results are available from the IBSR web. Thus, in Fig. 5 segmentation results for the axial plane and the ground truth segmentation for visual comparison. Moreover, in Fig. 6, results for the coronal plane are shown.

In Fig. 5 and Fig. 6, the segmentation results can be compared with the ground truth. Nevertheless, the use of an index which measure the similarity between each segment and its ground truth is convenient. This way we use the Average Overlap Metric, which is calculated from the Jaccard distance. Jaccard distance is a commonly used overlap measurement between two sets. This can be defined as shown in Equation 8. In this equation, $T_{P,AB}$ represents the number of pixels which are positive for a specific tissue in the set A (image A) and negative for the other (image B), $T_{P,BA}$ represents the number of pixels which are positive for the same tissue in the set B (image B) and positive for the other (image B). T_P is the number of pixels which are positive for the tissue for both, A and B images. However, the more commonly used similarity index is the Jaccard/Tanimoto coefficient. This is defined as $1 - J_d(A, B)$.

Fig. 5. Pixel Classification Process

(a) WM   (b) GM   (c) CSF
                       Ground Truth

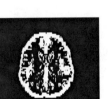 

(d) WM   (e) GM   (f) CSF
                    Segmentation Result

Fig. 6. Segmentation results and ground truth for the sagital plane, volume 100_23, slice 128:30:166. (a) White Matter, (b) Gray Matter and (c) Cerebrospinal Fluid.

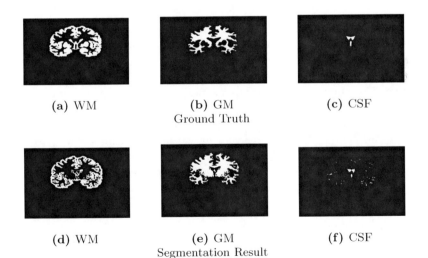

(a) WM   (b) GM   (c) CSF
                       Ground Truth

(d) WM   (e) GM   (f) CSF
                    Segmentation Result

Fig. 7. Segmentation results and ground truth for the coronal plane, volume 100_23, slice 128:30:166. (a) White Matter, (b) Gray Matter and (c) Cerebrospinal Fluid.

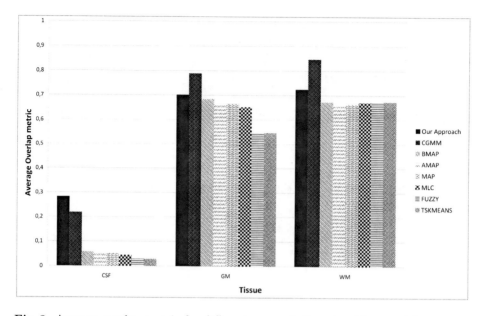

Fig. 8. Average overlap metric for different segmentation algorithms and for our approach for the 100_23 volume (Data from the IBSR web site)

$$J_d(A, B) = \frac{T_{P,AB} + T_{P,BA}}{T_P + T_{P,AB} + T_{P,BA}} \qquad (8)$$

This similarity measurement is widely used in the literature. Moreover, the IBSR web provides this data for different segmentation algorithms. Thus, we calculate the Jaccard coefficient for each pair of images, the ground truth segment and the automatically segmented by our algorithm.

In Fig. 7 we present the overlap average metric for the volume 100_23 on the IBSR database. At the same time, the overlap average comparison with other segmentation algorithms provided by the IBSR web site is depicted. As shown in Fig. 7, our proposal outperforms the segmentation techniques found in the IBSR database. Although there is a new segmentation proposal [19] which provides higher values for the overlap average metric, our algorithm outperforms it when classifying CSF pixels.

6   Conclusions

In this paper, we presented a segmentation method based on four basic stages: feature extraction, feature classification, training and segmentation. In order to select the more discriminant features which maximize the classifier performance, we use multiobjective optimization. The selected set of features is used to train a classifier consisting on a supervised version of SOM which uses an associative layer to build an effective classifier. Once the system has been trained with a percentage of the pixels present on a volume, it may classify pixels belonging to other volumes. The experimental results performed with volumes from the IBSR database shows that our proposal outperforms the segmentation algorithms published in the IBSR web page. However, there are more recent proposals which provide better results for WM and GM, but our algorithm clearly outperforms it when classifying CSF pixels.

References

1. Kapur, T., Grimson, W., Wells, I., Kikinis, R.: Segmentation of brain tissue from magnetic resonance images. Medical Image Analysis 1(2), 109–127 (1996)
2. Kennedy, D., Filipek, P., Caviness, V.: Anatomic segmentation and volumetric calculations in nuclear magnetic resonance imaging. IEEE Transactions on Medical Imaging 8(1), 1–7 (1989)
3. Smith, S., Brady, M., Zhang, Y.: Segmentation of brain images through a hidden Markov Random Field Model and the Expectation-Maximization Algorithm. IEEE Transactions on Medical Imaging 20(1) (2001)
4. Yang, Z., Laaksonen, J.: Interactive Retrieval in Facial Image Database Using Self-Organizing Maps. In: MVA (2005)
5. Tsai, Y., Chiang, I., Lee, Y., Liao, C., Wang, K.: Automatic MRI Meningioma Segmentation Using Estimation Maximization. In: Proceedings of the 27th IEEE Engineering in Medicine and Biology Annual Conference (2005)
6. Xie, J., Tsui, H.: Image Segmentation based on maximum-likelihood estimation and optimum entropy distribution (MLE-OED). Pattern Recognition Letters 25, 1133–1141 (2005)
7. Smith, S., Brady, M., Zhang, Y.: Segmentation of brain images through a hidden Markov Random Field Model and the Expectation-Maximization Algorithm. IEEE Transactions on Medical Imaging 20(1) (2001)
8. Wells, W., Grimson, W., Kikinis, R., Jolesz, F.: Adaptive segmentation of MRI data. IEEE Transactions on Medical Imaging 15(4), 429–442 (1996)
9. Mohamed, N., Ahmed, M., Farag, A.: Modified fuzzy c-mean in medical image segmentation. In: IEEE International Conference on Acoustics, Speech and Signal Processing (1999)
10. Parra, C., Iftekharuddin, K., Kozma, R.: Automated Brain Tumor Segmentation and Pattern recognition using AAN. In: Computational Intelligence, Robotics and Autonomous Systems
11. Sahoo, P., Soltani, S., Wong, A., Chen, Y.: A survey of thresholding techniques. Computer Vision, Graphics Image Process. 41, 233–260
12. Yang, Z., Laaksonen, J.: Interactive Retrieval in Facial Image Database Using Self-Organizing Maps. In: MVA (2005)
13. Güler, I., Demirhan, A., Karakis, R.: Interpretation of MR images using self-organizing maps and knowledge-based expert systems. Digital Signal Processing 19, 668–677 (2009)
14. Ong, S., Yeo, N., Lee, K., Venkatesh, Y., Cao, D.: Segmentation of color images using a two-stage self-organizing network. Image and Vision Computing 20, 279–289 (2002)
15. Alirezaie, J., Jernigan, M., Nahmias, C.: Automatic segmentation of cerebral MR images using artificial neural Networks. IEEE Transactions on Nuclear Science 45(4), 2174–2182 (1998)

16. Sun, W.: Segmentation method of MRI using fuzzy Gaussian basis neural network. Neural Information Processing 8(2), 19–24 (2005)
17. Fan, L., Tian, D.: A brain MR images segmentation method based on SOM neural network. In: IEEE International Conference on Bioinformatics and Biomedical Engineering (2007)
18. Kohonen, T.: Self-Organizing Maps, 3rd edn. Springer, Heidelberg (2001)
19. Haralick, R.M., Shanmugam, K., Dinstein, I.: Textural features for image classification. IEEE Transactions on Systems and Cybernet. 6, 610–621 (1973)
20. Greenspan, H., Ruf, A., Goldberger, J.: Constrained Gaussian Mixture Model Framework for Automatic Segmentation of MR Brain Images. IEEE Transactions on Medical Imaging 25(10), 1233–1245
21. Hodgson, M.E.: What Size Window for Image Classification? A Cognitive Perspective. Photogrammetric Engineering & Remote Sensing. American Society for Photogrammetry and Remote Sensing 64(8), 797–807

Effective Diagnosis of Alzheimer's Disease by Means of Distance Metric Learning and Random Forest

R. Chaves, J. Ramírez, J.M. Górriz, I. Illán, F. Segovia, and A. Olivares

University of Granada, Periodista Daniel Saucedo Aranda s/n,
18071, Granada, Spain
{rosach,javierrp,gorriz,illan
fsegovia,aolivares}@ugr.es
http://sipba.ugr.es/

Abstract. In this paper we present a novel classification method of SPECT images for the development of a computer aided diagnosis (CAD) system aiming to improve the early detection of the Alzheimer's Disease (AD). The system combines firstly template-based normalized mean square error (NMSE) features of tridimensional Regions of Interest (ROIs) t-test selected with secondly Kernel Principal Components Analysis (KPCA) to find the main features. Thirdly, aiming to separate examples from different classes (Controls and ATD) by a Large Margin Nearest Neighbors technique (LMNN), distance metric learning methods namely Mahalanobis and Euclidean distances are used. Moreover, the proposed system evaluates Random Forests (RF) classifier, yielding a 98.97% AD diagnosis accuracy, which reports clear improvements over existing techniques, for instance the Principal Component Analysis(PCA) or Normalized Minimum Squared Error (NMSE) evaluated with RF.

Keywords: SPECT Brain Imaging, Alzheimer's disease, Distance Metric Learning, Kernel Principal Components Analysis, Random Forest, Support Vector Machines.

1 Introduction

Alzheimer's Disease (AD) is the most common cause of dementia in the elderly and affects approximately 30 million individuals worldwide [2]. Its prevalence is expected to triple over the next 50 years due to the growth of the older population. To date there is no single test or biomarker that can predict whether a particular person will develop the disease. With the advent of several effective treatments of AD symptoms, current consensus statements have emphasized the need for early recognition.

SPECT (Single Positron Emission Computed Tomography) is a widely used technique to study the functional properties of the brain [3]. After the reconstruction and a proper normalization of the SPECT raw data, taken with Tc-99m

J.M. Ferrández et al. (Eds.): IWINAC 2011, Part II, LNCS 6687, pp. 59–67, 2011.

ethyl cysteinate dimer (ECD) as a tracer, one obtains an activation map displaying the local intensity of the regional cerebral blood flow (rCBF). Therefore, this technique is particularly applicable for the diagnosis of neuro-degenerative diseases like AD.

In order to improve the prediction accuracy especially in the early stage of the disease, when the patient could benefit most from drugs and treatments, computer aided diagnosis (CAD) tools are desirable. At this stage in the development of CAD systems, the main goal is to reproduce the knowledge of medical experts in the evaluation of a complete image database, i.e. distinguishing AD patients from controls, thus errors from single observer evaluation are avoided achieving a method for assisting the identification of early signs of AD.

In the context of *supervised* multivariate approaches, the classification is usually done by defining feature vectors representing the different SPECT images and training a classifier with a given set of known samples [4]. Firstly, the feature extraction consists of a combination of 3D cubic NMSE (Normalized Minimum Square Error) features over regions of interest (ROIs) which are inside a 3D mask and that are selected by a t-test with feature correlation weighting. Secondly, kernel principal component analysis (KPCA) is applied on the previous step extracted features as dimension reduction to a lower subspace.

Thirdly, we obtain a family of metrics over the KPCA eigenbrains by computing distance metric learning analysis. The goal of distance metric learning can be stated in two ways: to learn a linear transformation L, also called Euclidean Distance or to learn a Mahalanobis metric $M = L^T \cdot L$. In this paper, we show how to learn these metrics for kNN classification [5] being aimed at the organisation of the k-nearest neighbors to the same class, while examples from different classes are separated by a large margin. This technique is called *large margin nearest neighbor* (LMNN) classification.

Regarding the classifiers, Random Forest (RF) [6] was used. It is a tree-based method consisting of bootstrap aggregating for its training. Besides, the results are combined through a voting process in which the output is determined by a majority vote of the trees.

2   Matherial and Methods

2.1   Subjects and Preprocessing

Baseline SPECT data from 97 participants were collected from the Virgen de las Nieves hospital in Granada (Spain). The patients were injected with a gamma emitting ^{99m}Tc-ECD radiopharmeceutical and the SPECT raw data was acquired by a three head gamma camera Picker Prism 3000. A total of 180 projections were taken with a 2-degree angular resolution. The images of the brain cross sections were reconstructed from the projection data using the filtered backprojection (FBP) algorithm in combination with a Butterworth noise removal filter. The SPECT images are first spatially normalized using the SPM software, in order to ensure that voxels in different images refer to the same anatomical positions in the brain allowing us to compare the voxel intensities of different subjects. Then we normalize the intensities of the SPECT images with a method similar to [7]. After the spatial normalization, one obtains a $95 \times 69 \times 79$ voxel representation of each subject, where each voxel represents a brain volume of $2 \times 2 \times 2\,\mathrm{mm}^3$. The SPECT images were visually classified by experts of the Virgen de las Nieves hospital using 4 different labels to distinguish between different levels of the presence of typical characteristics for AD. The database consists of 43 NOR, 30 AD1, 20 AD2 and 4 AD3 patients.

3   Feature Extraction

In this article, we propose to apply a combination of VAF (Voxels as Features), NMSE and Kernel PCA. First of all, controls are averaged in a tridimensional image ($sm(x,y,z)$ as it is shown in figure 1.

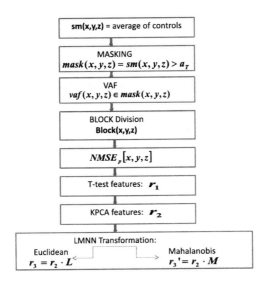

Fig. 1. Feature Extraction Process

In functional imaging, each voxel carries a grey level intensity $I(\mathbf{x}_j)$, which is related to the regional cerebral blood flow, glucose metabolism, etc. in the brain of a patient, depending on the image acquisition modality. Secondly, it is obtained a 3D $mask(x,y,z)$ that consists of all the voxels with $sm(x,y,z) > a_T$. The threshold a_T is equivalent to the 50% of the maximum Intensity in $sm(x,y,z)$.

Baseline VAF is a way of including in $vaf(x,y,z)$ all the voxels inside the obtained $mask(x,y,z)$ and considering them as features. Therefore, voxels outside the brain and poorly activated regions are excluded from this analysis. In this way, no explicit knowledge about the disease is needed, avoiding the inclusion of a priori information about the pathology into the system.

Each SPECT image is following divided into 3D $v \times v \times v$ cubic voxels defining Regions of Interest (ROIs), or **block(x,y,z)** centered in (x,y,z) coordinates which are inside of **vaf(x,y,z)**, as it is referenced in figure 1. Then, it is calculated the Normalized Minimum Squared Error or $NMSE_p$(x,y,z) for each subject and block. NMSE is given by equation 1 as:

$$NMSE_p(x,y,z) = \frac{\sum_{l,m,n=-v}^{v} [f(x-l, y-m, z-n) - g_p(x-l, y-m, z-n)]^2}{\sum_{l,m,n=-v}^{v} [f(x-l, y-m, z-n)]^2}$$

(1)

where f(x,y,z) is the mean voxel intensity at (x, y, z), and g_p(x, y, z) is the voxel intensity of the p subject. The most discriminant ROIs (r_1 in figure 1) are obtained with an absolute value two-sample t-test-based with pooled variance estimate on NMSE features as in [8]. The latter method, PCA is a multivariate approach often used in neuroimaging to significantly reduce the original high-dimensional space of the original brain images to a lower dimensional subspace [9] called r_2 in figure 1. PCA aims to find the projection directions that maximize the variance of a subspace, which is equivalent to finding the eigenvalues from the covariance matrix. PCA can be used in combination with the so-called kernel methods [10]. The basic idea of the kernel PCA method is to first preprocess the data by some non-linear mapping and then to apply the same linear PCA. In kernel PCA (KPCA) [10], each vector **x** is projected from the input space, $\mathbb{R}^n$, to a high-dimensional feature space, $\mathbb{R}^f$, by a non-linear mapping function: $\phi : \mathbb{R}^n \longrightarrow \mathbb{R}^f$ $f > n$. Note that the dimensionality of the feature space can be arbitrarily large. When we use the KPCA trick framework, the original LMNN can be immediately used as Kernel LMNN (KLMNN) as it is explained in [11].

3.1   Large Margin Nearest Neighbors (LMNN)

A training set of n labeled examples is denoted $\{(\boldsymbol{x_i}, y_i)\}$ $i = 1, ..., n$, with inputs $\boldsymbol{x_i} \epsilon \mathbb{R}^d$ and class labels y_i. We obtain a family of metrics over the training set by computing Euclidean distances after learning a linear transformation $\mathbf{L} : \mathbb{R}^d \to \mathbb{R}^d$ that optimizes kNN classification when squared distances are computed in this way:

$$D_L(\boldsymbol{x_i}, \boldsymbol{x_j}) = \|L(\boldsymbol{x_i} - \boldsymbol{x_j})\|^2$$

(2)

This equation 2 is commonly used to express squared distances in terms of the squared matrix. Using the eigenvalue decomposition, M can be decomposed into $M = L \cdot L^T$ in which $M \epsilon \mathbb{R}^*$ where $\mathbb{R}^*$ is $n \times n$ dimensional and positively semi-definite. Given two data points $x_1 \in \mathbb{R}^n$ and $x_2 \in \mathbb{R}^n$, the squared distances are denoted as Mahalanobis metrics and can be calculated in terms of M as follows [12]:

$$d_M(x_1, x_2) = \sqrt{(x_1 - x_2)^T \cdot M \cdot (x_1 - x_2)}$$

(3)

A Mahalanobis distance can be parametrized in terms of the matrix $\mathbf{L}$ or the matriz $\mathbf{M}$. The first is unconstrained, whereas the second must be positive semidefinite.

There is a cost function [5] that favors distance metric in which different labeled inputs maintain a large margin of distance, that is, the first term acts to *pull* target neighbors closer together penalizing large distances between each input and its target neighbors. Whereas, the other one acts to *push* differently labeled examples further apart. It penalizes small distances between differently labeled examples. Finally, we combine the two terms $\epsilon_{pull}(L)$ and $\epsilon_{push}(L)$ into a single loss function for distance metric learning. The two terms can have competing effects, to attract target neighbors on one hand and to repel impostors on the other. A weighting parameter $\mu\epsilon$ [0,1] that is selected 0.5 since it works well on practice, balances these goals.

$$\varepsilon(L) = (1 - \mu) \cdot \varepsilon_{pull} + \mu \cdot \varepsilon_{push}(L) \tag{4}$$

Large Margin principle [13] was used to design classification algorithms. In this work, its property of large margin separation was used as a previous step to the classification. In particular, matrices L and M representing both Euclidean and Mahalanobis distances respectively are multiplied by kernel pca features to obtain the new features r_3 and r'_3 which perform better results (see Experimental Section).

4   Random Forest Classification

Various ensemble classification methods have been proposed in recent years for improved classification accuracy [6]. In ensemble classification, several classifiers are trained and their results are combined through a voting process. Perhaps, the most widely used such methods are boosting and bagging. Boosting is based on sample re-weighting but bagging uses bootstrapping. The RF classifier [6] uses bagging, or bootstrap aggregating, to form an ensemble of classification and regression tree (CART)-like classifiers $h(x, T_k)$, k=1,..., where the Tk is bootstrap replica obtained by randomly selecting N observations out of N with replacement, where N is the dataset size, and x is an input pattern. For classification, each tree in the RF casts a unit vote for the most popular class at input x. The output of the classifier is determined by a majority vote of the trees. This method is not sensitive to noise or overtraining, as the resampling is not based on weighting. Furthermore, it is computationally more efficient than methods based on boosting and somewhat better than simple bagging [14].

5   Results

Several experiments were conducted to evaluate the combination of VAF, NMSE and PCA feature extraction aiming to posterior Distance Metric Learning to extend the margin between classes. The performance of the combination of Large Margin Nearest Neighbor-based feature extraction with Random Forest classifier was evaluated in depth as a tool for the early detection of the AD by means of Accuracy (Acc), Sensitivity (Sen) and Specificity (Spe), which were estimed by

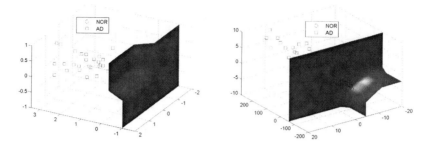

Fig. 2. Decision surfaces for Random Forest Classifier with LMNN methods using a)Euclidean (to the left) and b)Mahalanobis (to the right) distances

the kFold cross-validation. Firstly, it was experimentally proven that after the application of the tridimensional mask (sm), the best results were obtained with $r_1=100$ NMSE features in blocks of $5 \times 5 \times 5$ and t-test selected which were reduced to $r_2=10$ components using Gaussian Kernel PCA. A higher number of PCA components may lead to an increase in the computational cost of the CAD system, and a penalty in the Acc rate because no-discriminant features are being used too. Secondly, the LMNN-based space transformation is made, obtaining r_3 or r_3' depending respectively on the application of Euclidean or Mahalanobis techniques respectively.

As RF classification is concerned, the experiments considered both an increasing number of trees for a fixed number of folds (10) and also an increasing number of folds for a fixed number of trees which performed the best results for euclidean distance (13 trees) and mahalanobis distance (23 trees) as it is shown in table 1. Sen and Spe are defined as:

$$\text{Sensitivity} = \frac{TP}{TP+FN}; \quad \text{Specificity} = \frac{TN}{TN+FP}$$

respectively, where TP is the number of true positives: number of AD patients correctly classified; TN is the number of true negatives: number of controls correctly classified; FP is the number of false positives: number of controls classified as AD patient; FN is the number of false negatives: number of AD patients classified as control. Figure 2 shows decision surfaces of RF classifier when Euclidean and Mahalanobis distances are used. It is observed that when Mahalanobis space transformation is used, the separation between classes is higher than with Euclidean classes. In other words, this first is an effective way to obtain a large margin between classes NOR and AD.

In table 1, the performance of the Random Forest CAD system was further evaluated for the early detection of AD. The experiments considered an increasing number of features for designing the classifier using LMNN with both Euclidean and Mahalanobis distances. Performance in terms of Accuracy(%) was demonstrated to be independent of the number of trees. On the contrary, the fact of using Mahalanobis distance for the transformation space in the feature

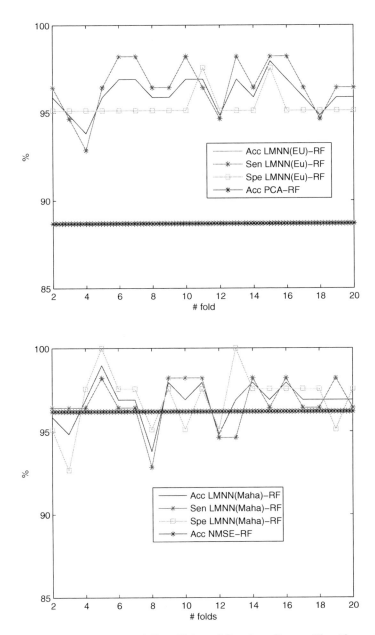

Fig. 3. Accuracy, Sensitivity and Specificity of Random Forest Classifier using in the feature extraction 100 NMSE features, Gaussian kernel PCA and LMNN methods with a)Euclidean (up) and b)Mahalanobis (down) distances for an increasing number of folds. Comparision with other reported methods: PCA-Random Forest compared with this work method using Euclidean distance and NMSE-Random Forest compared with this work method using Mahalanobis distance.

Table 1. Accuracy (%) with Random Forest classifier for an increasing number of trees using Euclidean and Mahalanobis learning distance metrics respectively

N trees	Euclidean	Mahalanobis
3	94.84	97.94
5	95.88	97.94
7	93.81	96.91
9	96.91	96.91
11	96.91	93.81
13	97.94	95.12
15	95.88	96.91
17	94.85	95.88
19	93.81	95.88
21	95.88	94.64
23	95.88	98.97
25	94.85	98.21

extraction reported better results in the majority of cases respectly to the Euclidean distance or linear transformation. In particular, the best results of Accuracy for Euclidean metric were obtained at 13 trees (97.94%) and for Mahalanobis at 23 trees (98.97%). In these experiments, 10-fold cross-validation was used since it is the most commonly used value.

The strategy to evaluate the method consists of differing, in terms of Acc, Sen, Spe the method which best discriminate Normal and AD subjects by means of the distance metric learning, that is, the highest margin which separates the two considered classes. In Figure 3, it is shown the Acc, Sen and Spe of Euclidean and Mahanobis metrics and other reported method such as PCA-Random Forest compared with this work method using Euclidean distance and as NMSE-Random Forest compared with this work method using Mahalanobis distance. In both comparisions, this work LMNN based, outperformed recent developed AD-CAD systems, yielding excellent results of classification. The best results of this LMNN-based Random Forest Classifier (Acc=98.97%, Sen=98.21% and Spe=100%) were reached at 5 folds for $r'_3=10$ Mahalanobis features.

6 Conclusions

Kernel Distance Metric Learning Methods were investigated for SPECT images Random Forest classified for the early AD's diagnosis, obtaining results of Acc of 98.97%, Sen of 98.21% and Spe of 100% for a LMNN (Mahalanobis)-based method, outperforming other recently reported methods including the NMSE-Random Forest or PCA-Random Forest. As regards feature extraction, the combination of NMSE t-test selection and then gaussian kernel PCA helped reach the best results of classification. It is remarkable that with the increasing number of trees selected in Random Forest there was no convergence but this independence favoured how to select the most adecuate number of trees to obtain excellent results of classification.

References

1. Evans, D., Funkenstein, H., Albert, M., Scherr, P., Cook, N., Chown, M., Hebert, L., Hennekens, C., Taylor, J.: Prevalence of Alzheimer's disease in a Community Population of older persons. Journal of the American Medical Association 262(18), 2551 (1989)
2. Petrella, J.R., Coleman, R.E., Doraiswamy, P.M.: Neuroimaging and Early Diagnosis of Alzheimer's Disease: A Look to the Future. Radiology 226, 315–336 (2003)
3. English, R.J., Childs, J.: SPECT: Single-Photon Emission Computed Tomography: A Primer. Society of Nuclear Medicine (1996)
4. Fung, G., Stoeckel, J.: SVM feature selection for classification of SPECT images of Alzheimer's disease using spatial information. Knowledge and Information Systems 11(2), 243–258 (2007)
5. Weinberger, K.Q., Blitzer, J., Saul, L.K.: Distance Metric Learning for Large Margin Nearest Neighbor Classification. Journal of Machine Learning Research 10, 207–244 (2009)
6. Breiman, L.: Random Forests. Machine Learning 45(1), 5–32 (2001)
7. Saxena, P., Pavel, D.G., Quintana, J.C., Horwitz, B.: An automatic threshold-based scaling method for enhancing the usefulness of tc-HMPAO SPECT in the diagnosis of alzheimer#146s disease. In: Wells, W.M., Colchester, A.C.F., Delp, S.L. (eds.) MICCAI 1998. LNCS, vol. 1496, pp. 623–630. Springer, Heidelberg (1998)
8. Chaves, R., Ramírez, J., Górriz, J.M., López, M., Salas-Gonzalez, D., Alvarez, I., Segovia, F.: SVM-based computer-aided diagnosis of the Alzheimer's disease using t-test NMSE feature selection with feature correlation weighting. Neuroscience Letters 461, 293–297 (2009)
9. Andersen, A.H., Gash, D.M., Avison, M.J.: Principal component analysis of the dynamic response measured by fMRI: a generalized linear systems framework. Journal of Magnetic Resonance Imaging 17, 795–815 (1999)
10. López, M., Ramírez, J., Górriz, J.M., Alvarez, I., Salas-Gonzalez, D., Segovia, F., Chaves, R.: SVM-based CAD system for early detection of the Alzheimer's disease using kernel PCA and LDA. Neuroscience Letters 464(3), 233–238 (2009)
11. Chatpatanasiri, R., Korsrilabutr, T., Tangchanachaianan, P., Kijsirikul, B.: A new kernelization framework for Mahalanobis distance learning algorithms. Neurocomputing 73, 1570–1579 (2010)
12. Xiang, S., Nie, F., Zhang, C.: Learning a Mahalanobis distance metric for data clustering and classification. Pattern Recognition 41, 3600–3612 (2008)
13. Tsochantaridis, I., Joachims, T., Hofmann, T., Altun, Y.: Large Margin Methods for Structured and Interdependent Output Variables. Journal of Machine Learning Research 6, 1453–1484 (2005)
14. Ramírez, J., Górriz, J.M., Chaves, R., López, M., Salas-Gonzalez, D., Alvarez, I., Segovia, F.: SPECT image classification using random forests. Electronic Letters 45(12) (2009)

Distance Metric Learning as Feature Reduction Technique for the Alzheimer's Disease Diagnosis

R. Chaves, J. Ramírez, J.M. Górriz, D. Salas-Gonzalez, and M. López

University of Granada, Periodista Daniel Saucedo Aranda s/n,
18071, Granada, Spain
{rosach,javierrp,gorriz
dsalas,miriamlp}@ugr.es
http://sipba.ugr.es/

Abstract. In this paper we present a novel classification method of SPECT images for the development of a computer aided diagnosis (CAD) system aiming to improve the early detection of the Alzheimer's Disease (AD). The system combines firstly template-based normalized mean square error (NMSE) features of tridimensional Regions of Interest (ROIs) t-test selected with secondly Large Margin Nearest Neighbors (LMNN), which is a distance metric technique aiming to separate examples from different classes (Controls and AD) by a Large Margin. LMNN uses a rectangular matrix (called RECT-LMNN) as an effective feature reduction technique. Moreover, the proposed system evaluates Support Vector Machine (SVM) classifier, yielding a 97.93% AD diagnosis accuracy, which reports clear improvements over existing techniques, for instance the Principal Component Analysis (PCA), Linear Discriminant Analysis (LDA) or Normalized Minimum Squared Error (NMSE) evaluated with SVM.

Keywords: SPECT Brain Imaging, Alzheimer's disease, Distance Metric Learning, feature reduction, Support Vector Machines.

1 Introduction

Alzheimer's Disease (AD) is the most common cause of dementia in the elderly and affects approximately 30 million individuals worldwide [1]. Its prevalence is expected to triple over the next 50 years due to the growth of the older population. To date there is no single test or biomarker that can predict whether a particular person will develop the disease. With the advent of several effective treatments of AD symptoms, current consensus statements have emphasized the need for early recognition.

SPECT (Single Positron Emission Computed Tomography) is a widely used technique to study the functional properties of the brain [2]. After the reconstruction and a proper normalization of the SPECT raw data, taken with Tc-99m ethyl cysteinate dimer (ECD) as a tracer, one obtains an activation map displaying the local intensity of the regional cerebral blood flow (rCBF). Therefore,

J.M. Ferrández et al. (Eds.): IWINAC 2011, Part II, LNCS 6687, pp. 68–76, 2011.

this technique is particularly applicable for the diagnosis of neuro-degenerative diseases like AD.

In order to improve the prediction accuracy especially in the early stage of the disease, when the patient could benefit most from drugs and treatments, computer aided diagnosis (CAD) tools are desirable. At this stage in the development of CAD systems, the main goal is to reproduce the knowledge of medical experts in the evaluation of a complete image database, i.e. distinguishing AD patients from controls, thus errors from single observer evaluation are avoided achieving a method for assisting the identification of early signs of AD.

In the context of *supervised* multivariate approaches, the classification is usually done by defining feature vectors representing the different SPECT images and training a classifier with a given set of known samples [3]. In this work, the feature extraction consists firstly of a combination of 3D cubic NMSE (Normalized Minimum Square Error) features over regions of interest (ROIs) which are inside a 3D mask and that are t-test selected with feature correlation weighting. Secondly, we obtain reduced features over the NMSE ones t-test selected by computing distance metric learning analysis.

In this paper, we show how to learn these metrics for kNN classification [4] being aimed at the organisation of the k-nearest neighbors to the same class, while examples from different classes are separated by a large margin. This technique is called *large margin nearest neighbor* (LMNN) classification. In particular, LMNN is a rectangular matrix (RECT-LMNN) whose transformation reduces the input features. Moreover, Support Vector Machines (SVMs) are trained on the features extracted from the neurological images to predict the class of the input (Normal or AD).

2   Matherial and Methods

2.1   Subjects and Preprocessing

Baseline SPECT data from 97 participants were collected from the Virgen de las Nieves hospital in Granada (Spain). The patients were injected with a gamma emitting ^{99m}Tc-ECD radiopharmeceutical and the SPECT raw data was acquired by a three head gamma camera Picker Prism 3000. A total of 180 projections were taken with a 2-degree angular resolution. The images of the brain cross sections were reconstructed from the projection data using the filtered backprojection (FBP) algorithm in combination with a Butterworth noise removal filter. The SPECT images are first spatially normalized using the SPM software, in order to ensure that voxels in different images refer to the same anatomical positions in the brain allowing us to compare the voxel intensities of different subjects. Then we normalize the intensities of the SPECT images with a method similar to [6]. After the spatial normalization, one obtains a $95 \times 69 \times 79$ voxel representation of each subject, where each voxel represents a brain volume of $2 \times 2 \times 2\,\mathrm{mm}^3$. The SPECT images were visually classified by experts of the Virgen de las Nieves hospital using 4 different labels to distinguish between different levels of the presence of typical characteristics for AD. The database consists of 43 NOR, 30 AD1, 20 AD2 and 4 AD3 patients.

3   Feature Extraction

In this article, we propose to apply a combination of VAF (Voxels as Features), NMSE and RECT-LMNN. First of all, controls are averaged in a tridimensional image (*sm(x,y,z)* as it is shown in figure 1. In functional imaging, each voxel carries a grey level intensity $I(\mathbf{x}_j)$, which is related to the regional cerebral blood flow, glucose metabolism, etc. in the brain of a patient, depending on the image acquisition modality. Secondly, it is obtained a 3D *mask(x,y,z)* that consists of all the voxels with *sm(x,y,z)*>a_T. The threshold a_T is equivalent to the 50% of the maximum Intensity in *sm(x,y,z)*. Baseline VAF is a way of including in *vaf(x,y,z)* all the voxels inside the obtained *mask(x,y,z)* and considering them as features. Therefore, voxels outside the brain and poorly activated regions are excluded from this analysis. In this way, no explicit knowledge about the disease is needed, avoiding the inclusion of a priori information about the pathology into the system. Each SPECT image is following divided into 3D $v \times v \times v$ cubic voxels defining Regions of Interest (ROIs), or *block(x,y,z)* centered in (x,y,z) coordinates which are inside of *vaf(x,y,z)*, as it is referenced in figure 1. Then, it is calculated the Normalized Minimum Squared Error for each subject or $NMSE_p$(x,y,z) and block centered at coordinates (x,y,z). NMSE is given by equation 1 as:

$$NMSE_p(x,y,z) = \frac{\sum_{l,m,n=-v}^{v} [f(x-l,y-m,z-n) - g_p(x-l,y-m,z-n)]^2}{\sum_{l,m,n=-v}^{v} [f(x-l,y-m,z-n)]^2}$$

(1)

where f(x,y,z) is the mean voxel intensity at (x, y, z), and g_p(x, y, z) is the voxel intensity of the p subject. The most discriminant ROIs (r_1 in figure 1)

Fig. 1. Feature Extraction Process

are obtained with an absolute value two-sample t-test-based with pooled variance estimate on NMSE features as in [10]. The latter method is RECT-LMNN transformation (L) as feature reduction technique.

3.1 Large Margin Nearest Neighbors (LMNN)

A training set of n labeled examples is denoted $\{(\boldsymbol{x_i}, y_i)\}$ $i = 1, ..., n$, with inputs $\boldsymbol{x_i} \epsilon \mathbb{R}^d$ and class labels y_i. We obtain a family of metrics over the training set by computing Euclidean distances after learning a linear transformation $\mathbf{L} : \mathbb{R}^d \to \mathbb{R}^d$ that optimizes kNN classification when squared distances are computed in this way:

$$D_L(\boldsymbol{x_i}, \boldsymbol{x_j}) = \|L(\boldsymbol{x_i} - \boldsymbol{x_j})\|^2 \tag{2}$$

This equation 2 is commonly used to express squared distances in terms of the squared matrix. Using the eigenvalue decomposition, M can be decomposed into $M = L \cdot L^T$ in which $M \epsilon \mathbb{R}^*$ where $\mathbb{R}^*$ is $n \times n$ dimensional and positively semidefinite. Given two data points $x_1 \in \mathbb{R}^n$ and $x_2 \in \mathbb{R}^n$, the squared distances are denoted as Mahalanobis metrics and can be calculated in terms of M as follows [12]:

$$d_M(x_1, x_2) = \sqrt{(x_1 - x_2)^T \cdot M \cdot (x_1 - x_2)} \tag{3}$$

A Mahalanobis distance can be parametrized in terms of the matrix $\mathbf{L}$ or the matriz $\mathbf{M}$. The first is unconstrained, whereas the second must be positive semidefinite.

There is a cost function [4] that favors distance metric in which different labeled inputs maintain a large margin of distance, that is, the first term acts to *pull* target neighbors closer together penalizing large distances between each input and its target neighbors. Whereas, the other one acts to *push* differently labeled examples further apart penalizing small distances between differently labeled examples. Finally, we combine the two terms $\epsilon_{pull}(L)$ and $\epsilon_{push}(L)$ into a single loss function for distance metric learning. The two terms can have competing effects, to attract target neighbors on one hand and to repel impostors on the other. A weighting parameter $\mu \epsilon$ [0,1] that is selected 0.5 since it works well on practice, balances these goals.

$$\varepsilon(L) = (1 - \mu) \cdot \varepsilon_{pull} + \mu \cdot \varepsilon_{push}(L) \tag{4}$$

Large Margin principle [13] was used to design classification algorithms. In this work, its property of large margin separation was used as a previous step to the classification.

In this approach, we explicitly parameterized the Mahalanobis distance metric as a low-rank matrix, writing $M = L \cdot L^T$, where L is a rectangular matrix (of size $r_1 \cdot r_2$) . To obtain the distance metric, we optimized in terms of the explicitly low-rank linear transformation L. The optimization over L is not convex unlike the original optimization over M, but a (possibly local) minimum can be computed by standard gradient-based methods. We call this approach RECT-LMNN [5]. In particular, matrix L is multiplied by r_1 to obtain the new features r_2 which perform better results (see Results Section).

4 Support Vector Machines Classification

Since their introduction in the late seventies, SVMs [9] marked the beginning of a new era in the learning from examples paradigm because they let to build reliable classifiers in very small sample size problems [8] and even may find nonlinear decision boundaries for small training sets. SVM [7] separates a set of binary-labeled training data by means of a maximal margin hyperplane, building a decision function $\mathbb{R}^N \to \{\pm 1\}$. The objective is to build a decision function f:$\mathbb{R}^N \to \{\pm 1\}$ using training data that is, N-dimensional patterns x_i and class labels y_i so that f will correctly classify new unseen examples (x, y): (x_1, y_1), (x_2, y_2), ... , (x_l, y_l).

Linear discriminant functions define decision hyperplanes in a multidimensional feature space: $g(x) = w^T \cdot x + w_0$ where w is the weight vector to be optimized that is orthogonal to the decision hyperplane and w_0 is the threshold. The optimization task consists of finding the unknown parameters w_i, i=1,...,N and w_0 that define the decision hyperplane. When no linear separation of the training data is possible, SVM can work effectively in combination with kernel techniques such as quadratic, polynomial or rbf, so that the hyperplane defining the SVM corresponds to a non-linear decision boundary in the input space [10].

5 Results

Several experiments were conducted to evaluate the combination of VAF, NMSE and Distance Metric Learning to extend the margin between classes. The performance of the combination of RECT-LMNN based feature extraction with Support Vector Machine classifier was evaluated in depth as a tool for the early detection of the AD in terms of Accuracy(Acc), Sensitivity (Sen) and Specificity (Spe), which were estimed by the kFold cross-validation.

Sen and Spe are defined as:

$$\text{Sensitivity} = \frac{TP}{TP+FN}; \quad \text{Specificity} = \frac{TN}{TN+FP}$$

respectively, where TP is the number of true positives: number of AD patients correctly classified; TN is the number of true negatives: number of controls correctly classified; FP is the number of false positives: number of controls classified as AD patients; FN is the number of false negatives: number of AD patients classified as controls. Firstly, it was experimentally proven that after the application of the tridimensional mask (sm), the best results were obtained with r_1=1000 NMSE features in blocks of $5 \times 5 \times 5$ and t-test selected which were reduced to r_2 components using RECT-LMNN transformation. A higher number of reduced components (r_2) may lead to an increase in the computational cost of the CAD system, and a penalty in the Acc rate because no-discriminant features are being used too as we can observe in table 1.

As SVM classification is concerned, the experiments considered both an increasing number of reduced RECT-LMNN features for a fixed number of folds (12) in table 1 and also an increasing number of folds for a fixed number of

Table 1. Accuracy (%) with Support Vector Machine classifier for an increasing number of feature-reduced using RECT-LMNN transformation

N features	Accuracy	Specificity	Sensitivity
3	92.7835	95.1220	91.0714
4	91.7526	92.6829	91.0714
5	95.8763	95.1220	96.4286
6	94.8454	92.6829	96.4286
7	95.8763	92.6829	98.2143
8	97.9381	97.5610	98.2143
9	96.9072	95.1220	98.2143
10	96.9072	95.1220	98.2143
11	96.9072	97.5610	96.4286
12	96.9072	95.1220	98.2143
13	95.8763	95.1220	96.4286
14	97.9381	97.5610	98.2143
15	95.8763	95.1220	96.4286
20	94.8454	95.1220	94.6429
25	92.7835	92.6829	92.8571

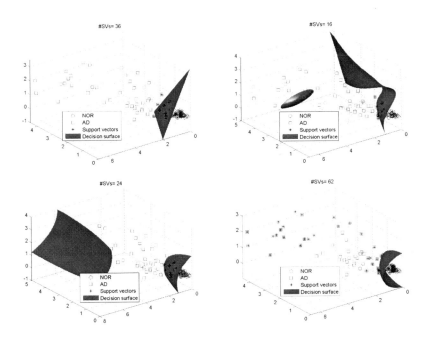

Fig. 2. Decision surfaces for Support Vector Machine Classifier with RECT-LMNN transformation using kernels: linear, polynomial, quadratic, rbf (from left to right, up to down respectively).

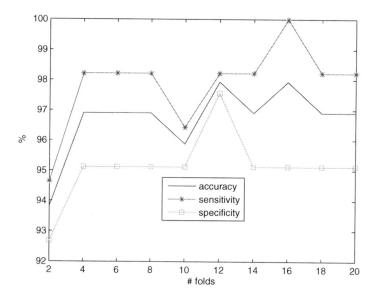

Fig. 3. Accuracy, Sensitivity and Specificity using Support Vector Machine Classifier with RECT-LMNN transformation for feature extraction using an increasing number of folds for the k-fold validation

Table 2. Accuracy (%) using RECT-LMNN and other reported feature extraction methods such as VAF, PCA, LDA/PCA evaluated with Support Vector Machines

Kernel	VAF	PCA	LDA/PCA	RECT-LMNN
Linear	85.71	89.01	90.11	97.94
Quadratic		87.91	89.01	95.88
Polynomial		85.71	90.11	97.94
rbf		89.01	90.11	95.88

features (8) which performed the best results in figure 3 for 12 folds. Furthermore, different kernel techniques were used (see figure 2 and table 2). Figure 2 shows decision surfaces of SVM classifier when linear, polynomial, quadratic and rbf kernels are used. It is observed that when polynomial and linear kernel-SVM space transformation is used, the separation between classes is higher than with quadratic and rbf kernels. These results match with those of the table 2 in which in addition, LMNN-RECT feature reduction method is compared with other reported methods such as VAF, PCA, LDA/PCA [11]. In other words, LMNN-RECT is an effective way to obtain a large margin between classes NOR and AD. Performance by means of Accuracy(%) was demonstrated to be independent of the number of folds up to 12. On the contrary, the fact of using linear or polynomial kernel for the transformation space in the feature extraction reported better results respectively to the others. In particular, the best results of

Accuracy for LMNN-RECT were obtained at 8 features using linear kernel as it is observed in table 1 and 2. In Figure 3, it is shown the Acc, Sen and Spe using the RECT-LMNN metric. The best results of this RECT-LMNN based Support Vector Machine Classifier (Acc=98.97%, Sen=98.21% and Spe=97.56%) were reached at 12 folds for r_2=8 LMNN-features.

6   Conclusions

Kernel Distance Metric Learning Methods were investigated for SPECT images SVM-based classification for the early AD's diagnosis, obtaining results of Acc of 97.94%, Sen of 98.21% and Spe of 97.56% for a RECT-LMNN based method, outperforming other recently reported methods including VAF, PCA, LDA/PCA. As regards feature extraction, the combination of NMSE t-test selection and RECT-LMNN space reduction helped reach the best results of classification. It is remarkable that increasing the number of folds there was no convergence but this independence favoured how to select the most adecuate number of features to obtain excellent results of classification. In particular, with 8 features, the best results were obtained. With a higher number of features, Acc rate is decreased.

References

1. Petrella, J.R., Coleman, R.E., Doraiswamy, P.M.: Neuroimaging and Early Diagnosis of Alzheimer's Disease: A Look to the Future. Radiology 226, 315–336 (2003)
2. English, R.J., Childs, J.: SPECT: Single-Photon Emission Computed Tomography: A Primer. Society of Nuclear Medicine (1996)
3. Fung, G., Stoeckel, J.: SVM feature selection for classification of SPECT images of Alzheimer's disease using spatial information. Knowledge and Information Systems 11(2), 243–258 (2007)
4. Weinberger, K.Q., Blitzer, J., Saul, L.K.: Distance Metric Learning for Large Margin Nearest Neighbor Classification. Journal of Machine Learning Research 10, 207–244 (2009)
5. Weinberger, K., Saul, L.K.: Fast Solvers and Efficient Implementations for Distance Metric Learning. In: Proceedings of the 25th International Conference on Machine Learning (2008)
6. Saxena, P., Pavel, F.G., Quintana, J.C., Horwitz, B.: An automatic threshold-based scaling method for enhancing the usefulness of tc-HMPAO SPECT in the diagnosis of alzheimer#146s disease. In: Wells, W.M., Colchester, A.C.F., Delp, S.L. (eds.) MICCAI 1998. LNCS, vol. 1496, pp. 623–630. Springer, Heidelberg (1998)
7. Burges, C.: A tutorial on support vector machines for pattern recognition. Data Mining and Knowledge Discovery 2(2), 121–167 (1998)
8. Duin, R.P.W.: Classifiers in almost empty spaces. In: International Conference on Pattern Recognition (ICPR), vol. 2 (2), pp. 4392–4395 (2000)
9. Vapnik, V.N.: Estimation of Dependences Based on Empirical Data. Springer, New York (1982)

10. Chaves, R., Ramírez, J., Górriz, J.M., López, M., Salas-Gonzalez, D., Alvarez, I., Segovia, F.: SVM-based computer-aided diagnosis of the Alzheimer's disease using t-test NMSE feature selection with feature correlation weighting. Neuroscience Letters 461, 293–297 (2009)
11. López, M., Ramíreza, J., Górriza, J.M., Alvareza, I., Salas-Gonzalez, D., Segovia, F., Chaves, R., Padilla, P., Gómez-Río, M.: Principal Component Analysis-Based Techniques and Supervised Classification Schemes for the Early Detection of the Alzheimer's Disease. Neurocomputing, doi:10.1016/j.neucom.2010.06.025
12. Xiang, S., Nie, F., Zhang, C.: Learning a Mahalanobis distance metric for data clustering and classification. Pattern Recognition 41, 3600–3612 (2008)
13. Tsochantaridis, I., Joachims, T., Hofmann, T., Altun, Y.: Large Margin Methods for Structured and Interdependent Output Variables. Journal of Machine Learning Research 6, 1453–1484 (2005)

Brain Status Data Analysis by Sliding EMD

A. Zeiler[1], R. Faltermeier[2], A. Brawanski[2], A.M. Tomé[3],
C.G. Puntonet[4], J.M. Górriz[5], and E.W. Lang[1]

[1] CIML Group, Biophysics, University of Regensburg, D-93040 Regensburg, Germany
angela.zeiler@biologie.uni-regensburg.de
[2] Neurosurgery, University Regensburg Medical Center, D-93040 Regensburg, Germany
[3] IEETA/DETI, Universidade de Aveiro, P-3810-193 Aveiro, Portugal
[4] DATC/ETSI, Universidad de Granada, E-18071 Granada, Spain
[5] DSTNC, Universidad de Granada, E-18071 Granada, Spain

Abstract. Biomedical signals are in general non-linear and non-stationary which renders them difficult to analyze with classical time series analysis techniques. *Empirical Mode Decomposition* (EMD) in conjunction with a Hilbert spectral transform, together called *Hilbert-Huang Transform*, is ideally suited to extract informative components which are characteristic of underlying biological or physiological processes. The method is fully adaptive and generates a complete set of orthogonal basis functions, called *Intrinsic Mode Functions* (IMFs), in a purely data-driven manner. Amplitude and frequency of IMFs may vary over time which renders them different from conventional basis systems and ideally suited to study non-linear and non-stationary time series. However, biomedical time series are often recorded over long time periods. This generates the need for efficient EMD algorithms which can analyze the data in real time. No such algorithms yet exist which are robust, efficient and easy to implement. The contribution shortly reviews the technique of EMD and related algorithms and develops an *on-line* variant, called *slidingEMD*, which is shown to perform well on large scale biomedical time series recorded during neuromonitoring.

1 Introduction

Recently an empirical nonlinear analysis tool for complex, non-stationary time series has been pioneered by N. E. Huang et al. [1]. It is commonly referred to as *Empirical Mode Decomposition* (EMD) and if combined with Hilbert spectral analysis it is called *Hilbert - Huang Transform* (HHT). It adaptively and locally decomposes any non-stationary time series in a sum of *Intrinsic Mode Functions* (IMF) which represent zero-mean amplitude and frequency modulated components. The EMD represents a fully data-driven, unsupervised signal decomposition and does not need any *a priori* defined basis system. EMD also satisfies the perfect reconstruction property, i.e. superimposing all extracted IMFs together with the residual slow trend reconstructs the original signal without information loss or distortion. The method is thus similar to the traditional Fourier or wavelet decompositions but the interpretation of IMFs is not similarly transparent [2]. It is still a challenging task to identify and/or combine extracted IMFs in a proper way to yield physically meaningful components. However, the empirical nature of EMD offers the advantage over other empirical signal decomposition

J.M. Ferrández et al. (Eds.): IWINAC 2011, Part II, LNCS 6687, pp. 77–86, 2011.
© Springer-Verlag Berlin Heidelberg 2011

techniques like *empirical matrix factorization* (EMF) of not being constrained by conditions which often only apply approximately. Especially with biomedical signals one often has only a rough idea about the underlying modes and mostly their number is unknown.

Furthermore, if biomedical time series are recorded over long time periods like, for example, in neuromonitoring, huge amounts of data are collected. Existing EMD algorithms then reveal two shortcomings: First, the analysis of such data has to wait until data collection is finished. Second, the huge amount of data does not allow their complete analysis, rather the recorded time series are segmented to render them tractable by current EMD techniques. These shortcomings generate the needs for an *on-line* variant of EMD algorithms which is able to analyze recorded data in real time. This contribution will review the technique of empirical mode decomposition and its recent extensions which include a recent suggestion of an *on-line* EMD algorithm [3] which, however, is neither robust nor efficient. Such an algorithm, called *slidingEMD*, is proposed here, its properties and applications to large scale brain status time series recorded during neuromonitoring are discussed as well.

2 Empirical Mode Decomposition

The EMD method was developed from the assumption that any non-stationary and non-linear time series consists of different simple intrinsic modes of oscillation. The essence of the method is to empirically identify these intrinsic oscillatory modes by their characteristic time scales in the data, and then decompose the data accordingly. Through a process called *sifting*, most of the *riding waves*, i.e. oscillations with no zero crossing between extrema, can be eliminated. The EMD algorithm thus considers signal oscillations at a very local level and separates the data into locally non-overlapping time scale components. It breaks down a signal $x(t)$ into its component IMFs obeying two properties:

1. An IMF has only one extremum between zero crossings, i.e. the number of local minima and maxima differs at most by one.
2. An IMF has a mean value of zero.

Note that the second condition implies that an IMF is stationary which simplifies its analysis. But an IMF may have amplitude modulation and also changing frequency.

The Standard EMD Algorithm. The sifting process can be summarized in the following algorithm. Decompose a data set $x(t)$ into IMFs $x_n(t)$ and a residuum $r(t)$ such that the signal can be represented as

$$x(t) = \sum_n x_n(t) + r(t) \tag{1}$$

Sifting then means the following steps illustrated for convenience in Fig. 1.

The average period of each IMF can be estimated by dividing twice the sample size $2 \cdot N$ by the number of zero crossings. Note also that the number of IMFs extracted from

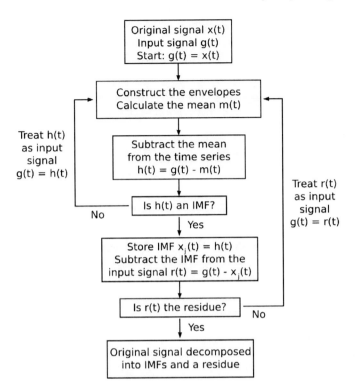

Fig. 1. Flow diagram of the EMD algorithm

a time series is roughly equal to $\log_2 N$. The sifting process separates the non-stationary time series data into locally non-overlapping intrinsic mode functions (IMFs). However, EMD is not a sub-band filtering technique with predefined waveforms like wavelets. Rather selection of modes corresponds to an automatic and adaptive time-variant filtering. Completeness of the decomposition process is automatically achieved by the algorithm as $x(t) = \sum_{n=1}^{i} x_n + r$ represents an identity. Further, the EMD algorithm produces *locally orthogonal* IMFs. Global orthogonality is not guaranteed as neighboring IMFs might have identical frequencies at different time points (typically in $< 1\%$ of the cases).

2.1   Recent Extensions of the Standard EMD

Plain EMD is applied to the full length signal which in view of limited resources like computer memory also limits the length of the time series to be dealt with. This is an especially serious problem with biomedical time series which often are recorded over very long time spans. A number of extensions to plain EMD have been proposed in recent years such as

- *Ensemble EMD* (EEMD) is a noise-assisted method to improve sifting [4], [5]. In practice EEMD works as follows:

- Add white noise to the data set
- Decompose the noisy data into IMFs
- Iterate these steps and at each iteration add white noise
- Calculate an ensemble average of the respective IMFs to yield the final result

  – *Local EMD* [3] pursues the idea to iterate the sifting process only in regions where the mean is still finite to finally meet the stopping criterion everywhere. Localization can be implemented via a weighting function $w(t)$ which is $w(t) = 1$ in regions where sifting is still necessary and decays to zero at the boundaries. This can be easily integrated into the EMD algorithm via

$$h_{j,n}(t) = h_{j,n-1}(t) - w_{j,n}(t)m_{j,n}(t) \qquad (2)$$

  This procedure essentially improves the sifting process and tries to avoid *over-sifting*.

  – *On-line EMD*: The application of EMD to biomedical time series is limited by the size of the working memory of the computer. Hence in practical applications only relatively short time series can be studied. However, many practical situations like continuous patient monitoring ask for an *on-line* processing of the recorded data. Recently, a blockwise processing, called *on-line EMD*, has been proposed [3]. The method is still in its infancy and needs yet to be developed to a robust and efficient *on-line* technique.

3   Sliding EMD

Biomedical data acquired during neuromonitoring accumulate to huge amounts of data when monitoring patients in intensive care units is extended over days. Up till now only short segments have been discussed. Analyzing larger data sets is hardly possible because of the computational load involved. Even more important, however, is the fact that data analysis has to wait until monitoring is finished. But an immediate *on-line* analysis of such time series data is of utmost importance. No proper EMD algorithm is available yet to achieve this goal. Following we will propose an *online* EMD algorithm which we call *slidingEMD*.

3.1   The Principle

In a first step, the recorded time series is split into segments which can be analyzed with EMD. Simply adding the IMFs extracted from the different segments together would induce boundary artifacts, however. This is illustrated in a simple example in Fig. 2 and Fig. 3, respectively.

Thus segmenting a time series into non-overlapping windows for further analysis leads to strong boundary artifacts which can be avoided when the segments m_i are shifted by k samples. Choosing the shift according to $\frac{m}{k} \notin \mathcal{N}$, however, leads to gaps between the windows and causes instable results. If, instead, the window size is a multiple of the step size, i.e. if $\frac{m}{k} \in \mathcal{N}$ holds, neighboring windows can be joined without having boundary artifacts. With this choice, every sample is represented equally often,

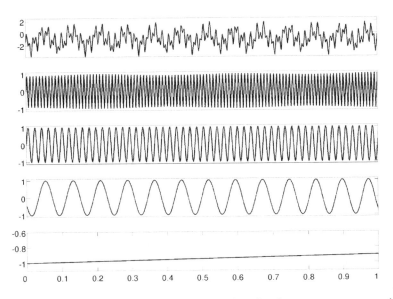

Fig. 2. Detail of a toy signal (top trace) and underlying signal components: sawtooth wave $(1.5708)^{-1}\arcsin\left(\sin(699\,x)\right)$ (second trace), a sinusoid $\sin(327\,x)$ (third trace), a cosine function with a time-dependent frequency $\cos\left(2\cdot(x+20)^2\right)$ (fourth trace) and a monotonic trend $0.1\,x-1$ (bottom trace) whereas $x=[0,20]$.

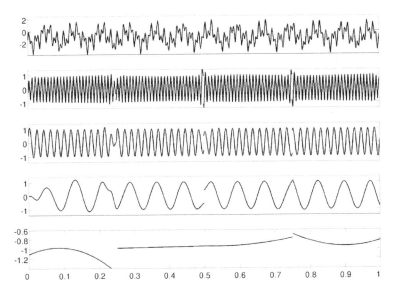

Fig. 3. EMD decomposition of the toy signal from Fig. 2. The time series has been segmented into 4 parts and decomposed with EMD. After joining the resulting IMFs together, boundary effects become clearly visible at $x=0.25, 0.5$ and $x=0.75$.

i.e. $n = \frac{m}{k}$ - times, in the overlapping windows for a later estimation of the corresponding mean sample value. If the conditions $\frac{m}{k} \in \mathcal{N}$ and $\frac{N}{m} \in \mathcal{N}$ hold with N the number of samples, the number of windows M is given by

$$M = \left(\frac{N - m}{k} \right) + 1 \tag{3}$$

A schematic illustration of the principle mode of function of the *slidingEMD* algorithm is shown in Fig. 4 below. The time series in every segment m_i is decomposed by EMD into j IMFs $c_{m_i j}(t)$ and a local residuum $r_{m_i}(t)$ according to

$$x_{m_i}(t) = \sum_j c_{m_i j}(t) + r_{m_i}(t) \tag{4}$$

whereby the number of sifting steps is kept equal in all segments. Resulting IMFs are collected in a matrix with corresponding sample points forming a column of the matrix with $n = \frac{m}{k}$ entries. Columns corresponding to the beginning or end of the time series are deficient, hence are omitted from further processing. This assures that all columns contain the same amount of information to estimate average IMF amplitudes at every time point in each segment. This finally yields

$$c_j(t) = \frac{1}{n} \sum_i^{i+n-1} c_{m_i j}(t), \quad r(t) = \frac{1}{n} \sum_i^{i+n-1} r_{m_i}(t), \quad i = \lfloor \frac{t - m}{k} \rfloor + 2 \tag{5}$$

Problems arise if a different number of IMFs results in different overlapping segments or if the same intrinsic frequency appears in different IMFs in different overlapping segments. Except for the stopping criterion any EMD algorithm can be applied but EEMD has proven to yield the best results so far.

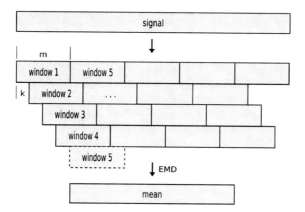

Fig. 4. Schema of the *slidingEMD* algorithm. The time series segments in the shifted windows are decomposed with EEMD. IMFs and the residuum are determined finally by mode amplitudes which are averaged over corresponding samples in all windows. m describes the window size and k the step size.

3.2   Stationarization of Brain Status Data with *slidingEMD*

The proposed algorithm *slidingEMD* can also be used to remove non-stationary components from the recorded signals. Which components should be separated into the residuum can be controlled by the segment size. The following applies very small step sizes only to achieve optimal results. In practical *on-line* applications such small step sizes are usually impractical. The investigations use ABP and ICP time series with $N = 50000$ samples, a variable segment size and step size $k = 2$. After decomposing the signal with *slidingEMD*, the residuum is subtracted from the original time series. Fig. 5 presents in its top trace results of the stationarization of an ICP time series using *slidingEMD* with a segment size $m = 2048$ and a step size $k = 2$. The related subfigures present the mean and standard deviation of the difference between the residuum and the original ICP time series for different step sizes and a segment size $m = 2048$. Both parameters fluctuate only slightly at small step sizes indicating that a high reconstruction quality could be achieved robustly.

To test for stationarity, mean values of the original time series and the difference of the original time series and the estimated residuum were followed over time by estimating them in subsequent subsegments of variable length s for segment sizes $m = 256, 512, 1024, 2048$. The resulting dependence of the standard deviations of the means on the number of subsegments s quickly saturates.

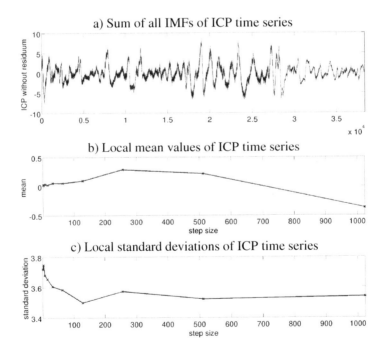

Fig. 5. a) Sum of all IMFs estimated from the ICP time series using *slidingEMD* with $m = 2048$ and $k = 2$. b) Mean and c) standard deviation of the stationarized ICP time series as function of the step size and a segment size $m = 2048$.

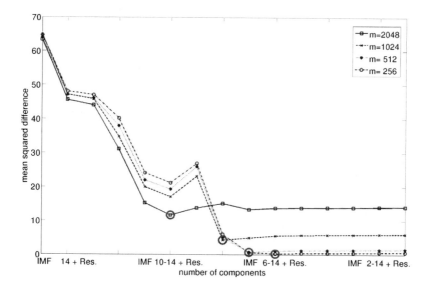

Fig. 6. Quality Q_{n+2} of the decomposition of an ICP time series with either *slidingEMD* or EMD for segment sizes $m = 256, 512, 1024, 2048$ and a step size $k = 2$ used within *slidingEMD*. Note that to the residuum estimated with EMD a variable amount of IMFs is added successively.

The results prove a substantial reduction of the non-stationary components in the difference signal. But weak stationarity is only reached if the number s of subsegments stays small in accord with results obtained earlier already. However, the amount of non-stationary signal parts removed with *slidingEMD* is much larger than when the residua are estimated with EMD and then subtracted from the original time series. Again a proper choice of the segment size is imperative not to remove informative oscillations from the recordings.

Comparative analysis of brain status data with EMD and *slidingEMD*. The following study compares the residua estimated with *slidingEMD* with the residua plus the sum of low frequency IMFs estimated with EMD. First the mean square difference between the residua estimated with *slidingEMD* and EMD is calculated according to

$$Q_1 = \frac{1}{N} \sum_{t=1}^{N} (r_s(t) - r_e(t))^2 \tag{6}$$

Here N denotes the total number of samples, $r_s(t)$ represents the residuum estimated with *slidingEMD* and $r_e(t)$ the corresponding residuum estimated with EMD. Next, the residuum $r_e(t)$ and the lowest frequency IMF is added and the sum is subtracted from the residuum $r_s(t)$. Next, the residuum $r_e(t)$ and the two lowest frequency IMFs are added and the sum is subtracted from the residuum $r_s(t)$. This process is iterated until the difference is negligible.

$$Q_{n+2} = \frac{1}{N} \sum_{t=1}^{N} \left(r_s(t) - \left(r_e(t) + \sum_{i=0}^{n} c_{14-i}(t) \right) \right)^2 \tag{7}$$

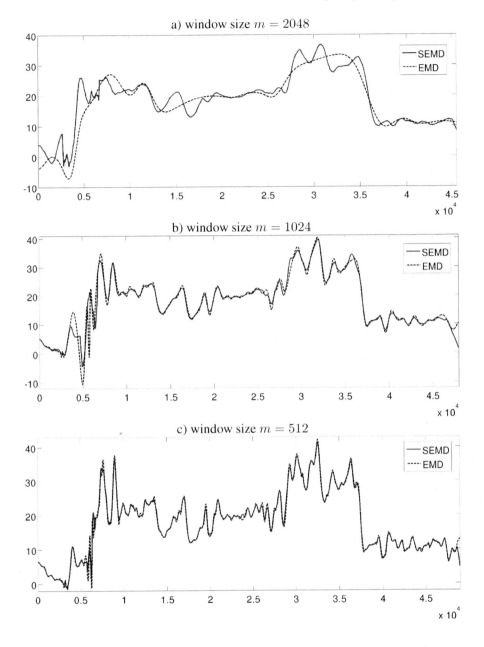

Fig. 7. Residuum $r_s(t)$ (full line) estimated with *slidingEMD* using a window size of a) $m = 2048$, b) $m = 1024$ and c) $m = 512$ samples and the sum of the respective residuum plus low frequency IMFs, i.e. $r_e(t) + IMF_i + \ldots + IMF_{14}$ where $a)i = 10, b)i = 8, c)i = 7$, estimated with **EMD** (broken line).

The result of this iteration ($n = \{0, 1, ..., 13\}$) is summarized in Fig. 6 in case of the ICP time series using segment sizes $m = 256, 512, 1024, 2048$ and a step size $k = 2$.

It becomes obvious that the residuum estimated with *slidingEMD* using a segment size $m = 2048$ and the residuum plus IMFs 10 to 14, all estimated from applying EMD, are very similar (see Fig. 7). Much the same holds true in the following constellations: segment size $m = 1024$ and $r_e(t) + IMF_8 + ... + IMF_{14}$, segment size $m = 512$ and $r_e(t) + IMF_7 + ... + IMF_{14}$ and segment size $m = 256$ and $r_e(t) + IMF_6 + ... + IMF_{14}$.

The results prove that *slidingEMD* yields a decomposition well in accord with standard EMD. However, *slidingEMD* offers the additional advantage of being a true *on-line* algorithm which, furthermore, is based on an ensemble of estimates. It thus provides a robust estimate of underlying intrinsic mode functions. It is also as flexible as standard EMD in detrending applications and allows to extract the stationary part of originally non-stationary biomedical time series data. For practical applications, hence, *slidingEMD* should be preferred.

References

1. Huang, N.E., Shen, Z., Long, S.R., Wu, M.L., Shih, H.H., Zheng, Q., Yen, N.C., Tung, C.C., Liu, H.H.: The empirical mode decomposition and Hilbert spectrum for nonlinear and nonstationary time series analysis. Proc. Roy. Soc. London A 454, 903–995 (1998)
2. Jánosi, I.M., Müller, R.: Empirical mode decomposition and correlation properties of long daily ozone records. Phys. Rev. E 71, 056126 (2005)
3. Rilling, G., Flandrin, P., Goncalès, P.: On empirical mode decomposition and its algorithms. In: Proc. 6th IEEE-EURASIP Workshop on Nonlinear Signal and Image Processing (2003)
4. Wu, Z., Huang, N.: Ensemble empirical mode decomposition: A noise assisted data analysis method. Technical report, Center for Ocean-Land-Atmosphere Studies, 193, 51 (2005)
5. Flandrin, P., Gonçalvès, P., Rilling, G.: Emd equivalent filter banks: From interpretation to application. In: Huang, N., Shen, S. (eds.) Hilbert-Huang Transform: Introduction and Application, pp. 67–87. World Scientific, Singapore (2005)

A Quantitative Study on Acupuncture Effects for Fighting Migraine Using SPECT Images

M. López[1], J. Ramírez[1], J. M. Górriz[1],
R. Chaves[1], and M. Gómez-Río[2]

[1] Dept. of Signal Theory, Networking and Communications
University of Granada, Spain
[2] Department of Nuclear Medicine
Hospital Universitario Virgen de las Nieves, Granada, Spain

Abstract. The aim of this paper is to quantitatively determine whether acupuncture, applied under real conditions of clinical practice in the area of primary healthcare, is effective for fighting migraine. This is done by evaluating SPECT images of migraine patients' brain in a context of image classification. Two different groups of patients are randomly collected and received *verum* and *sham* acupuncture, respectively. In order to make the image processing computationally efficient and solve the small sample size problem, an initial feature extraction step based on Principal Component Analysis is performed on the images. Differences among features extracted from pre– and post–acupuncture scans are quantified by means of Support Vector Machines for verum and sham modalities, and statistically reinforced by carrying out a statistical t–test. The conclusions of this work point at acupuncture as an effective method to fight migraine.

Keywords: SPECT, acupuncture, migraine, PCA, SVM, class separability.

1 Introduction

Acupuncture has been used as a therapy for several thousand years and is now being used as an alternative treatment for many medical conditions. Despite its popularity, there persists some controversy as to the differentiation between the specific and the nonspecific effects of acupuncture. Basic scientific studies on the mechanisms of acupuncture and well–designed randomized clinical trials that permit objective evaluation of this ancient science are urgently needed.

The development of imaging techniques, such as Single Photon Emission Computed Tomography (SPECT) and functional Magnetic Resonance Imaging (fMRI), have opened a "window" into the brain that allows us to gain an appreciation of the anatomy and physiological function involved during acupuncture in humans and animals non–invasively. Researchers such as Alavi [1] and Cho [2] were among the first to publish in this area. Their initial observations have subsequently been enhanced and further developed in relation to point specificity,

J.M. Ferrández et al. (Eds.): IWINAC 2011, Part II, LNCS 6687, pp. 87–95, 2011.
© Springer-Verlag Berlin Heidelberg 2011

effects in hearing, nausea and more generalized analgesic effects in both animal models and human experimental pain [3]. In this work we focus on migraine disorder, which is a chronic neurologic disease that can severely affect the patient's quality of life. Several authors have studied and analyzed the cerebral blood flow of migraine patients in SPECT imaging [4] and, in recent years, many randomized studies have been carried out to investigate the effectiveness of acupuncture as a treatment for migraine although it remains a controversial issue [5].

Up to now, acupuncture effects on brain activation function have been assessed by performing well–known approaches in neuroscience such as Statistical Parametric Mapping (SPM) [6]. However, this technique requires the number of observations (i.e. scans) to be greater than the number of components of the multivariate observation (i.e. voxels) and furthermore, final conclusions are assessed by experts in an subjective way. Our aim in this paper is to quantitatively determine the effects of the acupuncture on migraine patients in an automatic way. This quantitative approach used for SPECT image assessment is based on an analysis of the images in a classification frame, where Principal Component Analysis (PCA) is used for feature extraction and the evaluation of class separability is performed by evaluating the performance of Support Vector Machine (SVM)–based classifiers.

2   Database Description

2.1   Image Acquisition and Preprocesing

Tomographic radiopharmaceutical imaging such as SPECT provides in vivo three–dimensional maps of a pharmaceutical labeled with a gamma ray emitting radionuclide. The distribution of radionuclide concentrations are estimated from a set of projectional images acquired at many different angles around the patient. SPECT imaging techniques employ radioisotopes which decay emitting predominantly a single gamma photon. SPECT images used in this work were taken with a PRISM 3000 machine. All SPECT images were first co–registered to the SPM template [6] in order to accomplish the spatial normalization assumption: the same position in the volume coordinate system within different volumes corresponds to the same anatomical position. This makes it possible to do meaningful voxel wise comparisons between SPECT images. In addition SPECT imaging generates volumes that only give a relative measure of the blood flow. The blood flow measure is relative to the blood flow in other regions of the brain. Direct comparison of the voxel intensities, between images, even different acquisitions of the same subject, is thus not possible without normalization of the intensities.

After the normalization steps, a $68 \times 95 \times 79$ voxel–sized representation of each subject is obtained. For each patient, SPECT image is acquired as a baseline 20 minutes following the injection of ^{99m}Tc-ECD. After this initial image acquisition the patients are given the acupuncture session (verum or sham) under the procedure explained in [5] and finally, post image acquisition is acquired 30 minutes later the acupuncture session. Figures 1(a) and 1(b) show three consecutive

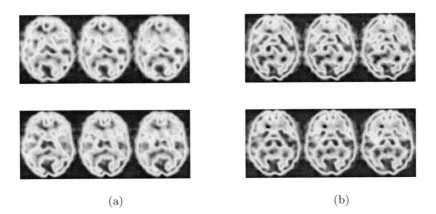

(a)   (b)

Fig. 1. Three slices from the pre–acupuncture (up) and post–acupuncture (down) images obtained from a patient that underwent (a) verum acupuncture and (b) sham acupuncture

slices along the transaxial axis of the pre–acupuncture and post–acupuncture sessions of two patients to whom verum and sham acupuncture has been applied, respectively. Visually it becomes a difficult task to determine the effects of the acupuncture, if any. We propose to categorize the images into groups and analyze class differences by quantifying class separability measures and statistical analyses, so that differences between the first and the second acquisition can determine if acupuncture produces the expected effects on the patients.

2.2   Demographic Details

Our database consists of 28 migraine patients, 14 subjected to individualized active or verum acupuncture and 14 with minimal or sham acupuncture. All pre–acupuncture acquisitions are considered as belonging to class ω_1 while label ω_2 is assigned to post–acupuncture acquisitions. On the other hand, migraine images are separated in group A or B if the acupuncture modality was verum or sham, respectively. Thus, each volume is included in one of these groups: $X_{\omega_1}^A$, $X_{\omega_2}^A$, $X_{\omega_1}^B$ and $X_{\omega_2}^B$. Apart from the samples of patients suffering from migraine,

Table 1. Demographic details of the dataset. Number of samples, percentage of males (M) and females (F), and age (mean/standard deviation) is given for each group.

	# Samples	Sex(M/F) %	Age
NORMAL	41	32.95/12.19	71.51/7.99
A	14	78.57/21.43	45.42/8.59
B	14	50/50	44.78/12.29

a set of controls consisting in 41 normal patients is available. A more detailed demographic description of the whole database is given in Table 1.

3   Feature Extraction and Classification

In order to set a method to measure the changes that acupuncture produces in the brain activation function of migraine patients, we deal with images establishing a classification frame. The idea is to quantify the effects of acupuncture by evaluating the discrimination power of a classifier when post–acupuncture scans are separated from pre–acupuncture acquisitions. To this purpose, two independent classification tasks are defined: separating $X_{\omega_1}^A$ from $X_{\omega_2}^A$ and separating $X_{\omega_1}^B$ from $X_{\omega_2}^B$.

PCA has shown to be a very efficient technique to extract features from brain images for classification purposes [7]. Furthermore, it consists of a very powerful technique to compress the information contained in the images, solving this way the so-called small sample size. PCA generates an orthonormal basis vector that maximizes the scatter of all the projected samples. After the preprocessing steps, the n voxels representing each subject are rearranged into a vector form. Let $\mathbf{X} = [\mathbf{x}_1, \mathbf{x}_2, ..., \mathbf{x}_N]$ be the sample set of these vectors, where N is the number of patients. After normalizing the vectors to unity norm and subtracting the grand mean, a new vector set $\mathbf{Y} = [\mathbf{y}_1, \mathbf{y}_2, ..., \mathbf{y}_N]$ is obtained, where each $\mathbf{y}_i$ represents an n–dimensional normalized vector, $\mathbf{y}_i = (y_{i1}, y_{i2}, ..., y_{in})^t, i = 1, 2, ..., N$. The covariance matrix of the normalized vectors set is defined as

$$\mathbf{\Sigma}_Y = \frac{1}{N} \sum_{i=1}^{N} \mathbf{y}_i \mathbf{y}_i^t = \frac{1}{N} \mathbf{Y}\mathbf{Y}^t \tag{1}$$

and the eigenvector and eigenvalue matrices $\mathbf{\Phi}$, $\mathbf{\Lambda}$ are computed as

$$\mathbf{\Sigma}_Y \mathbf{\Phi} = \mathbf{\Phi}\mathbf{\Lambda} \tag{2}$$

Note that $\mathbf{Y}\mathbf{Y}^t$ is an $n \times n$ matrix while $\mathbf{Y}^t\mathbf{Y}$ is an $N \times N$ matrix. If the sample size N is much smaller than the dimensionality n, then diagonalizing $\mathbf{Y}^t\mathbf{Y}$ instead of $\mathbf{Y}\mathbf{Y}^t$ reduces the computational complexity [8]

$$(\mathbf{Y}^t\mathbf{Y})\mathbf{\Psi} = \mathbf{\Psi}\mathbf{\Lambda}_1 \tag{3}$$

$$\mathbf{T} = \mathbf{Y}\mathbf{\Psi} \tag{4}$$

where $\mathbf{\Lambda}_1 = diag\{\lambda_1, \lambda_2, ..., \lambda_N\}$ and $\mathbf{T} = [\mathbf{\Phi}_1, \mathbf{\Phi}_2, ..., \mathbf{\Phi}_N]$.

The criterion to choose the most m discriminant eigenvectors is usually given by the value of the associated eigenvalues. However, for classification tasks is has been proved that the Fisher Discriminant Ratio (FDR) is more effective for choosing the more powerful projection vectors [7], since this criterion considers the class labels. The FDR is defined as

$$FDR = \frac{(\mu_1 - \mu_2)^2}{\sigma_1^2 + \sigma_2^2} \tag{5}$$

where μ_i and σ_i denote the i-th class within class mean value and variance, respectively. Thus, in the classification stage, the training set is used to compute the eigenvectors. Images are projected onto them giving rise to a set of PCA coefficients, and the FDR values are computed from the PCA coefficients. After that, the test sample is projected onto the eigenvectors and the obtained test coefficients are rearranged in decreasing order according to the previous FDR values obtained from the training set. Finally, the rearranged PCA coefficients are used to train and test an SVM-based classifier [9] with a Radial Basis Function (RBF) kernel, as done in [10,11].

4   Experiments and Results

4.1   Classification

In the context of image classification, separability of the classes can be measured quantitatively by training a classifier and determining its class separation power. Images are first transformed onto the PCA subspace, where each image is represented compactly by a few PCA coefficients, extracted from groups A and B separately. Figures 4.1 and 2(d) show the three first FDR-PCA coefficients used as features and represented as 3D points, as well as the 3D decision hyperplane defined by the SVM classifier with RBF kernel. Features corresponding to well–clustered classes will make the classifier perform better and vice versa, and this fact can be measured as a percentage of accuracy achieved by the classifier. In these terms, pre– and post–acupuncture features from group A are expected to be more easily separable than those from group B, since sham acupuncture has been applied to the latter set of patients.

Figures 2(a) and 2(b) show values of accuracy, sensibility (ability of the classifier to detect true positive samples) and specificity (ability of the classifier to detect true negative samples) obtained for each group when the number of PCA extracted coefficients increases. Due to the small number of samples that make each group up, leave-one-out cross validation strategy was used to train and test the classifier and to compute the averaged final values. For verum acupuncture, separation between classes is reached with up to 89.29% accuracy. On the contrary, the designed SVM performs like a random classifier for sham acupuncture modality, yielding a maximum accuracy peak of 64.29 %. This fact highlights the presence of acupuncture effects on patients by changing the brain perfusion pattern.

Besides the classification accuracy, sensibility and specificity rates, the distance from the sample features to the separation hyperplane reflects the separability of classes. Figure 4.1 shows the distance of the first, main PCA coefficient used for classification to the designed hyperplane by the SVM classifier. Distances to the hyperplane are in average greater for group A, which can be interpreted as higher effects of acupuncture when verum modality is used.

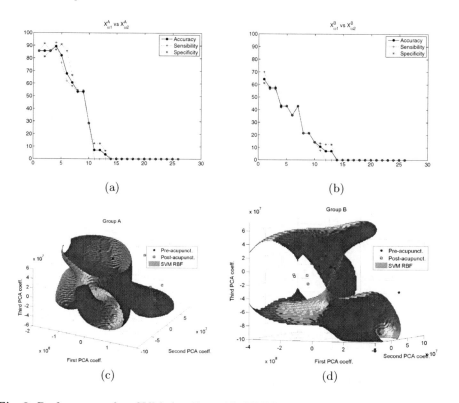

Fig. 2. Performance of an SVM classifier with RBF kernel when dealing with pre– and post–acupuncture samples of groups (a) A and (b) B as two different classes to be separated. The three main PCA coefficients and the designed separation surface are depicted in (c) and (d).

4.2   T–Test Analysis

The main problem we find when dealing with our database consists of having a small number of samples of patients affected by migraine in each group. T–test is one method for testing the degree of difference between two means in small sample. It uses T distribution theory to deduce the probability when difference happens, then judge whether the difference between two means is significant. Specifically, the comparison of two sample means deduces whether their representative population differs or not.

The statistic t is computed as

$$t = \frac{\overline{X}_1 - \overline{X}_2}{\sqrt{\frac{S_1^2}{N_1} + \frac{S_2^2}{N_2}}} \tag{6}$$

where $\overline{X}_1$ and $\overline{X}_2$ are the sample means, S_1 and S_2 are the sample standard deviation and N_1 and N_2 are the sample sizes. In this work, t–test will be used

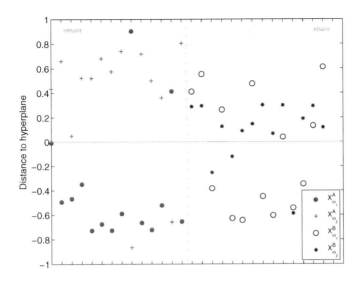

Fig. 3. Distance from each sample to the separation hyperplane

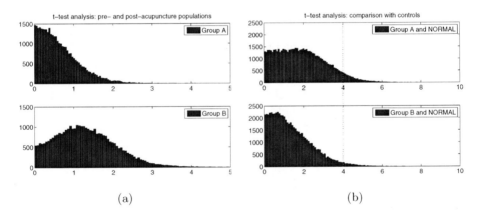

(a)   (b)

Fig. 4. Histogram of the absolute value of t values obtained for measuring differences between populations (a) pre– and post–acupuncture sessions for migraine patients and (b) post–acupuncture and control patients.

to provide statistical significance to the results obtained from the study of the populations, given the small number of available samples.

We analyze each group images statistically by means of the t–test in order to obtain an initial description of the classes. To this purpose, independent voxel-wise two sample t–tests are computed for groups $X_{\omega_1}^A$ and $X_{\omega_2}^A$ first, and groups $X_{\omega_1}^B$ and $X_{\omega_2}^B$. Figure 4 shows the histograms of the absolute value of the t tests performed to compare populations. Higher values of t mean greater differences between both scans under study. The first analysis (4(a)) reveals a lager number of voxels with high t value for group B, which can be interpreted as lower differences between pre– and post–acupuncture brain activation patterns, that is, sham acupuncture produced low effects compared to verum modality.

Next matter to be solved is related to the effects produced by acupuncture in migraine patients. Is acupuncture effective for fighting migraine or does it just produce activation in brain with no specific analgesic effects? We then analyze post–acupuncture scans of groups A and B and compare each one with the control population represented by 41 NORMAL samples. Results of the t–test are now represented in Fig. 4(b), where it can be seen that post–acupuncture t–test distribution for group A is significantly higher than for group B, i. e., brain activity represented by each single voxel intensity value is "nearer" to controls' pattern for group A, to whom verum acupuncture was applied.

5   Conclusions

This work shows a quantitative method to evaluate acupuncture effects on migraine patients by using SPECT brain images, in order to prove and quantify the analgesic effects of acupuncture in an objective way. Verum and sham acupuncture modalities are applied to patients suffering from migraine, and pre– and post–acupuncture image perfusion patterns are analyzed in a image classification context. This is done by means of PCA as feature extraction technique and a SVM-based classifier. The quantification of the effects of acupuncture can be measured in terms of capability of a classifier to separate pre– and post–acupuncture acquisitions. The SVM classifier yields high accuracy results for verum modality while the classifier responds to sham acupuncture samples as a random classifier. The acupuncture effects can be quantitatively assessed by computing the distance from each sample to the separation hyperplane. On the other hand, a dataset of controls and statistical t–test are used to prove that changes produced by verum acupuncture make the brain perfusion pattern statistically more similar to the normal pattern when verum modality is applied.

Acknowledgments. This work was partly supported by the Spanish Government under the PETRI DENCLASES (PET2006-0253), TEC2008-02113, NAPOLEON (TEC2007-68030-C02-01) projects and the Consejería de Innovación, Ciencia y Empresa (Junta de Andalucía, Spain) under the Excellence Projects TIC-02566 and TIC-4530. The SPECT database was provided by "Virgen de las Nieves" hospital in Granada (Spain).

References

1. Alavi, A., Lariccia, P., Sadek, A., Newberg, A., Lee, L., Reich, H., Lattanand, C., Mozley, P.: Neuroimaging of acupuncture in patients with chronic pain. The Journal of Alternative and Complementary Medicine 3(1), S41–S53 (1997)
2. Cho, Z., Chung, S., Jones, J., Park, J., Park, H., Lee, H., Wong, E., Min, B.: New findings of the correlation between acupoints and corresponding brain cortices using functional mri. Proceedings of the National Academy of Sciences 95, 2670–2673 (1998)

3. Lewitha, G.T., White, P., Pariente, J.: Investigating acupuncture using brain imaging techniques: The current state of play. Advance Access Publication (2005)

4. Battistella, P.A., Ruffilli, R., Pozza, F.D., Pitassi, I., Casara, G., Boniver, C., Bendagli, A., Condini, A.: 99mTc HM-PAO SPECT in pediatric migraine. Headache: The Journal of Head and Face Pain 30(10), 646–649 (1990)

5. Vas, J., Modesto, M., Méndez, C., Perea-Milla, E., Aguilar, I., Carrasco-Lozano, J., Faus, V., Martos, F.: Effectiveness of acupuncture, special dressings and simple, low-adherence dressings for healing venous leg ulcers in primary healthcare: study protocol for a cluster-randomized open-labeled trial. BMC Complementary and Alternative Medicine 8(29) (2008)

6. Friston, K.J., Ashburner, J., Kiebel, S.J., Nichols, T.E., Penny, W.D.: Statistical Parametric Mapping: The Analysis of Functional Brain Images. Academic Press, London (2007)

7. López, M., Ramírez, J., Górriz, J.M., Salas-Gonzalez, D., Álvarez, I., Segovia, F., Puntonet, C.G.: Automatic tool for the Alzheimer's disease diagnosis using PCA and Bayesian classification rules. IET Electronics Letters 45(8), 389–391 (2009)

8. Turk, M., Pentland, A.: Eigenfaces for recognition. Journal of Congnitive Neuroscience 3(1), 71–86 (1991)

9. Burges, C.J.C.: A tutorial on Support Vector Machines for pattern recognition. Data Mining and Knowledge Discovery 2(2), 121–167 (1998)

10. Ramírez, J., Górriz, J., Romero, A., Lassl, A., Salas-Gonzalez, D., López, M., Río, M.G.: Computer aided diagnosis of Alzheimer type dementia combining support vector machines and discriminant set of features. Information Sciences (2009) (accepted)

11. López, M., Ramírez, J., Górriz, J., Álvarez, I., Salas-Gonzalez, D., Segovia, F., Chaves, R., Padilla, P., Gómez-Río, M.: The Alzheimer's Disease Neuroimaging Initiative: Principal component analysis-based techniques and supervised classification schemes for the early detection of alzheimer's disease. Neurocomputing (2010) (accepted)

High Resolution Segmentation of CSF on Phase Contrast MRI

Elsa Fernández[1], Manuel Graña[1], and Jorge Villanúa[2]

[1] Grupo de Inteligencia Computacional, Universidad del País Vasco, Spain
[2] Osatek, Hospital Donostia,
Paseo Dr. Beguiristain 109, 20014 San Sebastián, Spain

Abstract. Dynamic velocity-encoded Phase-contrast MRI (PC-MRI) techniques are being used increasingly to quantify pulsatile flows for a variety of flow clinical application. A method for igh resolution segmentation of cerebrospinal fluid (CSF) velocity is described. The method works on PC-MRI with high temporal and spatial resolution. It has been applied in this paper to the CSF flow at the Aqueduct of Sylvius (AS). The approach first selects the regions with high flow applying a threshold on the coefficient of variation of the image pixels velocity profiles. The AS corresponds to the most central detected region. We perform a lattice independent component analysis (LICA) on this small region, so that the image abundances provide the high resolution segmentation of the CSF flow at the AS. Long term goal of our work is to use this detection and segmentation to take some measurements and evaluate the changes in patients with suspected Idiopathic Normal Pressure Hydrocephalus (iNPH).

1 Introduction

The syndrome of progressive mental deterioration and neurologic disturbances including psychomotor retardation, gait unsteadiness, and incontinence of urine associated with hydrocephalus, in the setting of normal CSF pressure on lumbar puncture and the absence of papilledema, was coined normal pressure hydrocephalus (NPH) in 1965[1]. To this day, the exact cause of the idiopathic form of NPH (iNPH) remains unclear. However, it is generally agreed on that on the basis of NPH, there is an abnormal absorption of CSF with a consequent accumulation of CSF in the ventricular space. For at least the last 20 years, the therapy for NPH has been diversion of CSF via a ventriculoperitoneal shunt (VPS)[3,2].

Phase-Contrast Magnetic Resonance Imaging (PC-MRI) is a well-known method of obtaining additional velocity measurements that can be used for analysis of the blood flow and tissue motion. It has been extensively used for cardiovascular flow measurement [10]. Also, PC-MRI has been proposed to quantify the flow of CSF in the Aqueduct of Sylvius (AS)[9,14]. The speed of CSF flow in AS has been studied as a predictor for the positive evolution of VPS. The stroke volume (SV), defined as the mean volume passing through the aqueduct

J.M. Ferrández et al. (Eds.): IWINAC 2011, Part II, LNCS 6687, pp. 96–103, 2011.

during both systole and diastole, can be calculated based on PC-MRI. A SV greater than or equal to 5ml serves as a selection criterion for patients with good probabilities of improvement after VPS [4]. High resolution segmentation of the flow at the AS may be useful to study the evolution of iNPH patients. In this paper we show results of the application of LICA [6] to this task. The approach has two steps. First, it performs a detection of high velocity regions by computing the coefficient of variation of each PC-MRI pixel velocity profile and selecting those above a threshold set as the 95% percentile of the distribution of the coefficients of variation. The central of such regions corresponds to the AS. We focus the LICA analysis on the AS pixel velocity profiles, obtaining a high resolution segmentation given by the abundance images computed by LICA. Actually, the performed detection allows to find automatically the image pixel with the highest peak velocity. The LICA segmentation gives qualitatively different results for healthy and iNPH subjects, allowing to localize some flow effects in the AS region.

The contents of the paper is as follows: Section 2 gives a brief review of PC-MRI. Section 3 describes the segmentation methods. Section 4 gives some experimental results. Finally, section 5 gives our conclusions.

2   Basics of Phase-Contrast Imaging

Phase contrast techniques derive contrast between flowing blood and stationary tissues from the magnetization phase. The phase of the magnetization from the stationary spins is zero and the phase of the magnetization from the moving spins is non-zero. The phase is a measure of how far the magnetization precess from the time it is tipped into the transverse plane until the time it is detected. The data acquired with phase contrast techniques can be processed to produce phase difference, complex difference, and magnitude images. Phase contrast MRI is based on the property that a uniform motion of tissue in a magnetic field gradient produces a change in the MR signal phase, Φ. This change is proportional to the velocity of the tissue, v. The MR signal from a volume element accumulates the phase[11].

$$\Phi(r, T) = \gamma B_0 T + \gamma v \cdot \int_0^T G(r, t) t \, dt$$

$$= \gamma B_0 T + \gamma v \cdot \overline{G},$$

during time T , where B_0 is a static magnetic field, γ the gyro-magnetic ratio and G(r, t) is the magnetic field gradient. Notice that G is exactly the first moment of G(r, t) with respect to time. If the field gradient is altered between two consecutive recordings, then by subtracting the resulting phases

$$\Phi_1 - \Phi_2 = \gamma v \cdot (\overline{G}_1 - \overline{G}_2),$$

the velocity in the (G1 − G2)-direction is implicitly given. In this way a desired velocity component can be calculated for every volume element simultaneously. To construct the velocity vector in 3D, the natural way is to apply appropriate gradients to produce the x-, y- and z-components respectively.

3   Segmentation Methods

Segmentation is done in two steps. First the AS region in the PC-MRI image is located and the velocity profiles of the pixels lying in this region are selected for further processing. To perform the AS segmentation, we compute the coefficient of variation of the velocity profiles of the PC-MRI pixels. We compute the coefficient of variation histogram to estimate its distribution over the image. Then we select the 95% percentile as the threshold to detect the regions where some actual flow is happening. Second, we apply Lattice Independent Component Analysis (LICA) to obtain a high resolution segmentation of the AS region on the basis of the velocity profiles.

LICA is based on the Lattice Independence property discovered when dealing with noise robustness in Morphological Associative Memories [12]. Works on finding lattice independent sources (aka endmembers) for linear unmixing started on hyperspectral image processing [8,13]. Since then, it has been also proposed for functional MRI analysis [7,6]. It is a data dimension reduction proposed in the framework of Lattice Computing [5].

Under the Linear Mixing Model (LMM) the design matrix is composed of endmembers which define a convex region covering the measured data. The linear coefficients are known as fractional abundance coefficients that give the contribution of each endmember to the observed data:

$$\mathbf{y} = \sum_{i=1}^{M} a_i \mathbf{s}_i + \mathbf{w} = \mathbf{S}\mathbf{a} + \mathbf{w}, \tag{1}$$

where $\mathbf{y}$ is the d-dimension measured vector, $\mathbf{S}$ is the $d \times M$ matrix whose columns are the d-dimension endmembers $\mathbf{s}_i, i = 1, .., M$, $\mathbf{a}$ is the M-dimension abundance vector, and $\mathbf{w}$ is the d-dimension additive observation noise vector. Under this generative model, two constraints on the abundance coefficients hold. First, to be physically meaningful, all abundance coefficients must be non-negative $a_i \geq 0, i = 1, .., M$, because the negative contribution is not possible in the physical sense. Second, to account for the entire composition, they must be fully additive $\sum_{i=1}^{M} a_i = 1$. As a side effect, there is a saturation condition $a_i \leq 1, i = 1, .., M$, because no isolate endmember can account for more than the observed material. From a geometrical point of view, these restrictions mean that we expect the endmembers in $\mathbf{S}$ to be an Affine Independent set of points, and that the convex region defined by them covers *all* the data points.

The LICA approach assumes the LMM as expressed in equation 1. Moreover, the equivalence between Affine Independence and Strong Lattice Independence [?] is used to induce from the data the endmembers that compose the matrix $\mathbf{S}$. Briefly, LICA consists of two steps:

1. Use an Endmember Induction Algorithm (EIA) to induce from the data a set of Strongly Lattice Independent vectors. In our works we use the algorithm described in [8,7]. These vectors are taken as a set of affine independent vectors that forms the matrix **S** of equation 1.
2. Apply the Full Constrained Least Squares estimation to obtain the abundance vector according to the conditions for LMM. Abundance images provide the desired segmentation results.

The advantages of this approach are (1) that we are not imposing statistical assumptions to find the sources, (2) that the algorithm is one-pass and very fast because it only uses lattice operators and addition, (3) that it is unsupervised and incremental, and (4) that, depending on the EIA used, it can be tuned to detect the number of endmembers by adjusting a noise-filtering related parameter.

4   Experimental Results

The data used for this experiments consist of three volumes (256x256x18) of PC-MRI velocity data of the brain. Figure 1(a) illustrated the location of the capture of the PC-MRI data, synchronized with heart rate. Figure 1(b) shows a typical PC-MRI image obtained from a healthy subject, which will be the first subject from now on. The hyperintense blob at the image center corresponds to the AS. The results of the PC-MRI segmentation of the three subjects based on the coefficient of variation of the velocity profiles are given in figure 2. The central blob, signaled by an arrow in each image, corresponds to the AS while other blobs correspond to nearby blood vessels. Extracting the velocity profiles of the PC-MRI pixels corresponding to the As region and plotting them we obtain the plots in figure 3. The velocity encoding is positive, with a value of 2000 corresponding to zero velocity. Below and above that value, corresponds to flow in one or other sense: from or towards the head. Clinical practice at the Hospital Donostia consists in the manual inspection of the PC-MRI image looking for the pixel with the highest velocity value. This information can be easily extracted from the plots in figure 3. Notice that for the range of velocities for the first subject is smalled than for the other two, which are patients of iNPH. Figure 4 shows the plots of the endmembers obtained by the EIA algorithm from [8] with nominal parameters. The first subject has a more uniform flow (figure 3(a)) the EIA only obtains one endmember. For the other two subjects it gets five endmembers. The endmember with smaller variation around the 2000 encoding value corresponds to a "normal" flow pixel, the others corresponding to high speed flows inside the AS region. Also, some phase changes can be appreciated in the endmembers of figures 4(b) and 4(c). Computing the abundance images gives the images of figure 5. The first subject (figure 5(a)) has only one abundance image which covers all the AS region with a positive value, meaning that an homogeneous flow is found. The second and third subjects have a big percentage of the AS region with an uniform flow, detected in the first abundance image of each (top-leftmost image). The other abundances correspond to pixels with specific flow patterns, which are mostly out of phase relative to the main flow. These flow inhomogeneities can be of interest for further studies.

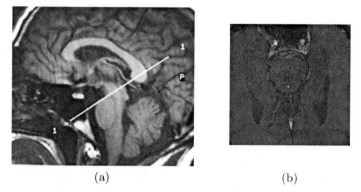

(a)   (b)

Fig. 1. Localization of the PC-MRI capture with the AS in the image center

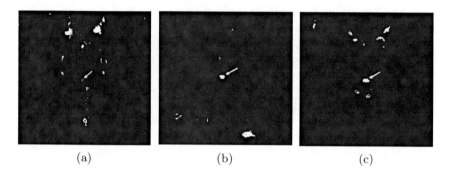

(a)   (b)   (c)

Fig. 2. Detection of high velocity regions based on the coefficient of variation in the three volumes considered

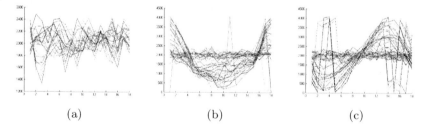

(a)   (b)   (c)

Fig. 3. Plots of the velocity profiles of the pixels in the segmented AS for each subject

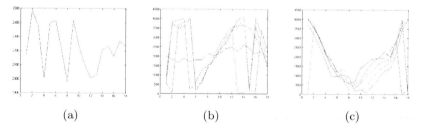

(a)   (b)   (c)

Fig. 4. Plot of the endmember velocity profiles found in each subject in the AS region

(a)

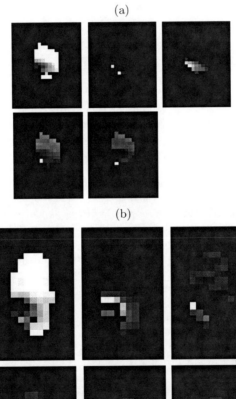

(b)

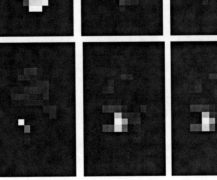

(c)

Fig. 5. Abundance images found by LICA in each case

5 Conclusions

In this paper we present an approach for the automatic segmentation of the Aqueduct of Sylvius from PC-MRI data and further high resolution segmentation of the flow patterns inside the AS region. The AS detection is based on the coefficient of variation of the PC-MRI pixel velocity profiles. The high resolution segmentation is obtained by an application of the LICA approach. We obtain a good identification of the AS region in the image. The high resolution segmentation identifies various flow patterns easily. The long term research work will address the use of LICA segmentation results to monitor the evolution of iNPH patients.

References

1. Adams, R.D., Fisher, C.M., Hakim, S., Ojemann, R.G., Sweet, W.H.: Symptomatic occult hydrocephalus with "Normal" cerebrospinal-fluid pressure. a treatable syndrome. The New England Journal of Medicine 273, 117–126 (1965); PMID: 14303656
2. Bergsneider, M., McL Black, P., Klinge, P., Marmarou, A., Relkin, N.: Surgical management of idiopathic normal-pressure hydrocephalus. Neurosurgery 57(3 suppl.), 29–39 (2005); PMID: 16160427
3. Black, P.M.: Idiopathic normal-pressure hydrocephalus. results of shunting in 62 patients. Journal of Neurosurgery 52(3), 371–377 (1980);PMID: 7359191
4. Bradley, W.G., Scalzo, D., Queralt, J., Nitz, W.N., Atkinson, D.J., Wong, P.: Normal-pressure hydrocephalus: evaluation with cerebrospinal fluid flow measurements at MR imaging. Radiology 198(2), 523–529 (1996)
5. Graña, M.: A brief review of lattice computing. In: IEEE International Conference on Fuzzy Systems, FUZZ-IEEE 2008, IEEE World Congress on Computational Intelligence, pp. 1777–1781 (June 2008)
6. Graña, M., Chyzhyk, D., García-Sebastián, M., Hernández, C.: Lattice independent component analysis for functional magnetic resonance imaging. Information Sciences (2010) (in Press) Corrected Proof
7. Graña, M., Manhaes-Savio, A., García-Sebastián, M., Fernandez, E.: A lattice computing approach for on-line fmri analysis. Image and Vision Computing 28(7), 1155–1161 (2010)
8. Graña, M., Villaverde, I., Maldonado, J.O., Hernandez, C.: Two lattice computing approaches for the unsupervised segmentation of hyperspectral images. Neurocomputing 72(10-12), 2111–2120 (2009)
9. Lee, J.H., Lee, H.K., Kim, J.K., Kim, H.J., Park, J.K., Choi, C.G.: CSF flow quantification of the cerebral aqueduct in normal volunteers using phase contrast cine MR imaging. CSF flow quantification of the cerebral aqueduct in normal volunteers using phase contrast cine MR imaging 5(2), 81–86 (2004); PMID: 15235231 PMCID: 2698144
10. Lotz, J., Meier, C., Leppert, A., Galanski, M.: Cardiovascular flow measurement with Phase-Contrast MR imaging: Basic facts and implementation1. Radiographics 22(3), 651–671 (2002)

11. Pelc, N.J., Herfkens, R.J., Shimakawa, A., Enzmann, D.R.: Phase contrast cine magnetic resonance imaging. Magnetic Resonance Quarterly 7(4), 229–254 (1991); PMID: 1790111
12. Ritter, G.X., Sussner, P., Diaz de Leon, J.L.: Morphological associative memories. IEEE Transactions on Neural Networks 9(2), 281–293 (1998)
13. Ritter, G.X., Urcid, G., Schmalz, M.S.: Autonomous single-pass endmember approximation using lattice auto-associative memories. Neurocomputing 72(10-12), 2101–2110 (2009)
14. Thomsen, C., Stahlberg, F., Stubgaard, M., Nordell, B.: Fourier analysis of cerebrospinal fluid flow velocities: MR imaging study. the scandinavian flow group. Radiology 177(3), 659–665 (1990); PMID: 2243965

Exploration of LICA Detections in Resting State fMRI

Darya Chyzhyk[1], Ann K. Shinn[2], and Manuel Graña[1]

[1] Computational Intelligence Group
Dept. CCIA, UPV/EHU, Apdo. 649, 20080 San Sebastian, Spain
www.ehu.es/ccwintco
[2] McLean Hospital, Belmont, Massachusetts; Harvard Medical School,
Boston, Massachusetts, US

Abstract. Lattice Independent Component Analysis (LICA) approach consists of a detection of lattice independent vectors (endmembers) that are used as a basis for a linear decomposition of the data (unmixing). In this paper we explore the network detections obtained with LICA in resting state fMRI data from healthy controls and schizophrenic patients. We compare with the findings of a standard Independent Component Analysis (ICA) algorithm. We do not find agreement between LICA and ICA. When comparing findings on a control versus a schizophrenic patient, the results from LICA show greater negative correlations than ICA, pointing to a greater potential for discrimination and construction of specific classifiers.

1 Introduction

Many works on fMRI analysis are based on the Independent Component Analysis (ICA) [18]. The approaches to solve the ICA problem obtain both the independent sources and the linear unmixing matrix. These approaches are unsupervised because no *a priori* information about the sources or the mixing process is included, hence the alternative name of Blind Deconvolution. Spatial sources in fMRI correspond to the activations localizations in the brain volume, while temporal sources correspond to statistical independent patterns of BOLD signal intensities. The main advantage of ICA is that it does not impose a priori assumptions on the selection of observations, thus avoiding "double dipping" effects biasing the results. We have used the FastICA algorithm implementation available at [1].

We have proposed Lattice Independent Component Analysis (LICA) [7] a Lattice Computing [6] approach that we call that consists of two steps. First it selects Strong Lattice Independent (SLI) vectors from the input dataset using an incremental algorithm, the Incremental Endmember Induction Algorithm (IEIA) [11]. Second, because of the conjectured equivalence between SLI and Affine Independence [9], it performs the linear unmixing of the input dataset based on these endmembers. We assume that the data is generated as a convex combination of a set of endmembers which are the vertices of a convex polytope

J.M. Ferrández et al. (Eds.): IWINAC 2011, Part II, LNCS 6687, pp. 104–111, 2011.

covering some region of the input data. This assumption is similar to the linear mixture assumed by the ICA approach, however we do not impose any probabilistic assumption on the data. Endmembers correspond to the ICA's temporal independent sources. LICA is unsupervised, as ICA, and it does not impose any a priori assumption on the data. We have applied LICA to fMRI activation detection [8,7], and Voxel Based Morphometry of structural MRI [2]. In this paper we explore its application to resting state fMRI.

The outline of the paper is as follows: Section 2 reviews resting state fMRI. Section 3 overviews the LICA. Section 4 presents results of LICA versus ICA on small case study. Section 5 provides some conclusions.

2   Resting State fMRI Background

Resting state fMRI data has been used to study the connectivity of brain activations [4,13,17]. The assumption is that temporal correlation of low frequency oscillations in diverse areas of the brain reveal their functional relations. When no explicit cognitive task is being performed, the connections discovered are assumed as some kind of brain fingerprint, the so-called default-mode network. Caution must be taken on the confounding effects of the ambient noise, the respiratory and cardiac cycles. One strong reason for resting state fMRI experiments is that they do not impose constraints on the cognitive abilities of the subjects. For instance in pediatric applications, such as the study of brain maturation [14], there is no single cognitive task which is appropriate across the aging population. Several machine learning and data mining approaches have been taken: hierarchical clustering [3], independent component analysis (ICA) [5,16], fractional amplitude of low frequency analysis [23], multivariate pattern analysis (MVPA) [14,15]. Graph analysis has been suggested [17] as a tool to study the connectivity structure of the brain. Resting state fMRI has being found useful for performing studies on brain evolution based on the variations in activity of the default mode network [14], depression (using regional homogeneity measures) [20], Alzheimer's Disease [10], and schizophrenia.

Schizophrenia is a severe psychiatric disease that is characterized by delusions and hallucinations, loss of emotion and disrupted thinking. Functional disconnection between brain regions is suspected to cause these symptoms, because of known aberrant effects on gray and white matter in brain regions that overlap with the default mode network. Resting state fMRI studies [12,21,22] have indicated aberrant default mode functional connectivity in schizophrenic patients. These studies suggest an important role for the default mode network in the pathophysiology of schizophrenia. Functional disconnectivity in schizophrenia could be expressed in altered connectivity of specific functional connections and/or functional networks, but it could also be related to a changed organization of the functional brain network. Resting state studies for schizophrenia patients with auditory hallucinations have also been performed [19] showing reduced connectivity.

3   The Lattice Independent Component Analysis

The linear mixing model can be expressed as follows: $\mathbf{x} = \sum_{i=1}^{M} a_i \mathbf{e}_i + \mathbf{w} = \mathbf{Ea} + \mathbf{w}$, where $\mathbf{x}$ is the d-dimension pattern vector corresponding to the fMRI voxel time series vector, $\mathbf{E}$ is a $d \times M$ matrix whose columns are the d-dimensional vectors, when these vectors are the vertices of a convex region covering the data they are called endmembers $\mathbf{e}_i, i = 1, .., M$, $\mathbf{a}$ is the M-dimension vector of linear mixing coefficients, which correspond to fractional abundances in the convex case, and $\mathbf{w}$ is the d-dimension additive observation noise vector. The linear mixing model is subjected to two constraints on the abundance coefficients when the data points fall into a simplex whose vertices are the endmembers, all abundance coefficients must be non-negative $a_i \geq 0, i = 1, .., M$ and normalized to unity summation $\sum_{i=1}^{M} a_i = 1$. Under this circumstance, we expect that the vectors in $\mathbf{E}$ are affinely independent and that the convex region defined by them includes *all* the data points. Once the endmembers have been determined the unmixing process is the computation of the matrix inversion that gives the coordinates of the point relative to the convex region vertices. The simplest approach is the unconstrained least squared error (LSE) estimation given by: $\hat{\mathbf{a}} = \left(\mathbf{E}^T \mathbf{E}\right)^{-1} \mathbf{E}^T \mathbf{x}$. Even when the vectors in $\mathbf{E}$ are affinely independent, the coefficients that result from this estimation do not necessarily fulfill the non-negativity and unity normalization. Ensuring both conditions is a complex problem.

The *Lattice Independent Component Analysis* (LICA) is defined by the following steps:

1. Induce from the given data a set of Strongly Lattice Independent vectors. In this paper we apply the Incremental Endmember Induction Algorithm (IEIA) [11,7]. These vectors are taken as a set of affine independent vectors. The advantages of this approach are (1) that we are not imposing statistical assumptions, (2) that the algorithm is one-pass and very fast because it only uses comparisons and addition, (3) that it is unsupervised and incremental, and (4) that it detects naturally the number of endmembers.
2. Apply the unconstrained least squares estimation to obtain the mixing matrix. The detection results are based on the analysis of the coefficients of this matrix. Therefore, the approach is a combination of linear and lattice computing: a linear component analysis where the components have been discovered by non-linear, lattice theory based, algorithms.

4   A Resting State fMRI Case Study

The results shown in this section are explorations over resting state fMRI data obtained from a healthy control subject and an schizophrenia patient with auditory hallucinations selected from an on-going study in the McLean Hospital. Details of image acquisition and demographic information will be given elsewhere. For each subject we have 240 BOLD volumes and one T1-weighted anatomical image. The functional images were coregistered to the T1-weighted anatomical image. The data preprocessing begins with the skull extraction usind the BET tool from

FSL (http://www.fmrib.ox.ac.uk/fsl/). Further preprocessing, including slice timing, head motion correction (a least squares approach and a 6-parameter spatial transformation), smoothing and spatial normalization to the Montreal Neurological Institute (MNI) template (resampling voxel size = 3.5 mm × 3.5 mm × 3.5 mm), were conducted using the Statistical Parametric Mapping (SPM8, http://www.fil.ion.ucl.ac.uk/spm) package. Results of LICA and ICA have been computed in Matlab. Visualization of results has been done in FSL. Because both ICA and LICA are unsupervised in the sense that the pattern searched is not predefined, they suffer from the identifiability problem: we do not know beforehand which of the discovered independent sources/endmembers will correspond to a significant brain connection. Therefore, results need a careful assessment by the medical expert. We will not give here any neurological conclusion.

The application of LICA with nominal parameters give 8 endmembers. Accordingly we have computed fastICA setting the number of independent sources to 8. We compute the LICA abundance distributions. For each endmember, we set the 95% percentile of its abundance distribution as the threshold for the detection of the corresponding endmember in the abundance volume. We do the

Table 1. Pearson's Correlation coefficients between ICA and LICA source/endmember detections for the schizophrenia patient

LICA	ICA							
	#1	#2	#3	#4	#5	#6	#7	#8
#1	0.02	0	-0.04	-0.02	0.02	0.03	0.01	0.01
#2	0.03	0.08	-0.1	-0.04	0	0.01	-0.33	0
#3	-0.01	**0.36**	0.01	-0.07	-0.01	-0.02	0.13	-0.01
#4	0.03	0	-0.03	-0.11	0	-0.01	0	-0.01
#5	-0.03	-0.02	**0.16**	-0.01	-0.01	-0.11	-0.02	**0.46**
#6	0.38	-0.03	0.01	-0.13	**0.17**	0	-0.01	0.01
#7	0	-0.02	0.06	-0.02	-0.01	-0.02	-0.02	-0.01
#8	0.25	0.01	-0.22	0.04	-0.52	0.05	0.02	-0.05

Table 2. Pearson's Correlation coefficients between ICA and LICA source/endmember detections for the healthy control

LICA	ICA							
	#1	#2	#3	#4	#5	#6	#7	#8
#1	0.04	-0.02	-0.04	0.05	-0.06	-0.07	0.03	0.01
#2	-0.02	0.02	0	-0.08	-0.03	-0.03	**0.15**	-0.01
#3	**0.22**	-0.05	**0.13**	0.06	0.01	0.08	-0.03	0.07
#4	0.05	-0.22	0.06	-0.09	0.06	0.08	0.08	-0.03
#5	0.03	0.07	-0.07	**0.12**	**0.14**	-0.13	0.04	-0.01
#6	0.04	0	0.05	-0.06	-0.1	0.02	-0.05	-0.03
#7	0.08	**0.1**	0	0.03	-0.03	-0.02	0.09	0.03
#8	-0.02	-0.04	0.02	0.04	-0.05	-0.07	0.07	0.03

same with the ICA mixture distributions. To explore the agreement between ICA and LICA detections, we have computed the Pearson's correlation between the abundance/mixing volumes of each source/endmember, shown in table 1 for the schizophrenia patient and in table 2 for the control subject. In both cases, agreement between detections of LICA and ICA is low. The best correlation is ICA #8 versus LICA #5 for the schizophrenia patient. For a visual assessment of the agreement between both detection analysis, we show in figure 1 the detections obtained by both algorithms applying the 95% percentile on their respective mixing and abundance coefficients. In this figure, the detection found by LICA is highlighted in blue and the detection found by ICA in red. Overlapping voxels appear in a shade of magenta. It can be appreciated that the LICA detections appear as more compact clusters. Some spurious detections are shown in the sorroundings of the brain due to the diffusion produced by the smoothing filter. From these results is clear that we can not use ICA to validate the findings of LICA.

For further comparison, we have computed the correlations intra-algorithm of the patient versus the control data, meaning that we compute the correlations of the abundance/mixing volumes obtained by the LICA/ICA on the patient and the control data. The aim is to get an idea of the ability of each approach to produce discriminant features. If we find negative correlations of high magnitude then we can say that the corresponding approach has a great potential to generate features that discriminate patients from controls. Table 3 shows the correlations between the LICA abundances obtained from the patient and the control subject. Table 4 shows the same information for the mixing coefficients of ICA. In these tables we are interested in finding the most negatively correlated detections, implying complementary detections. We have highlighted in bold the negative correlations below -0.15. We show in figure 2 the detections with greatest negative correlation between patient and control for both LICA and ICA. In this figure, red corresponds to the patient volume detection, blue corresponds to the control volume detection. Notice again that LICA detections produce more compact clusters. The greatest discrimination is obtained by LICA.

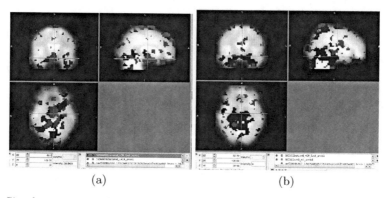

(a)   (b)

Fig. 1. Simultaneous visualization of the best correlated detection results from LICA and ICA from tables 1 and 2. Red corresponds to ICA detection, Blue to LICA detection. (a) Patient, (b) Control.

Table 3. Correlation between patient and control detections obtained by LICA

control	patient							
	#1	#2	#3	#4	#5	#6	#7	#8
#1	-0.17	-0.04	-0.03	0.24	0.08	0.08	0.09	0.01
#2	0.02	**-0.21**	0.04	0.1	0.15	0.09	0.02	-0.09
#3	**-0.32**	0.05	-0.05	0.14	0.24	0.13	0.13	0.15
#4	0.01	0.15	0.08	0.05	-0.03	0.02	0.05	0.17
#5	-0.14	-0.13	**-0.18**	0.13	0.14	0.11	0.12	-0.04
#6	0.01	-0.06	0.11	0.02	0.02	0.02	-0.02	-0.02
#7	0.06	-0.11	-0.05	-0.15	-0.05	-0.05	-0.12	0.03
#8	**-0.32**	-0.19	-0.02	0.23	0.22	0.2	0.05	0.02

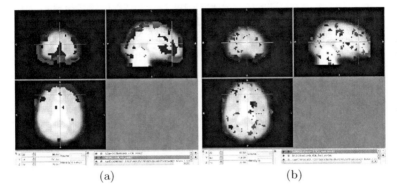

(a)   (b)

Fig. 2. Findings in the patient versus the control. Greatest negative correlated detections (a) found by LICA, (b) found by ICA.

Table 4. Correlation between patient and control detections obtained by ICA

control	patient							
	#1	#2	#3	#4	#5	#6	#7	#8
#1	0.41	0.01	-0.15	0.01	**-0.18**	0.02	0.02	-0.04
#2	-0.12	0.02	0.06	-0.04	0.08	-0.01	0.01	0.05
#3	0.02	-0.02	**-0.24**	-0.02	0.01	0.01	0	0.03
#4	0.03	0	0	0	0.02	0.02	0.02	0
#5	0.04	-0.01	0.06	-0.03	-0.05	-0.01	0.01	0.36
#6	0.04	0.07	-0.05	0.01	0	0	**-0.25**	0
#7	0.03	0	-0.02	0	-0.01	0.06	0	-0.03
#8	0.02	0.03	-0.01	0	-0.02	0	0.01	0

5 Summary and Conclusions

We are exploring the application fo Lattice Independent Component Analysis (LICA) to resting state fMRI. We present results on selected subjects from a study under way in the McLean Hospital. We compare LICA and ICA findings in the form of detections based on the thresholding of the abundance images and mixing matrices. Both LICA and ICA are unsupervised approaches, so they do not force *a priori* assumptions on the localization of the findings, which must be interpreted after the analysis, risking to obtain results not in agreement with the expectations of the analysis. LICA detections are less sparse than those of ICA, but the medical assessment of findings is being carried out actually. The main quantitative conclusion of this study is that there is little agreement between LICA and ICA on this data. Moreover, when we consider the correlation of findings by LICA or ICA on the control versus the schizophrenic patient, we find that the LICA results show greater negative correlation than the results of ICA. We interpret this result as pointing to a greater capability to produce features for discrimination between control and patients based on resting state fMRI data. Anyway, we can not use ICA results as a validation reference, so validation of LICA results must rest on the medical expert assessment of its findings.

References

1. http://www.cis.hut.fi/projects/ica/fastica/
2. Chyzyk, D., Termenon, M., Savio, A.: A Comparison of VBM Results by SPM, ICA and LICA. In: Corchado, E., Graña Romay, M., Manhaes Savio, A. (eds.) HAIS 2010. LNCS, vol. 6077, pp. 429–435. Springer, Heidelberg (2010)
3. Cordes, D., Haughton, V., Carew, J.D., Arfanakis, K., Maravilla, K.: Hierarchical clustering to measure connectivity in fmri resting-state data. Magnetic Resonance Imaging 20(4), 305–317 (2002)
4. Craddock, R.C., Holtzheimer III, P.E., Hu, X.P., Mayberg, H.S.: Disease state prediction from resting state functional connectivity. Magnetic Resonance in Medicine 62, 1619–1628 (2009)
5. Demirci, O., Stevens, M.C., Andreasen, N.C., Michael, A., Liu, J., White, T., Pearlson, G.D., Clark, V.P., Calhoun, V.D.: Investigation of relationships between fMRI brain networks in the spectral domain using ICA and granger causality reveals distinct differences between schizophrenia patients and healthy controls. NeuroImage 46(2), 419–431 (2009)
6. Graña, M.: A brief review of lattice computing. In: Proc. WCCI, pp. 1777–1781 (2008)
7. Graña, M., Chyzyk, D., García-Sebastián, M., Hernández, C.: Lattice independent component analysis for functional magnetic resonance imaging. Information Sciences (2010) (in press)
8. Graña, M., Savio, A.M., Garcia-Sebastian, M., Fernandez, E.: A lattice computing approach for on-line fmri analysis. Image and Vision Computing (2009) (in press)
9. Schmalz, M.S., Ritter, G.X., Urcid, G.: Autonomous single-pass endmember approximation using lattice auto-associative memories. Neurocomputing 72(10-12), 2101–2110 (2009)

10. Liu, Y., Wang, K., Yu, C., He, Y., Zhou, Y., Liang, M., Wang, L., Jiang, T.: Regional homogeneity, functional connectivity and imaging markers of alzheimer's disease: A review of resting-state fmri studies. Neuropsychologia 46(6), 1648–1656 (2008); Neuroimaging of Early Alzheimer's Disease
11. Maldonado, J.O., Hernandez, C., Graña, M., Villaverde, I.: Two lattice computing approaches for the unsupervised segmentation of hyperspectral images. Neurocomputing 72(10-12), 2111–2120 (2009)
12. Mingoia, G., Wagner, G., Langbein, K., Scherpiet, S., Schloesser, R., Gaser, C., Sauer, H., Nenadic, I.: Altered default-mode network activity in schizophrenia: A resting state fmri study. Schizophrenia Research 117(2-3), 355–356 (2010); 2nd Biennial Schizophrenia International Research Conference
13. Northoff, G., Duncan, N.W., Hayes, D.J.: The brain and its resting state activity–experimental and methodological implications. Progress in Neurobiology 92(4), 593–600 (2010)
14. Dosenbach, N.U.F., et al.: Prediction of individual brain maturity using fmri. Science 329, 1358–1361 (2010)
15. Pereira, F., Mitchell, T., Botvinick, M.: Machine learning classifiers and fMRI: A tutorial overview. NeuroImage 45(supplement 1), S199–S209 (2009); Mathematics in Brain Imaging
16. Remes, J.J., Starck, T., Nikkinen, J., Ollila, E., Beckmann, C.F., Tervonen, O., Kiviniemi, V., Silven, O.: Effects of repeatability measures on results of fmri sica: A study on simulated and real resting-state effects. NeuroImage (2010) (in Press) Corrected proof
17. van den Heuvel, M.P., Pol, H.E.H.: Exploring the brain network: A review on resting-state fmri functional connectivity. European Neuropsychopharmacology 20(8), 519–534 (2010)
18. Calhoun, T.V.D., Adali, T.: Unmixing fmri with independent component analysis. IEEE Engineering in Medicine and Biology Magazine 25(2), 79–90 (2006)
19. Vercammen, A., Knegtering, H., den Boer, J.A., Liemburg, E.J., Aleman, A.: Auditory hallucinations in schizophrenia are associated with reduced functional connectivity of the temporo-parietal area. Biological Psychiatry 67(10), 912–918 (2010); Anhedonia in Schizophrenia
20. Yao, Z., Wang, L., Lu, Q., Liu, H., Teng, G.: Regional homogeneity in depression and its relationship with separate depressive symptom clusters: A resting-state fmri study. Journal of Affective Disorders 115(3), 430–438 (2009)
21. Zhou, Y., Liang, M., Jiang, T., Tian, L., Liu, Y., Liu, Z., Liu, H., Kuang, F.: Functional dysconnectivity of the dorsolateral prefrontal cortex in first-episode schizophrenia using resting-state fmri. Neuroscience Letters 417(3), 297–302 (2007)
22. Zhou, Y., Shu, N., Liu, Y., Song, M., Hao, Y., Liu, H., Yu, C., Liu, Z., Jiang, T.: Altered resting-state functional connectivity and anatomical connectivity of hippocampus in schizophrenia. Schizophrenia Research 100(1-3), 120–132 (2008)
23. Zou, Q.-H., Zhu, C.-Z., Yang, Y., Zuo, X.-N., Long, X.-Y., Cao, Q.-J., Wang, Y.-F., Zang, Y.-F.: An improved approach to detection of amplitude of low-frequency fluctuation (alff) for resting-state fmri: Fractional alff. Journal of Neuroscience Methods 172(1), 137–141 (2008)

FreeSurfer Automatic Brain Segmentation Adaptation to Medial Temporal Lobe Structures: Volumetric Assessment and Diagnosis of Mild Cognitive Impairment

R. Insausti[1], M. Rincón[2], E. Díaz-López[2], E. Artacho-Pérula[1], F. Mansilla[1],
J. Florensa[1], C. González-Moreno[3], J. Álvarez-Linera[4], S. García[5],
H. Peraita[5], E. Pais[6], and A.M. Insausti[6]

[1] Human Neuroanatomy Laboratory, School of Medicine,
University of Castilla-La Mancha, Albacete, Spain
Ricardo.Insausti@uclm.es
[2] Dept. Inteligencia Artificial. E.T.S.I. Informática,
Universidad Nacional de Educación a Distancia, Madrid, Spain
[3] DEIMOS Space S.L.U., Ronda de Poniente, 19, Edificio Fiteni VI,
2-2ª 28760 Tres Cantos, Madrid, Spain
[4] Unidad de investigación del Proyecto Alzheimer; Fundación CIEN y Fundación
Reina Sofía, Comunidad Autónoma de Madrid
[5] Facultad de Psicología, Universidad Nacional de Educación a Distancia,
Madrid, Spain
[6] Departamento de Ciencias de la Salud, Campus de Tudela,
Universidad Pública de Navarra, Pamplona, Spain

Abstract. Alzheimer's disease is a prevalent and progressive neurode-generative disease that often starts clinically as a memory deficit. Specif-ically, the hippocampal formation (HF) and medial part of the temporal lobe (MTL) are severely affected. Those structures are at the core of the neural system responsible for encoding and retrieval of the memory for facts and events (episodic memory) which is dependent on the HF and MTL. Clinical lesions as well as experimental evidence point that the HF (hippocampus plus entorhinal cortex) and the adjacent cortex in the MTL, are the regions critical for normal episodic memory func-tion. Structural MRI studies can be processed by FreeSurfer to obtain an automatic segmentation of many brain structures. We wanted to ex-plore the advantages of complementing the automatic segmentation of FreeSurfer with a manual segmentation of the HF and MTL to obtain a more accurate evaluation of these memory centers.

We examined a library of cases in which neuroanatomical delimita-tion of the extent of the HF and MTL was made in 48 control and 16 AD brains, and the knowledge provided was applied to 7 cases (2 con-trols and 5 MCI patients) in which 3T MRI scans were obtained at two time points, one year and a half apart. Our results show that volumetric values were preserved in controls as well as non amnestic MCI patients, while the amnestic type (the more often to develop full AD) showed a volume decrease in the HF and MTL structures. The methodology still

J.M. Ferrández et al. (Eds.): IWINAC 2011, Part II, LNCS 6687, pp. 112–119, 2011.
© Springer-Verlag Berlin Heidelberg 2011

needs further development to a full automatization, but it seems to be promising enough for early detection of volume changes in patients at risk of developing AD.

1 Introduction

FreeSurfer is a free program available on the Web that offers an advanced tool for the delimitation of human brain neuroanatomical structures. Mostly based on local histogram analysis and atlas-based information, it is capable of separating small portions of the human brain gray matter that form the cortical mantle and subcortical brain centers identifiable macroscopically. Those brain nuclei can be identified through labels assimilated to well known anatomical structures as recognized in brain atlases. Based on this analysis of brain structures, it is possible to calculate the volume of a segmented structure. The outcome of FreeSurfer, albeit providing in many cases a rough estimate, and not free of limitations such as overestimations [14], is fast compared to the manual segmentations for similar structures. In this way, compared with pathological cases, opens the possibility of further insight into the pathology and differential diagnosis of different neurological and psychiatric diseases.

Among the neuropsychiatric syndromes that affect humans, in particular at old age, is Alzheimer's disease. This prevalent disease is highly incapacitating, and whose ultimate diagnosis is made after postmortem examination of the brain. From the appearance of the first symptoms to the death of the patient, there is progressive deterioration of intellectual and functional abilities that lead to the prostration of the person, with the subsequent burden to caregivers and social cost.

Although much research is being devoted to the molecular mechanisms of the cellular pathology, AD is characterized by a common final outcome which is neuronal death and loss of interaction between neurons belonging to different neural networks trough the loss synaptic junctions, particularly in specific regions that support basic functions such as memory for facts and events. The neuronal loss manifests itself as atrophy that can be evidenced at relatively early stages on structural magnetic resonance imaging (MRI), as well as other neuroimaging procedures [9].

In this regard, programs such as FreeSurfer offer the possibility of assessing the volumes of the centers involved in memory processing, and in particular those located in the medial aspect of the human temporal lobe. The medial temporal lobe (MTL) contains the hippocampal formation (HF), necessary for the processing of everyday's events that are stored as specific memories, which can be stored as long-term memory. However, the HF, in the normal human brain, is a component of a more extended network that comprise cortical as well as subcortical centers, and which, by means of neuroanatomical and functional connections, make possible the processing of information that ultimately, will be stored in a distributed way as long term memories [15].

The neuropathology of AD starts at specific portions of the MTL such as the entorhinal cortex (EC), from where it affects other parts of the HF and in a stepwise mode, spreads to association cortex, [1,3]. Therefore, the EC results most severely affected. The pathology extends in such a way that it follows a progression that reflects the known connectivity of the EC, especially with the hippocampus and cerebral cortex [1,7,8]. The cell loss in the HF has a profound impact on memory processing networks, and, as a result, the HF and the neocortex become disconnected [4,16] leading to memory loss, a hallmark symptom in AD patients. The cell and the ensuing axonal arbor loss produces a decrease on the volume of MTL structures perfectly visible on MRI images, even at a relatively early stage of the disease [5,11,17,10,9].

The automatic segmentation of brain structures by the FreeSurfer provides a useful tool for the evaluation of volume changes in well defined structures. However, a tendency to overestimation has been reported in well defined structures such as the hippocampus [14], while the cortex surrounding the hippocampus (the EC and the cortices comprising the parahippocampal region. The parahippocampal region is a cortical ribbon that surrounds the HF in the MTL along its anterior-posterior axis, and is formed by the temporopolar (TPC), perirhinal (PRC) and posterior parahippocampal cortex (PPH). Experimental studies in nonhuman primates show that the EC is a gateway of cortical information to the hippocampus, and that the parahippocampal region provides two-thirds of all the cortical input to the EC. This neuroanatomical substrate provides the basis of the cortical dialog between the HF and the neocortex. FreeSurfer, although able to segment the hippocampus, is not able to segment neither the EC nor the cortices of the parahippocampal region, and these remain included among the cortical ribbon of the temporal lobe. The MRI segmentation of the EC and cortices of the parahippocampal region has been manually traced [5] in the volumetric assessment of the MTL in AD [11], showing the usefulness of the method, although it is time consuming.

In this report we propose an elaboration of the MTL segmentation by complementing the automatic segmentation of FreeSurfer to define more accurately volume changes in a small sample of patients affected with mild cognitive impairment (MCI) which is in many instances a start form of AD that can progress to a full AD [17]. Our method, based on the anatomical features of the human HF and MTL aims at providing first a more accurate evaluation of memory centers, and second to set the basis towards an automatic procedure to segment this part of the brain, based on the declarative and procedural knowledge involved.

2   Methods

We have used a series of 48 control brains ranging from 12 to 110 years of age, in which a cytoarchitectonic microscopic analysis was performed to determine the extent of the EC and the parahippocampal region along the MTL. The descriptive data of this series was applied to a further series of 16 cases with AD, in which a similar analysis was carried out. In this way, we determined the mean distances, relationships of the cortical areas to the sulci, and relative distances to several landmarks along the anterior-posterior length of the MTL, from TPC to PPH, at the end of the HF. The criteria for the MRI segmentation of the EC, TPC, PRC and PPH have been reported elsewhere [5,6]. The results of this analysis led to the construction of a table (Table 1) in which the first and last section for each of the constituents of the cortical ribbon of the MTL. The thickness and spacing of each one of the sections resulted in a distance in mm. that was annotated. This series was compared to a series of 6 brains in which ex-vivo MRI at 3T was obtained. Once the MRI series was obtained, the brains were serially sectioned, processed for cytoarchitectonic evaluation and analysis of the congruence between the MRI images and the neuroanatomical sections. An example of this procedure has been reported previously [2].

In this context we examined a series of 8 cases in subjects (5 males, 3 females) that were evaluated with a battery of neuropsychological tests and characterized as MCI (5 cases) or controls (2 cases) used in another study [13]. MRI scans were obtained in a GE Healthcare scan at 3T. Scans were repeated 18 months after in order to assess volumetric changes.

All MRI scans were processed with FreeSurfer and two neuroanatomists (RI and AI) who were blind to the neuropsychological state of the patients. The FreeSurfer automatic segmentation was first examined to determine the longitudinal extent of the MTL cortical structures and fulfill the information in Table 1. This was done independently by the two neuroanatomists (RI and AI), and criteria were compared. Thereafter, 3D Slicer editing tools extended with our own semiautomatic algorithms in each of the MRI sections were used to label distinct areas in different colors in each of the MRI sections. After completion, the volumes of TPC, PRC, PPH, EC, and hippocampus were calculated.

Table 1. Table used as a template in which the number of the section corresponding to the first and last section of the MTL cortical structures is annotated

SERIES
CASE IDENTIFICATION

Section #	Beginning TPC	End TPC	Comments
	Beginning PRC	End PRC	
	Limen insulae		
	Beginning EC	End EC	
	Beginning Head Hippocampus	End Head Hippocampus	
	Gyrus. Intralimbicus		
	Beginning PPH	End PPH	

3   Results

The examination of the MRI scans revealed a close correspondence with anatomical landmarks that can be easily identified to set limits to the MTL cortical constituents. Results in a normal brain are presented in Figure 1.

The main landmarks are the first section in which the temporal pole (TPC) is displayed (Fig. 1 a, b). The temporal lobe increases in size rapidly as the superior and middle temporal gyri are included in the plane of section. The limen insulae is very easy to detect, as it is the point where the frontal and temporal lobes meet (Fig. 1 c). It usually takes place about 2.5 cm from the temporal pole. PRC continues the TPC along the MTL at a variable distance from the limen insulae, in relation to the variability of the collateral sulcus, which is a frontier with the EC. The EC starts shortly after, and it extends for about 2 cm (Fig. 1 c). The medial bank of the collateral sulcus makes the transition with PRC. The hippocampus starts as the hippocampal head (Fig. 1 d), usually about 1 cm. posterior to the EC start.

The hippocampus presents at this level a number of convolutions (Fig. 1 e, f) that characterize this region, and ends in a second critical landmark that is the gyrus intralimbicus (Fig. 1 f), which corresponds to the end of the hippocampal head. Approximately 2 mm behind the gyrus intralimbicus the end of the EC can be placed, and the caudal limit of PRC is 2 mm. after the EC. The PRC is replaced by the PPH, that lasts as far as the fimbria and fornix can be seen in continuation one to the other. The gyrus intralimbicus defines the beginning of the body of the hippocampus (Fig. 1 g), and the limit between body and tail of the hippocampus (Fig. 1 h) was set approximately midway between the gyrus intralimbicus and the hippocampal termination under the splenium of the corpus callosum (Fig. 1 i).

All these landmarks were corroborated at the neuroanatomical examination of the cases from our archive and the post-mortem MRI images.

The analysis of the eight cases in which two MRI scans were obtained one year and a half later were confirmatory of the initial classification of controls or MCI patients (Fig. 2). Controls did not show any significant change in the volumes of the MTL structures. MCI patients were classified as belonging to the amnestic or non-amnestic type. The non-amnestic MCI cases did not present significant variation in the volume of MTL structures. MCI amnestic patients showed a slight decrease in all MTL structures.

4   Discussion

Our methodology suggests that this important region of the human brain, heavily involved in memory processing can be segmented manually according to neuroanatomical criteria that take advantage of the peculiar morphology of the human HF and the surrounding cortex that makes up the parahippocampal region. Both the HF and the parahippocampal region are heavily damaged in AD and present abundant neuropathological lesions (neuritic plaques and neurofibrillary tangles) which are pathognomonic of AD [1].

The main interest in the segmentation of the constituents of the MTL is that it may provide the basis for the development of algorithms that might lead to the automatic segmentation of the MTL and faster and more reproducible results in the clinical setting by combining statistically significant data samples with

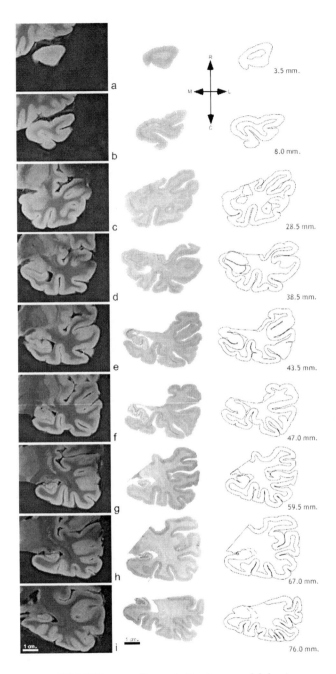

Fig. 1. Post-mortem 3T MRI scans through the temporal lobe in a control case. Adjacent are the actual neuroanatomical series on which the limits of the HF and MTL were set. On the right hand side, line drawings of the sections with the distances from the temporal pole are depicted.

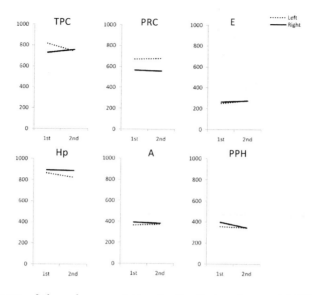

Fig. 2. Histogram of the volume variation for the first and second MRI, in which the preservation of the volumes along the series can be noticed

the procedural knowledge used by the expert while tracing the contours. It is noteworthy to point out that the early diagnosis of AD is still in its infancy, although accurate volumetric assessment of critical brain regions such as the MTL may provide a basis for a more precise follow-up of patients that still have moderate symptoms of cognitive decline.

Acknowledgements

Supported by projects PI 59/2007 FISCAM (Junta de Comunidades de Castilla-la Mancha, Spain), TSI-020110-2009-362 from the Spanish Ministry of Industry, Tourism and Trade and SEJ2007-63325 from the Spanish Ministery of Science and Innovation (MICINN).

References

1. Arnold, S.E., Hyman, B.T., Flory, J., Damasio, A.R., Van Hoesen, G.W.: The topographical and neuroanatomical distribution of neurofibrillary tangles and neu-ritic plaques in the cerebral cortex of patients with Alzheimer's disease. Cereb. Cortex 1, 103–116 (1991)
2. Blaizot, X., Mansilla, F., Insausti, A.M., Constans, J.M., Salinas-Alamán, A., Pró-Sistiaga, P., Mohedano-Moriano, A., Insausti, R.: The human parahippocampal region: I. Temporal pole cytoarchitectonic and MRI correlation. Cer. Cortex 20, 2198–2212 (2010)
3. Braak, H., Braak, E.: Staging of Alzheimer's disease-related neurofibrillary changes. Neurobiol. Aging 16, 271–278 (1995)

4. Hyman, B.T., Van Hoesen, G.W., Damasio, A.R., Barnes, C.L.: Alzheimer's disease: cell-specific pathology isolates the hippocampal formation. Science 225, 1168–1170 (1984)
5. Insausti, R., Juottonen, K., Soininen, H., Insausti, A.M., Partanen, K., Vainio, P., Laakso, M.P., Pitkanen, A.: MR volumetric analysis of the human entorhinal, perirhinal, and temporopolar cortices. Amer J. Neuroradiol. 19, 656–671 (1998)
6. Insausti, R., Insausti, A.M., Mansilla, F., Abizanda, P., Artacho-Pérula, E., Arroyo-Jimenez, M.M., Martinez-Marcos, A., Marcos, P., Muñoz-Lopez, M.: The human parahippocampal gyrus. Anatomical and MRI correlates. In: 33th Annual Meeting of Society for Neuroscience New Orleans, USA (November 2003)
7. Insausti, R., Amaral, D.G.: The Human Hippocampal Formation. In: Paxinos, G., Mai, J. (eds.) The Human Nervous System, 2nd edn., pp. 871–912. Academic Press, San Diego (2004)
8. Insausti, R., Amaral, D.G.: Entorhinal cortex of the monkey: IV. Topographical and laminar organization of cortical afferents. J. Comp. Neurol. 20, 608–641 (2008)
9. Jack, C.R., Knopman, D.S., Jagust, W.J., Shaw, L.M., Aisen, P.S., Weiner, M.W., Petersen, R.C., Trojanowski, J.Q.: Hypothetical model of dynamic biomarkers of the Alzheimer's. pathological cascade. Lancet Neurol. 9, 119–128 (2010)
10. Jack, C.R., Lowe, V.J., Senjem, M.L., Weigand, S.D., Kemp, B.J., Shiung, M.M., Knopman, D.S., Boeve, B.F., Klunk, W.E., Mathis, C.A., Petersen, R.C.: 11C PiB and structural MRI provide complementary information in imaging of Alzheimer's disease and amnestic mild cognitive impairment. Brain 131, 665–680 (2007)
11. Juottonen, K., Laakso, M.P., Insausti, R., Lehtovirta, M., Pitkänen, A., Partanen, K., Soininen, H.: Volumes of the entorhinal and perirhinal cortices in Alzheimer's disease. Neurobiol. Aging 19, 15–22 (1998)
12. Malykhina, N.V., Bouchard, T.P., Ogilvie, C.J., Coupland, N.J., Seres, P., Camicioli, R.: Three-dimensional volumetric analysis and reconstruction of amygdala and hippocampal head, body and tail. Psychiat. Res. Neuroimaging 155, 155–165 (2007)
13. Peraita, H., Hernández-Tamames, J.A., Dobato, J.L., Díaz, C., Seco de Herrera, A.G., García, S., Linera, J.A.: Neuropsychological and imaging (MR) biomarkers in the early detection of Mild Cognitive Impairment (MCI). Póster. In: 3nd World congress on Controversies in Neurology (CONy), Pragu, October 8-11 (2009)
14. Sánchez-Benavides, G., Gómez-Ansón, B., Sainz, A., Vives, Y., Delfino, M., Peña-Casanova, J.: Manual validatiion of FreeSurfer's automated hippocampal segmetation in normal aging, mild cognitive impairment and Alzheimer disease subjects. Psychiatry Dis. Neuroimaging 181, 219–225 (2010)
15. Squire, L.R., Stark, C.E.L., Clark, R.E.: The Medial Temporal Lobe. Annu. Rev. Neurosci. 27, 279–306 (2004)
16. Stoub, T.R., de Toledo-Morrell, L., Stebbins, G.T., Leurgans, S., Bennett, D.A., Shah, R.C.: Hippocampal disconnection contributes to memory dysfunction in individuals at risk for Alzheimer's disease. Proc. Natl. Acad. Sci. U. S. A 103, 10041–10045 (2006)
17. Whitwell, J.L., Przybelski, S.A., Weigand, S.D., Knopman, D.S., Boeve, B.F., Petersen, R.C., Jack, C.R.: 3D maps from multiple MRI illustrate changing atrophy patterns as subjects progress from mild cognitive impairment to Alzheimer's disease. Brain 130, 1777–1786 (2008)

Alzheimer Disease Classification on Diffusion Weighted Imaging Features

M. Termenon[1], A. Besga[2], J. Echeveste[3], A. Gonzalez-Pinto[2], and M. Graña[1]

[1] Grupo de Inteligencia Computacional, UPV/EHU
www.ehu.es/ccwintco
[2] Unidad de Investigación en Psiquiatría del Hospital de Santiago Apostol,
Vitoria-Gasteiz
[3] Departamento de Resonancia Magnética, Osatek-Vitoria

Abstract. An on-going study in Hospital de Santiago Apostol collects anatomical T1-weighted MRI volumes and Diffusion Weighted Imaging (DWI) data of control and Alzheimer's Disease patients. The aim of this paper is to obtain discriminant features from scalar measures of DWI data, the Fractional Anisotropy (FA) and Mean Diffusivity (MD) volumes, and to train and test classifiers able to discriminate AD patients from controls on the basis of features selected from the FA or MD volumes. In this study, separate classifiers were trained and tested on FA and MD data. Feature selection is done according to the Pearson's correlation between voxel values across subjects and the control variable giving the subject class (1 for AD patients, 0 for controls). Some of the tested classifiers reach very high accuracy with this simple feature selection process. Those results point to the validity of DWI data as a image-marker for AD.

1 Introduction

Alzheimer's Disease (AD) is the most common form of dementia in elderly people. This degenerative disorder presents a cognitive and behavioral impairment that interferes with the daily life of the individual and its social network (family), with a high economical and psychological cost. The diagnosis of AD can be done after the exclusion of other forms of dementia but a definitive diagnosis can only be made after a post-mortem study of brain tissue. This is one of the reasons why early diagnosis based on Magnetic Resonance Imaging (MRI) is a current research hot topic in the neurosciences.

Diffusion Weighted Imaging (DWI) provides a measure of the integrity of the White Matter (WM) fibers measuring the movements of the water molecules inside the brain. DWI measures the motion of water molecules in several directions. This information can be used to provide structural information *in vivo* [7,1] through the computation of diffusion tensors, the so called Diffusion Tensor Imaging (DTI). Scalar measures of diffusion computed from DTI are fractional anisotropy (FA) and mean diffusivity (MD), which give information about the magnitude of the diffusion process at each voxel, though they do not give direction. Multiple WM abnormalities have been found in DTI studies about AD

J.M. Ferrández et al. (Eds.): IWINAC 2011, Part II, LNCS 6687, pp. 120–127, 2011.

such as differences in the splenium of the corpus callosum and temporal lobe between AD patients and controls [9], temporal lobe WM atrophy with WM microscopic damage in the thalamic radiations and in the corpus callosum [8]. Some studies on FA show effects in the bilateral posterior cingulate gyri and bilateral superior longitudinal fascicles [6]. However, there have been few whole-brain voxel-wise studies in which no a priori hypothesis was made regarding anatomical localization of white matter abnormalities [3].

The present paper will focus on the application of Machine Learning (ML) algorithms for the computer aided diagnosis (CAD) of Alzheimer Disease, on the basis of feature vectors extracted from DTI scalar measures FA and MD. Previous works have performed the feature extraction on T1-weighted MRI anatomical data using Voxel Based Morphometry (VBM) or sophisticated wraper feature selection methods[5,4]. In this paper, feature selection is made on the basis of the Pearson's correlation between the voxel value across the sample subjects and the control variable giving the subject class (1 for AD patients, 0 for controls). We select as features, the sites of voxels showing an absolute correlation value above some percentile of its empirical distribution. Previous to the feature extraction process, a careful registration of the FA and MD volumes is performed, including affine and non-linear registrations, to ensure that the same voxel site corresponds to the same anatomical location across subjects. Classifier building algorithms tested include Support Vector Machines (SVM), Relevance Vector Machines (RVM) and 1-NN (Nearest Neighbour) classifiers. Results on the collected subjects indicate that FA and MD data are useful to discriminate between AD patients and controls.

Section 2 describes the materials available for this study and the acquisition protocol. Section 3 gives a summary of the image processing pipeline followed to register the MRI and DWI volumes, it explains the feature extraction process and it summarizes the classification algorithms used for this study. Section 4 shows the results of our computational experiments. Section 5 gives our final comments and conclusions.

2   Materials

Thirty five men and women (aged 60-89), twenty controls and fifteen patients, from Hospital de Santiago Apostol (Vitoria-Gasteiz), were the subjects of this study. Structural MRI and DTI data were used for this experiment. Patients also include two cases of very mild to mild AD.

MR scanning was performed on a 1.5 Tesla scanner (Magnetom Avanto, Siemens). Study protocol consists of 3D T1 acquisition (isometric 1x1x1mm, 176 slices, TR=1900ms, TE=337ms and FOV=256/76%), a 3D Flair sequence (isometric 1x1x1mm, 176 slices, TR=5000ms, TE=333ms and FOV=260/87.5%) and diffusion tensor sequence (slice thickness = 5mm, 19 slices, TR=2700ms, TE=88ms, matrix 120/100, 3 average, b=1000 and 30 tensor directions). All the studies were performed at the same scanner and there was no modification of the acquisition parameters of the different sequences.

Algorithm 1. T1 and DWI data processing pipeline to obtain corrected FA and MD.

1. Convert DICOM to nifti
2. Skull stripping T1 volumes
3. Affine registration of T1 skull stripped volumes to template MNI152.
4. Correct DWI scans.
5. To obtain skull stripped brain masks for each DWI corrected scans.
6. To apply the diffusion tensor analysis.
7. Rigid registration 6DoF of FA and MD volumes to T1 affine registered volumes, resulting of Step3.

3   Methods

3.1   Image Processing

First step was converting DICOM images to nifti format. Second, T1-weighted sMRI volumes are skull stripped, and affine registered to the Montreal Neurological Institute (MNI152) standard template.

To obtain these FA and MD volumes, we apply the DWI correction for eddy currents, we extract the brain mask of the volumes, compute the DTI [2], and finally the FA and MD scalar measures. We perform spatial normalization of the data to ensure the correspondence between voxel sites and anatomical features across all subjects. Algorithm (1) summarizes the image processing.

3.2   Feature Extraction

We process the spatially normalized FA and MD maps independently. Considering each voxel site independently, we compose a vector with the FA or MD intensities at the voxel site across all the subjects. We compute Pearson's correlation coefficient between this vector and the control variable, where Control=0 and Patients=1, obtaining two independent volumes, one for FA and other for MD, of correlation values at each voxel. Then, we select a threshold corresponding to a percentile of the absolute correlation distribution, retaining the voxel sites with absolute value of correlation above this threshold. Finally, for each percentile selected we compose two feature vector for each subject, one extracted from the FA data and other from MD data. We show in figures 1 and 2 the localization of the voxel sites for the feature extraction of FA and MD data, respectively, when applying a 99,5% percentile overlayed to MNI152 template.

Voxel sites selected to build the feature vectors, were localized in many different regions of the brain [1], being different for FA and MD maps. For the FA case, most significant differences were found in the thalamus, temporal lobe and

---

[1] This specification of the voxel locations were obtained with the "atlasquery" tool from FMRIB's FSL (http://www.fmrib.ox.ac.uk/fsl/) using the "MNI Structural Atlas" and the "JHU White-Matter Tractography Atlas".

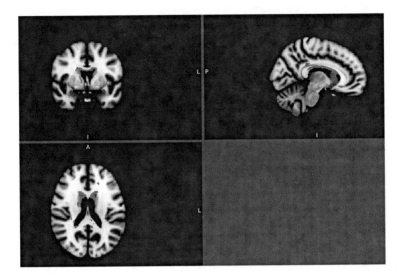

Fig. 1. Voxel sites for FA features selected with a 99,5% percentile on the correlation distribution

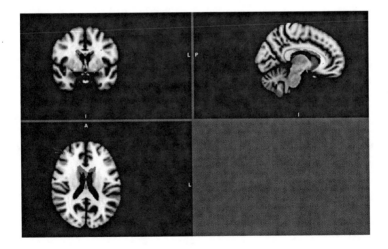

Fig. 2. Voxel sites for MD features selected with a 99,5% percentile on the correlation distribution

corpus callosum. In white matter, we found discriminant voxel values in the cingulum gyrus, anterior thalamic radiation, corticospinal tract and uncinate fasciculum. In the MD maps, there were also findings in the inferior fronto-occipital fasciculus.

3.3   Classification

Support Vector Machines: The Support Vector Machines (SVM) [12] approach is a pattern recognition technique based on the statistical learning theory. Its training principle consist of finding the optimal linear hyperplane that minimize the expected classification error. The classification approach works to solve the following optimization problem:

$$\min_{\mathbf{w},b,\boldsymbol{\xi}} \frac{1}{2}\mathbf{w}^T\mathbf{w} + C\sum_{i=1}^{l}\xi_i, \tag{1}$$

subject to

$$y_i(\mathbf{w}^T\phi(\mathbf{x}_i)+b) \geq (1-\xi_i),\ \xi_i \geq 0,\ i=1,2,\ldots,n. \tag{2}$$

The minimization problem is solved via its dual optimization problem:

$$\min_{\alpha} \frac{1}{2}\boldsymbol{\alpha}^T Q\boldsymbol{\alpha} - \mathbf{e}^T\boldsymbol{\alpha}, \tag{3}$$

subject to

$$\mathbf{y}^T\boldsymbol{\alpha} = 0,\ 0 \leq \alpha_i \leq C,\ i=1,\ldots,l. \tag{4}$$

Where $\mathbf{e}$ is a vector of all ones, $C > 0$ is the upper bound on the error and Q is an $l \times l$ positive semidefinite matrix. The Q elements are based on a kernel function, $K(\mathbf{x}_i, \mathbf{x}_j)$, that describes the behavior of the support vectors. The chosen kernel function results in different kinds of SVM with different performance levels, and the choice of the appropriate kernel for a specific application is a difficult task. In this study we only needed to use a linear kernel, defined as:

$$K(\mathbf{x}_i, \mathbf{x}_j) = 1 + \mathbf{x}_i^T\mathbf{x}_j. \tag{5}$$

This kernel shows good performance for linearly separable data.

Relevance Vector Machines: The Relevance Vector Machine (RVM) is a Bayesian sparse kernel technique for classification and regression. It is a model of identical functional form to the Support Vector Machine but embedded in a probabilistic Bayesian framework. It utilizes fewer basis functions while offering a number of additional advantages [10] like the benefits of probabilistic predictions, automatic estimation of 'nuisance' parameters and the facility to use arbitrary basis functions.

In the classification case, given a data set of input-target pairs $\{\mathbf{x}_n, t_n\}_{n=1}^{N}$ where $\mathbf{x}_n \in \mathbb{R}^d$ and $t_n \in \{0, 1\}$, a RVM classifier output is the posterior probability of membership of one of the classes given the input $\mathbf{x}$. Predictions are based on a linear function $y(x)$ defined over the input space:

$$y(\mathbf{x}; \mathbf{w}) = \sum_{i=1}^{N} w_i K(\mathbf{x}, \mathbf{x_i}) + w_0, \tag{6}$$

where $K(\mathbf{x}, \mathbf{x_i})$ is the *kernel* function that defines a basis function for each sample in the training set. Linear model is generalized by applying the logistic sigmoid link function $\sigma(y) = 1/(1+e^{-y})$ to $y(\mathbf{x})$. To avoid overfitting, it is imposed an additional constraint on the parameters by defining an explicit prior probability distribution over them, in this case, a zero-mean Gaussian prior distribution over $\mathbf{w}$:

$$p(\mathbf{w}|\boldsymbol{\alpha}) = \prod_{i=0}^{N} \mathcal{N}\left(w_i|0, \alpha_i^{-1}\right), \tag{7}$$

with $\boldsymbol{\alpha}$ a vector of $N+1$ hyperparameters, having an individual hyperparameter associated independently with each weight, moderating the strength of the prior. These parameters α_i respond to a zero-mean Gaussian distribution with variance λ_i^{-1}, forcing them to be concentrated around zero. The few training vectors that have finite α value are called Relevance Vectors (RV) and they are the only samples that contribute to the decision function $y(x)$.

4   Computational Experiments Results

We evaluated the performance of the classifiers built using a 10-fold cross-validation methodology, repeating the experiment 50 times. To quantify the results, we measured the Accuracy, defined as the ratio of the number of test volumes correctly classified to the total of tested volumes, $Accuracy = \frac{TP+TN}{N}$, the $Sensitivity = \frac{TP}{TP+FN}$ and the $Specificity = \frac{TN}{TN+FP}$. True Positives (TP) are the number of patient volumes correctly classified; True Negatives (TN) are the number of control volumes correctly classified; False Positives (FP) are the number of control volumes classified as diseased patients; False Negatives (FN) are the number of diseased patient volumes classified as control subjects and N, the total of subjects. We have labelled patients as class 1 and controls as class 0. As the image assessment is an additional finding meant to support other diagnostic information sources, there is a specific need for high sensitivity and specificity systems, thus these performance measures were preferred above others in our ongoing works. For all the classifiers, we have determined the optimal values of the classifier parameters via independent grid searches performed at each cross-validation fold.

The average results of 10-fold cross-validation tests computed on the features are presented in table 1. Tested approaches are the RVM [11] and SVM with linear kernels and the nearest neighbor (1-NN). Best results are achieved with linear SVM classifier, resulting in a very high values for all the cases of features extracted.

Table 1. Classification results for the FA and MD volumes

	Pcr 99.50%		Pcr 99.90%		Pcr 99.95%		Pcr 99.99%	
1NN	**FA**	**MD**	**FA**	**MD**	**FA**	**MD**	**FA**	**MD**
Acc.	0.67	0.74	0.72	0.80	0.85	0.81	0.94	0.75
Sens.	0.68	0.78	0.69	0.81	0.86	0.81	0.87	0.75
Spec.	0.67	0.70	0.75	0.79	0.84	0.80	0.99	0.75
SVM	**FA**	**MD**	**FA**	**MD**	**FA**	**MD**	**FA**	**MD**
Acc.	0.96	0.99	0.96	0.99	0.97	0.98	0.99	0.94
Sens.	0.98	0.99	0.97	0.99	0.99	0.95	0.98	0.90
Spec.	0.95	0.98	0.95	0.99	0.96	0.99	0.99	0.97
RVM	**FA**	**MD**	**FA**	**MD**	**FA**	**MD**	**FA**	**MD**
Acc.	0.83	0.71	0.89	0.63	0.89	0.63	0.91	0.57
Sens.	0.78	0.67	0.88	0.65	0.88	0.67	0.89	0.67
Spec.	0.87	0.73	0.87	0.63	0.89	0.60	0.91	0.53

5   Conclusion

The aim of this paper was to test the hypothesis that features extracted from DTI images of Alzheimer patients and control subjects could be differentiated using classification techniques based on Machine Learning. The way to build the feature vectors has been the direct selection of voxels from the DTI-derived FA and MD scalar valued volumes that show a high correlation with the control variable that labels the subjects. The selected voxels correspond to findings reported in the medical literature. Surprisingly, all the classifiers obtain very good results, except for the RVM classifier with Mean Diffusivity features. We think that appropriate pre-processing of the data is of paramount importance and can not be disregarded trusting that ensuing statistical or machine learning processes may cope with the errors introduced by lack of appropriate data normalization. Therefore, our main conclusion is that the proposed feature extraction is very effective providing a good discrimination between AD patients that can easily be exploited by the classifier construction algorithms. The main limitation of this study is that the results come from a small database. Therefore, more extensive testing will be needed to confirm our conclusions.

References

1. Basser, P.J., Mattiello, J., LeBihan, D.: MR diffusion tensor spectroscopy and imaging. Biophysical Journal 66(1), 259–267 (1994); PMID: 8130344 PMCID: 1275686
2. Behrens, T.E.J., Woolrich, M.W., Jenkinson, M., Johansen-Berg, H., Nunes, R.G., Clare, S., Matthews, P.M., Brady, J.M., Smith, S.M.: Characterization and propagation of uncertainty in diffusion-weighted MR imaging. Magnetic Resonance in Medicine 50(5), 1077–1088 (2003)
3. Douaud, G., Jbabdi, S., Behrens, T.E.J., Menke, R.A., Gass, A., Monsch, A.U., Rao, A., Whitcher, B., Kindlmann, G., Matthews, P.M., Smith, S.: DTI measures in crossing-fibre areas: Increased diffusion anisotropy reveals early white matter alteration in MCI and mild alzheimer's disease. NeuroImage (in press) Uncorrected Proof

4. Fan, Y., Shen, D., Gur, R.C., Gur, R.E., Davatzikos, C.: COMPARE: classification of morphological patterns using adaptive regional elements. IEEE Transactions on Medical Imaging 26(1), 93–105 (2007); PMID: 17243588
5. García-Sebastián, M., Savio, A., Graña, M., Villanúa, J.: On the use of morphometry based features for alzheimer's disease detection on MRI. In: Cabestany, J., Sandoval, F., Prieto, A., Corchado, J.M. (eds.) IWANN 2009. LNCS, vol. 5517, pp. 957–964. Springer, Heidelberg (2009)
6. Parente, D.B., Gasparetto, E.L., da Cruz, L.C.H., Domingues, R.C., Baptista, A.C., Carvalho, A.C.P., Domingues, R.C.: Potential role of diffusion tensor MRI in the differential diagnosis of mild cognitive impairment and alzheimer's disease. Am. J. Roentgenol. 190(5), 1369–1374 (2008)
7. Pierpaoli, C., Jezzard, P., Basser, P.J., Barnett, A., Di Chiro, G.: Diffusion tensor MR imaging of the human brain. Radiology 201(3), 637–648 (1996)
8. Serra, L., Cercignani, M., Lenzi, D., Perri, R., Fadda, L., Caltagirone, C., Macaluso, E., Bozzali, M.: Grey and white matter changes at different stages of alzheimer's disease. Journal of Alzheimer's Disease: JAD 19(1), 147–159 (2010); PMID: 20061634
9. Stahl, R., Dietrich, O., Teipel, S.J., Hampel, H., Reiser, M.F., Schoenberg, S.O.: White matter damage in alzheimer disease and mild cognitive impairment: Assessment with Diffusion-Tensor MR imaging and parallel imaging techniques1. Radiology 243(2), 483–492 (2007)
10. Tipping, M.E.: Sparse bayesian learning and the relevance vector machine. Journal of Machine Learning Research 1(3), 211–244 (2001)
11. Tipping, M.E., Faul, A., Thomson Avenue, J.J., Thomson Avenue, J.J.: Fast marginal likelihood maximisation for sparse bayesian models. In: Proceedings of the Ninth International Workshop on Artificial Intelligence and Statistics, pp. 3–6 (2003)
12. Vapnik, V.N.: Statistical Learning Theory. Wiley Interscience, Hoboken (1998)

Future Applications with Diffusion Tensor Imaging

T. García-Saiz[1], M. Rincón[1], and A. Lundervold[2]

[1] Dept. Inteligencia Artificial. E.T.S.I. Informática,
Universidad Nacional de Educación a Distancia, Madrid, Spain
{tomasgs,mrincon}@dia.uned.es
[2] Department of Biomedicine, University of Bergen, Bergen, Norway

Abstract. In this paper we present which are the operations, that all programs that work with Diffusion Tensor Imaging (DTI), should run to execute to transform the initial set of images to calculate the ellipsoids to be used for the definition of the fiber tracks. At this time we do not care to check the quality of the results of different programs, because our interest is to seek difference options for using the information that is calculated in the intermediate steps, because now the programs only are using the information for the computation of fiber tracks.

The second idea of the paper is the comparison of DTI with Magnetic Resonance Imaging (MRI).

1 What Are the Diffusion Tensor Imaging (DTI)?

The use of DTI is being directed specifically to the construction of the fiber tracks of the white matter. As a starting point for analysis of brain connectivity. But the focus of efforts in this direction, is causing not use all information available in the DTI.

In DTI[1,2,3,4] images we measure is the amount of water molecules, therefore the cerebrospinal fluid have the highest values of the image, and there won't be a big difference between the white and gray matter.

What would the interest of using such images? The interest is found in the behavior of liquids to be affected by magnetic fields, as liquids, e.g. water, moves in the direction of the magnetic field[4], and the only limitation to be fluid in their movement come associated with the container that contains it. Furthermore fluids behave the same way regardless of the direction of the magnetic field.

Extrapolating this information to the brain, we could study the behavior of water molecules contained in the cerebrospinal fluid inside the skull, or study the movement of water molecules within the different substances that form the brain. As expected the CSF would move in the same way regardless of the direction of the magnetic field, and the only limitations that would be the skull. However, as the brain's white matter has a definite directionality, when studying the movement of water molecules within it, observe how it is possible that these molecules are moving more in some directions than others, as can see in the

J.M. Ferrández et al. (Eds.): IWINAC 2011, Part II, LNCS 6687, pp. 128–135, 2011.

figure (1). The direction in which the molecules move more freely may be that corresponding to the direction of the axons and the directions in which they can move less correspond to the limits of the white matter, and that as expected the water molecules contained in the white matter may not cross the membrane that defines it, and therefore can not leave it.

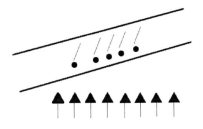

Fig. 1. Movement of water molecules, influenced by a magnetic field

2   Definition of a Tensor

A tensor is a function that represents as the forces are modified, it's visible in the figure (2), for it the tensor must break our force into its three component because, we are working in three dimensions, and we must give the result back into three dimensions, and we have to indicate which is the resultant force. The way of representing this information would be a square matrix of size 3.

It is also interesting to note that a tensor matrix can be used to define an ellipsoid. Also from the point of view of the ellipsoid, this will be defined with an orthonormal basis, which will be the eigenvectors of the tensor, and their three corresponding eigenvalues. The three eigenvalues we will have sorted from highest to lowest, the eigenvalue of greater value express the direction of the major axis of the ellipsoid.

What is the interest of using an ellipsoid? As previously mentioned the movement of water molecules contained in the white matter are limited by their

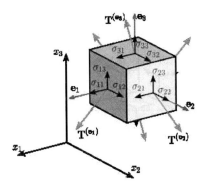

Fig. 2. Tensor

structures, and the resulting movement is close to an ellipsoid. For this reason we are interested in the ellipsoid, and to represent the structure of white matter at a point in the image.

3 Apparent Diffusion Coefficient

To calculate the tensor image initially need to calculate the apparent diffusion coefficient (ADC)[3,4]. This coefficient is calculated using the equation

$$S = S_0 e^{-b(ADC)} \tag{1}$$

for this calculation we need to have two images, S and S_0.

The image S_0 corresponds to a low-intensity magnetic field to prevent the movement of water molecules, and the image S will correspond with a higher magnetic field stimulation, to obtain a displacement of water molecules. Therefore the difference in intensity between the image S_0 and the image S will indicate how many water molecules have been displaced.

This movement of water molecules will be influenced by the direction and intensity of the magnetic field. We are interested in the direction of magnetic field for the study of white matter, because the movement of water molecules is maximum when the magnetic field is parallel to the axons of white matter. In the formula we have presented is not present magnetic field direction, which will be introduced later, now will study the ways that could vary the intensity of magnetic field.

The magnetic field strength is reflected in the formula in the parameter b, which is calculated with the formula

$$b = \gamma^2 G^2 \delta^2 \left(\Delta - \frac{\delta}{3} \right) \tag{2}$$

where

- γ is the gyromagnetic ration, which is specific to each machine.
- G is the intensity of magnetic field.
- δ is the length of the pulsed gradients.
- Δ is the time between pulsed.

The figure (3) shows these parameters.

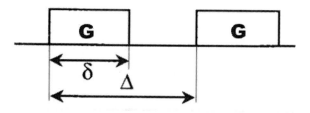

Fig. 3. Parameters of magnetic field

The first thing to check is the sign of the parameter b. The only component that could be negative parameter would be $\Delta - \frac{\delta}{3}$, but as Δ corresponds to the interval between pulses, and δ is the size of the pulse, it follows that $\Delta > \delta$, and therefore $\Delta - \frac{\delta}{3} > 0$. The rest of the components to be squared will always be positive, and therefore the sign of the parameter b is always positive.

In order to increase the value of the parameter b, we can work with the parameters $\{G, \delta, \Delta\}$, so the parameter b will increase if we increase the field strength, G, to extend the pulse magnetic field, δ, or space pulses, Δ. Any of these effects should lead to a reduction in the intensity of the image S.

Now, when the configuration of our system, with his parameter b, is defined we can study the result that we can expect from the ADC parameter.

Remember that all elements of the formula (1) are know less the ADC parameter. For this reason it is necessary to clear the parameter using the logarithms and obtain the formula

$$LnS = LnS_0 - b(ADC) \tag{3}$$

now, it's necessary to pass the $Ln\, S_0$ on the other side of the equation we have

$$LnS - LnS_0 = -b(ADC) \tag{4}$$

which can be written as

$$Ln(S/S_0) = -b(ADC) \tag{5}$$

Therefore to know what will happen to the ADC parameter we study the $Ln(S/S_0)$, since the parameter b is already fixed, and has a positive value.

As the values of the images will always be greater than zero, the division of $\frac{S}{S_0}$ be greater than zero too, it is possible to calculate the logarithm of the division in all situation, and the result of this logarithm it's represent in the equation (6)

$$\begin{cases} S > S_0 \Longrightarrow \frac{S}{S_0} > 1 \Longrightarrow Ln\left(\frac{S}{S_0}\right) > 0 \\ S = S_0 \Longrightarrow \frac{S}{S_0} = 1 \Longrightarrow Ln\left(\frac{S}{S_0}\right) = 0 \\ S < S_0 \Longrightarrow 0 < \frac{S}{S_0} < 1 \Longrightarrow Ln\left(\frac{S}{S_0}\right) < 0 \end{cases} \tag{6}$$

Therefore the sign of the ADC parameter will be the opposite of the logarithm and hence, the value of the ADC parameter it is represented in the equation (7)

$$\begin{cases} Ln\left(\frac{S}{S_0}\right) < 0 \Longleftrightarrow ADC > 0 \\ Ln\left(\frac{S}{S_0}\right) = 0 \Longleftrightarrow ADC = 0 \\ Ln\left(\frac{S}{S_0}\right) > 0 \Longleftrightarrow ADC < 0 \end{cases} \tag{7}$$

As result, we ensure that:

- If the image has a value below zero image, $S < S_0$, the coefficient $ADC > 0$, and we interpret it as there is an increase in water molecules.
- If the image has the same value as the image zero, $S = S_0$, the coefficient $ADC = 0$, and we interpret it as there is no change the number of water molecules.
- If the image has a value greater than zero image, $S > S_0$, the coefficient $ADC < 0$, and we interpret it as there is an increase in water molecules.

The reducing or increasing of the number of water molecules does not involve the destruction or generation of them, the idea is that the water molecules can move. As mentioned before water molecules would move freely in all direction with the only possible limitation of the vessel containing them.

So if on the same image position by acting multiple magnetic field have the same value for the parameter b, and the only difference is the direction of it, and the results are different, these differences are due to any limitations the possibility of movement of water molecules. From the point of view of a brain imaging should speak to the molecules of water contained in different substances can not leave these substances, and therefore should be to move through it. And so if we study an area of the image corresponding to the white matter we will see how water molecules can move freely in certain address and less so in others. The direction in which they can move more freely correspond with the directionality of white matter.

3.1   Build Tensor Matrix

As just presented the magnetic field direction is important to know perfectly the movement of water molecules, and in this way have information about the directionality of white matter, and thus building the fiber tracks of it. But the unique formula we have presented so far have not incorporated the directionality, the equation (1) is independent of direction. So to really build the tensor, we have to extend this formula to enter the magnetic field direction, and the formula will be using is the same equation for each of the directions of magnetic field and therefore the equation we would

$$S_j = S_0 e^{-b(ADC_j)} \tag{8}$$

where S_j is the image in the j-th direction[4]. As a result we have a parameter ADC_j for each of the directions of magnetic field that will indicate the movement of water molecules, regardless of the intensity of magnetic field which is the parameter b of the original equation (1).

Now we can to build the matrix tensor, D, from the parameter ADC_j, now we will use the vectors $\hat{g}_j$, which are the gradient direction[4], which define the direction of the magnetic field, using the equation

$$ADC_j = \hat{g}_j^T D \hat{g}_j \tag{9}$$

3.2   Build Ellipsoid from Tensor Matrix

Using Ce equation (9) have defined the tensor matrix for each points of an image, and it is possible to obtain the eigenvalues and eigenvectors of the ellipsoid defined by the tensor matrix. The reason to obtain the eigenvectors are because they define the axes of the ellipsoid. And it is interesting to know that the largest eigenvalues correspond to the major axis of ellipsoid, and the mayor propagation of the water molecules.

To build the eigenvalues and eigenvectors of the tensor matrix, that form an orthonormal basis i.e. they are perpendicular to each other and unit vector, just have to diagonalize the matrix[4].

The interpretation of the eigenvalues and eigenvectors [4] have to be the directions of movement of water molecules. If the point of the image that we are studying is located within the white matter should consider that the eigenvectors associated to eigenvalues greater than express the direction of axons in the white matter, while the other will bring us closer to the limits. Once we found the eigenvector indicates the direction of white matter may be used for construction of the fiber tracks.

4   Differences between DTI and MRI

The interest of this paper is to differentiate the MRI of the DTI in terms of the amount of information that we get from both. So the biggest difference we find is that while the MRI images are a 3D image, the DTI are multiple 3D images, each of these images correspond to one of the directions of magnetic field that we have used in the study. All these images are needed to build the tensor matrix, and therefore the eigenvectors to be used in the construction of the fiber traks.

Now we will present different problems where the use of the information contained in the DTI images is beneficial for the simplification or automation of the problem.

5   Building Planes Defined in the Images

Suppose that it is interesting to work with a plane contained within our image that is defined from the white matter. In the case of working with MRI images the only way we will define that plane with the information you give us the expert on it, however when looking for the same plane within the DTI may use the information we have regarding for the structure of white matter. E.g. it would be easy to calculate perpendicular planes to the white matter, visible in the figure (5), as we have that the first eigenvalue of the tensor matrix, and therefore in this case would not be necessary to inject information about the image.

Fig. 4. Perpendicular plane to fiber tracks.

6   Removal of the Skull

The removal of the skull in MRI is based on the Image registration that we make of that image with respect to our template, and to use only one thresholding with respect to the intensity of the image generates too many errors. But when we work with DTI we can extend the information used for disposal of the skull, because the cerebrospinal fluid has a high ADC value in its coefficient, right side of the figure (5), is therefore easy to segment, the skull itself, has a low ADC coefficient, the elements outside the skull tend to have a negative ADC, left side of the figure (5), with all this information is simple the construction of a mask that eliminates the skull of the images using only their own information to them, and this is a way to minimize errors that might commit.

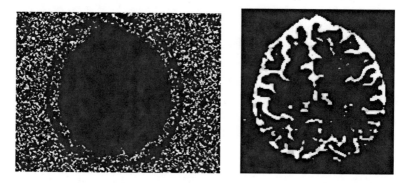

Fig. 5. ADC Images

7   Conclusions

The use of DTI and not only to build the fiber tracks, would be an advance in comparison with the MRI, in a MRI only have one layer of information, only have one 3d image, but in a DTI have a layer for each direction of the magnetic field, and in this case it is possible to combine all layers to obtain a new information in each pixels of the images.

Acknowledgments. The information of this paper was obtained during a stay in the Department of Biomedicine of university of Bergen, supported by a grant from Iceland, Liechtenstein and Norway through the EEA Financial Mechanism. Supported and coordinated by Universidad Complutense de Madrid.

References

1. Basser, P.J., Mattiello, J., LeBihan, D.: Mr diffusion tensor spectroscopy and imaging (January 1994)
2. Le, B.D., Mangin, J.F., Poupon, C., Clark, C.A., Pappata, S., Molko, N.: Diffusion tensor imaging: concepts and applications. J. Magn. Reson. Imaging 13(4) (2001)
3. Melhem, E.R., Mori, S., Mukundan, G., Kraut, M.A., Pomper, M.G., van Zijl, P.: Diffusion tensor mr imaging of the brain and white matter tractography. American Journal of Roentgenology 178(1), 3 (2002)
4. Zhang, F., Hancock, E.R., Goodlett, C., Gerig, G.: Probabilistic white matter fiber tracking using particle filtering and von mises-fisher sampling. Medical Image Analysis 13(1), 5–18 (2009)

Monitoring Neurological Disease in Phonation

Pedro Gómez-Vilda[1], Roberto Fernández-Baíllo[1],
José Manuel Ferrández-Vicente[2], Victoria Rodellar-Biarge[1],
Agustín Álvarez-Marquina[1], Luis Miguel Mazaira-Fernández[1],
Rafael Martínez-Olalla[1], and Cristina Muñoz-Mulas[1]

[1]Grupo de Informática Aplicada al Tratamiento de Señal e Imagen,
Facultad de Informática, Universidad Politécnica de Madrid,
Campus de Montegancedo, s/n, 28660 Madrid
[2]Dpto. Electrónica, Tecnología de Computadoras,
Univ. Politécnica de Cartagena,
30202, Cartagena
pedro@pino.datsi.fi.upm.es

Abstract. It is well known that many neurological diseases leave a fingerprint in voice and speech production. The dramatic impact of these pathologies in life quality is a growing concert. Many techniques have been designed for the detection, diagnose and monitoring the neurological disease. Most of them are costly or difficult to extend to primary services. The present paper shows that some neurological diseases can be traced a the level of voice production. The detection procedure would be based on a simple voice test. The availability of advanced tools and methodologies to monitor the organic pathology of voice would facilitate the implantation of these tests. The paper hypothesizes some of the underlying mechanisms affecting the production of voice and presents a general description of the methodological foundations for the voice analysis system which can estimate correlates to the neurological disease. A case of study is presented from spasmodic dysphonia to illustrate the possibilities of the methodology to monitor other neurological problems as well.

1 Introduction

During the last years a methodology to accurately detect voice pathology of organic origin has been developed [4]. This methodology supports early pathology detection and monitoring may be extended to other diseases of neurological nature if these leave alterations or correlates in voice production [2]. This objective could be of strategic interest towards research and innovation in e-health services, as according to the World Health Organization an estimate of 6.8 million people die yearly from diseases of neurological origin [10]. The highest incidence is due to Parkinson, epilepsy and Alzheimer, among others. The largest cause of disability and impairment among population above 65 is due to this group of diseases. Besides its enormous social impact the neurological diseases present also a strong economical burden, being responsible of at least 5 % of hospitalizations.

J.M. Ferrández et al. (Eds.): IWINAC 2011, Part II, LNCS 6687, pp. 136–147, 2011.

In the western world at least one out of 20 persons demanded a neurological disease consultation from the medical services during 2005. The resources devoted to their treatment were of around 139 billion euros in the Euro Zone during that same year. The evolution of the age pyramid and the increment in the number of elder people in the next years all over the first world will mean a larger impact of neurological diseases. It is important to remark that the diagnosis of these illnesses is rather complex because some of their symptoms may be related with other diseases as well. The attention to the patient with neurological disease will demand larger and more specialized sanitary resources to allow a more accurate and early diagnose. Early detection of these diseases is considered a key factor contributing to efficient attention and treatment. This early warning can be made possible from correlates of the neuromotor functions, as handwriting, eye movement, walking, as well as from electro-encephalography (EEG), functional Magnetic Resonance (fMR), etc. [6]. Voice alterations are good correlates due to the non-invasive character and simple methodology of voice inspection, which consists in obtaining samples involving common phonation styles. The rhythm, prosody, fluentness and other related concepts have been largely used by neurologists to evaluate and grade the extent and severity of the neurological pathology [9]. In this sense the advances in the reconstruction of the glottal excitation and the biomechanical correlates found in this signal, including vocal fold tension patterns provided by the organic pathology detection methodology already developed [2] may be applied in the early detection of the neurological disease, as the voice production apparatus, and especially the larynx, is largely dependent on the muscular tone granted by complex neuromotor circuits and terminals. Tremor and spasm as correlates of dystonia are common symptoms appearing relatively often in a list of non-organic dysphonias. The paper is organized as follows: the relationship between the neurological disease and voice and speech production is reviewed in section 2. In section 3 the methodology used in voice pathology detection and grading is briefly presented. In section 4 the application of this methodology to the detection and grading of a kind of neurological alteration of voice known as spasmodic dysphonia which presents neuromuscular alterations to the control of vocal folds is presented, and some results are given from a study case, accompanied by a brief discussion. Conclusions and future lines are presented in section 5.

2   Neurological Diseases and Voice Production

Voice pathologies are disorders of the phonatory system of different origins which manifest distinct symptoms. Classically two main groups are distinguished: organic and neurological. The first group is related with alterations found in the physiological structures of the larynx (nodules, polyps, edemae, sulci, cysts, granulomae, papilomae, carcinomae, etc). The second group does not show clear organic alterations in the larynx, and have to see more with neuromuscular diseases. Their origin has to be found in the central neural system (cortical areas and centres related with language production) or within neural transmission (brain stem, vagus nerve, larynx muscle innervation, etc.). According to the symptoms shown many of them can be associated with certain neurodegenerative diseases as Parkinson, Alzheimer, Permanent Non-Fluent Apraxia, etc. Spasmodic dysphonia is one of them, which involves involuntary transient muscle spasms which interfere with voice production resulting in voice breaks either within vowels in the adductor form of the disorder or during vowel onsets after voiceless consonants in the abductor form. The adductor dystonia is due to an over-stressing spasm, which presses both vocal folds above the average, resulting in over-stressed glottal cycles during the episode (sharper and more abrupt). On the contrary, abductor dystonia is manifested as a decay of the muscular tone resulting in an opening of the passage between vocal folds, which eventually may stop vibrating. Spasmodic dysphonia may present different origins (see Fig.1), of which the neurological type differentiates from the psychologic and idiopatic variants in that it is not reversible by voice therapy [7]. This makes it a good study model as it may convey information on neurological underlying malfunction.

The common treatment for neurologic dysphonia is either by botulinum toxin infiltration in the muscle (temporal effects) or by surgical disruption of

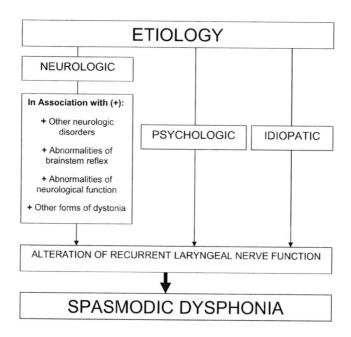

Fig. 1. Classical differentiation of spasmodic dysphonia attending to the underlying cause

innervations (permanent effects). Most of the times a clear diagnostic differentiating among the different variants is rather complex [7]. One of the objectives of the present study is to introduce possible inspection tools based on acoustic analysis of voicing speech by means of estimates of the biomechanical correlates of the vocal folds, which could be used in further discriminating the different origins of this set of dysphonias. Spasmodic dysphonia has been selected as a target because it shows clear biomechanical correlates as it will be presented in the sequel. This approach could open the study of other pathologies of neurological origin as well [3].

3   Detection and Grading Dysphonic Voice

Once the glottal source has been reconstructed different parameters are obtained from its power spectral density envelope and timely evolution. These are estimated for a large database including normophonic and dysphonic male and female subjects [8]. The normophonic set is carefully modelled on the most sensitive parameters by Gaussian Mixtures as in Fig.3. This allows estimating a normophonic condition probability from sample $\mathbf{y}_t$ from subject t with respect to the normophonic model Γ $_{Mm,f}$ (m: male, f: female populations) as:

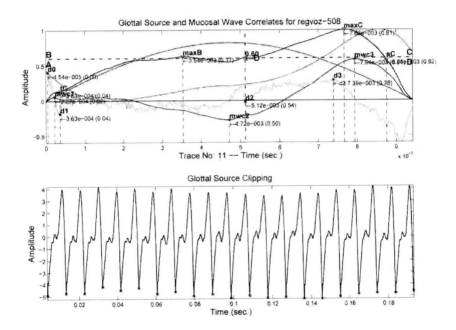

Fig. 2. Glottal Source Template (normophonic male voice). Top: Time Domain Parameterization. Bottom: Phonation Cycle Clipping. The classical Liljencrants-Fant (LF) pattern [5] may be clearly appreciated as a recovery phase (0-maxB), a closed resting phase (maxB-O), an opening phase (O-maxC) and a closing phase (maxC-0).

$$p(\mathbf{y}_t/\Gamma_{Mm,f}) = \sum_j w_j \left\{ \frac{1}{(2\pi)^{r_{m,f}/2}\,|C_{Mm,f}|^{1/2}} e^{-1/2(y_t - \Psi_{Mm,f})^T C_{Mm,F}^{-1}(y_t - \Psi_{Mm,f})} \right\}_j$$

(1)

where $\Psi_{Mm,f}$ and $C_{Mm,f}$ are the mean vector and covariance matrix of the population distribution. Once this probability has been calculated for all the gaussians j in the mixture weighted by the mixing vector w_j a likelihood ratio (Λ) may be defined the normophonic to dysphonic conditional probabilities:

$$\Lambda(\mathbf{y}_t) = \log[p(\mathbf{y}_t/\Gamma_{Mm,f})] - \log[1 - p(\mathbf{y}_t/\Gamma_{Nm,f})]$$

(2)

The log-likelihood ratio is contrasted against a threshold resulting in a decision of normophony vs pathology as:

$$L_t = \begin{cases} N; \Lambda(\mathbf{y}_t) \geq \vartheta \\ D; \Lambda(\mathbf{y}_t) < \vartheta \end{cases}$$

(3)

where N stands for normophonic and D for dysphonic. If a subject's voice is labelled as dysphonic the grade is of most importance to decide on how to proceed as most pathologies manifest as mild dysphonias which may go unnoticed to the general practitioner (GP).

The cases marked by green circles are the normophonic set, distributed accordingly with the three most relevant biometrical parameters (22 and 23 are singularity marks on the glottal source power spectral density, whereas 45 is a biomechanical unbalance, for a complete explanation see [5]). Subject names are printed near each mark. The red diamonds are the dysphonic set. The blue stars are samples from the same subjects before (T1636 and T1637) and after treatment (T2636 and T2637). A prototype gaussian is tentatively placed on the normophonic distribution centroid. Mahalanobis distances from a given sample to the respective gaussian centroid are used to grade dysphonia. This grade is well correlated with subjective estimations of the grade in GRBAS scale [4]. The detection Receiver Operator Curves after pathology detection from cross-validation tests are shown in Fig.4.

The accuracy of the detection process is based on a trade-off between the sensitivity (ratio of detected dysphonic subjects to total number of dysphonic subjects) and the specificity (ratio of detected normophonics to the total number of normophonic subjects). This balance is given by the point in the DET curve where the percentage of false dysphonic and normophonic subjects become equal, which is around 3 %. The voice analysis methodology would then consist in obtaining a sample of voice from the subject, estimating the spectral and temporal correlates on the glottal source, contrasting them against the normophonic/pathologic database, estimating the dysphonic grade and deriving the case and its anamnesis to specialized services [2].

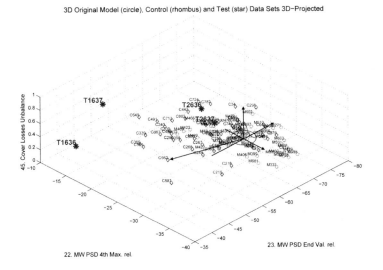

Fig. 3. Estimating the distance of a given subject sample to the normophonic model set as an index of normophonic vs dysphonic voice (female set). Green: samples from normophonic subjects. Red: samples from dysphonic subjects. Blue: Pre-post treatment cases: subjects before treatment (T1636 and T1637) and the same subjects after treatment (T2636 and T2637).

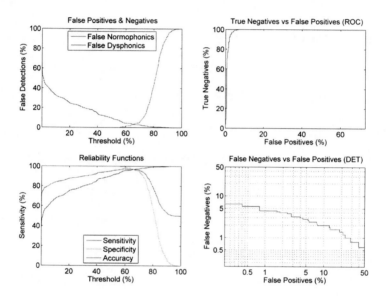

Fig. 4. ROC and DET curves from cross-validation of GLOTTEX for pathology detection in a female data set of normophonic and dysphonic cases. Top left: False detection cases as a function of the threshold. Top right: Associate Receiver Operator Characteristic (ROC) curve. Bottom left: Sensitivity, Specificity and Accuracy as a function of the threshold. Bottom right: Detection-Error Trade-off (DET) curve. The equal tax point is around 3 %.

4   Study Case: Abductor Spasmodic Dysphonia

Spasmodic dysphonia is a neurological disease characterized by the inability to maintain constant the phonation tension in vowels and other voiced sounds. This is due to strong involuntary changes in the neuromuscular activity of the larynx, resulting in fluctuations of the vocal fold tenseness. Pitch does not change much, but the amplitude of voice suffers large changes as seen in Fig.5.

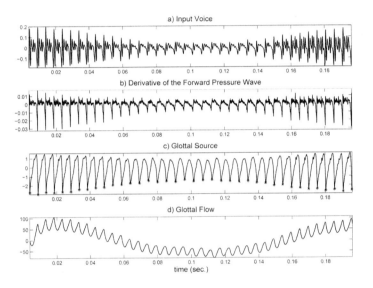

Fig. 5. Episode of spasmodic dysphonia of abductor type. a) Voice trace. b) Residual after vocal tract removal. c) Glottal Source (estimate of supraglottal pressure wave). d) Glottal flow (the fluctuation due to integration effects). All magnitudes are in relative units.

This tendency would be reflected in a smooth decaying and recovering of the shimmer (distortion parameter measuring the relative change in amplitude between neighbour phonation cycles). This case was graded as pathologic level 1 in subjective assessment, and 2 in objective assessment [4]. The detailed results of the detection process may be seen in Fig.6. The upper template gives the glottal source (black line), which is related to the supraglottal presure, raising from a minimum to a maximum following a glottal cycle. The opening point is given by (O). It may be seen that the glottal source in this case is rather different than the characteristic LF pattern in Fig.2, as the plateau observable in it before the opening does not exist. The closed phase (AO) is shorter than in a normal phonation cycle (34 % of the total phonation cycle), the open interval (OD) is much larger than expected (66 %). The lower template shows the evolution of the glottal arches at the beginning of the dystonic episode, where the LF plateau is still perceptible as a hunchback, and at the middle of the episode, where that mark is not present.

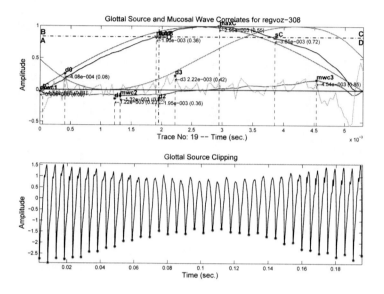

Fig. 6. Close-up view of the glottal source in the middle of the dystonic episode (upper template, black line), corresponding to a cycle in the middle of the lower template

This means that the extreme glottal cycles are almost normal, and the middle ones are under-stressed. The most important characteristic of the methodology used (see [5]) is the possibility of estimating the biomechanical parameters of the vocal fold given as the dynamic mass, losses and stiffness correlates of the fold body and cover. The templates in Fig.7 show the estimates of the dynamic mass of the vocal fold body for each phonation cycle (upper left). This parameter experiences a 13 % decay from normal to dystonic phonation. The boxplot in the upper right shows the median, first and third quartiles illustrating the resulting statistical spread. The median dynamic mass is around 13.5 mg. The middle left and right templates show similar estimates for the losses (due to viscosity and turbulence), which experience an increment of about 12 % during the episode. The bottom left and right give the corresponding estimates for the vocal fold body stiffness, measuring the strain experienced by the average vocal fold structure along its extension, and directly related with muscular tension.

It may be seen that the tension decays in around 30 % during the episode, to recover immediately after. The question which now arises is how biomechanical parameter estimations, and especially tension, may help in monitoring the neurological disease. The answer must be modulated depending on the specific task to fulfil. The classical ones in organic pathology are detection, grading, early warning, pre- and post-treatment monitoring, among others.

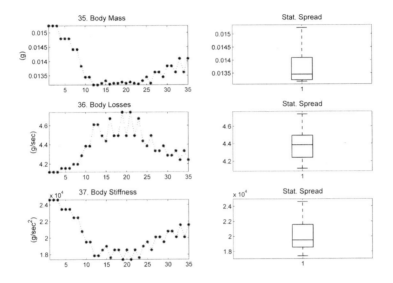

Fig. 7. Monitoring an episode of spasmodic dysphonia with GLOTTEX (female voice). Top two templates: Glottal Source in detail and clipping (compare them with those from normal phonation in Fig.2). Bottom three templates (right): Estimates of vocal fold body mass, losses and stiffness. Left: statistical dispersion boxplots. The dystonic correlate of the disease is evident in the decay appreciated in vocal fold tension (stiffness) during the episode.

Pre- post-monitoring is by far the most complex task to track the grades obtained before and after treatment (surgical, pharmaceutical or voice therapy), relative to a historical discourse. The key point is to establish intra- and inter-subject distributions of the most discriminating parameters. As the historical account is very important for pre-post monitoring intra-subject distributions may have more relevance than in organic pathological studies. This is the case of spasmodic dysphonia, showing important intra-speaker distribution changes, indicating that biomechanical parameter unbalance may play a very special role. These questions are subjects of further study.

5 Conclusions

Through the present work several important facts have been presented, the relevance of which is to be emphasized. On one hand the importance of the neurological disease has been presented. Among different diseases of this kind spasmodic dysphonia has been selected as a study target because, having in most cases a neurologic origin it might show correlates with vocal fold biomechanics under the working hypothesis. The classical methodology to detect the dysphonic condition was then proposed as a possible candidate to characterize spasmodic dysphonia using the biomechanical parameters of the vocal fold. This was confirmed by the estimation of the mass, losses and stiffness of the fold body. Especially this last one showed a clear correspondence with the abductor-type of pathology monitored. Of course, this case of study will need a confirmation against a wide set of cases including adduction and both male and female subjects. Besides, the neurologic origin of each case studied has to be carefully asserted, and the dysphonic grade must evaluated both subjectively as by the GRBAS index as well as by the dysphonic index proposed in [4]. This would allow to give this study the statistical validation required to consolidate the conclusions commented, defining the next steps in the study. The most interesting fact is that this methodology may open ways to characterize and quantify the degree of spasmodic dysphonia, and by extension to other neurological diseases, helping in early warning and monitoring.

Acknowledgements

This work is being funded by grants TEC2009-14123-C04-03 from Plan Nacional de I+D+i, Ministry of Science and Technology of Spain and CCG06-UPM/TIC-0028 from CAM/UPM. Special thanks are due to the support and invaluable help of the doctors at the ORL service in HU "Gregorio Marañón" of Madrid, especially to Drs. Scola and Ramírez without whose encouragement and push forward this project would not have been possible.

References

1. Bobadilla, J., Gómez, P., Godino, J.I.: Mapaci: A Real Time e-Health Application to Assist Throat Complaint Patients. In: Proc. of ICIW 2007 (2007)
2. Colton, R.H., Kasper, J.K., Leonard, R.: Understanding voice problems, pp. 107–151. Williams and Wilkins, Baltimore (1990)
3. Das, R.: A comparison of multiple classification methods for diagnosis of Parkinson disease. Expert Systems with Applications 37, 1568–1572 (2010)
4. Gómez, P., Fernández, R., Rodellar, V., Godino, J.I.: Evaluating the grade of voice pathology from the glottal-source spectral profile. In: Proc of AVFA 2009, Madrid, pp. 161–164 (April 2009)
5. Gómez, P., Fernández-Baíllo, R., Rodellar, V., Nieto, V., Álvarez, A., Mazaira, L.M., Martínez, R., Godino, J.I.: Glottal Source Biometrical Signature for Voice Pathology Detection. Speech Communication 51, 759–781 (2009)

6. Kloppel, S., Stonnington, C.M., Chu, C., Draganski, B., Scahill, R.I., Rohrer, J.D., Fox, N.C., Jack, C.R., Ashburner, J., Frackowiak, R.S.J.: Automatic classification of MR scans in Alzheimer's disease. Brain 131(3), 681–689 (2008)
7. Ludlow, C. L. Spasmodic Dysphonia is a Neurological Disorder. National Spasmodic Dysphonia Association, www.dysphonia.org (retrieved 2.1.2011)
8. Project MAPACI, http://www.mapaci.com
9. Rosenfeld, D.B.: Neurolaryngology. Journal of Ear Nose and Throat 66(8), 323–326 (1987)
10. World Health Organization, http://www.who.int/healthinfo

Group Formation for Minimizing Bullying Probability. A Proposal Based on Genetic Algorithms

L. Pedro Salcedo[1], M. Angélica Pinninghoff J.[2], and A. Ricardo Contreras[2]

[1] Research and Educational Informatics Department
[2] Department of Computer Science
University of Concepción, Chile
{psalcedo,mpinning,rcontrer}@udec.cl

Abstract. Bullying is a problem that needs to be considered in the early stages of group formation. Unfortunately, as far as we are aware, there is not known procedure helping teachers to cope with this problem. It has been established that, in a certain group, a specific configuration in the students distribution affects the behavior among them. Based on this fact, we propose the use of genetic algorithms for helping in students distribution in a classroom, taking into account elements like leadership traits among other features. The sociogram is a technique that teachers have been using for years for supporting group formation. The sociogram is a sociometric diagram representing the pattern of relationships among individuals in a group, usually expressed in terms of which persons they prefer to associate with. This work combines the concepts of genetic algorithms and sociograms, that can be easily represented by means of relationships graphs. A set of tests is applied to the students to collect relevant data, and results can be validated with the help of specialists.

Keywords: Bullying, Genetic algorithms, Sociograms.

1 Introduction

Bullying is currently a serious phenomena that affects negatively the school environment when it cannot be avoided or even detected. The problem can be treated from different perspectives; however a single strategy cannot always guarantee a successful action. There are some experiences that use intelligent techniques for group formation in learning collaborative environments [1]. Based on this idea we suggest that an analogous reasoning may help in determining groups in a classroom, in such a way that the potential negative effect due to bullying is minimized.

The definition of school bullying includes several key elements: physical, verbal, or psychological attack or intimidation that is intended to cause fear. Distress or harm to the victim; bullying can be generalized as an imbalance of power (psychological or physical), where the dominant child (or children) opress the less powerful ones. It has also been noticed that whenever bullying occurs, the

J.M. Ferrández et al. (Eds.): IWINAC 2011, Part II, LNCS 6687, pp. 148–156, 2011.

incident tends to repeat for the same children over a prolonged period. Many school-based intervention programs have been devised and implemented in an attempt to reduce school bullying. Depending on the programm the focus can be on bullies, victims, peers, teachers, or even on the institution. It has been observed that many of the programs have been based on commonsense ideas about what might reduce bullying rather than on empirically- supported theories of why children bully, why children become victims, or why bullying events occur [2].

What makes normal conflicts different from bullying is that in normal conflicts there is an adaptation between peers of the same social status, while in the latter a children is repeatedely being exposed over time, to negative actions from the part of one or more students who are perceived as stronger. Bullying can include among others: verbal abuse, hitting, kicking, beating, destroying others belongings or blackmail.

A range of intervention approaches have been proposed in the literature. There has also been some proposals involving school policies and legislation; however we believe they are unlikely to make any difference to bullying. We believe changes have to start at home with appropriate parenting, no tolerance for sibling bullying, training of teachers and consistent implementation of rules to deal with bullying in school. In particular, positive modeling and teaching alternatives to reaching high peer status is required. The above can include collaborative working and compassionate leadership while being allowed to compete in other settings (sport, music) or rewarding support to other students, befriending and peer counseling schemes [7].

In the process of searching for a mechanism that helps to avoid bullying, one of the more important references on the topic reports 26 studies of school-based interventions. These reports included 10 curriculum based, 10 whole school interventions, 4 social skills groups, 1 regarding the mentoring of bullied children, and 1 regarding the increased availability of social worker support in the school. The outcomes were bullying, victimization, aggression, and how schools responded to violence. Some other outcomes often related to bullying (e.g. academic achievement, how safe children felt, levels of self-esteem, and knowledge and attitudes about bullying) were among those used in some of the studies [6].

In this work we propose that by generating particular classroom configurations it is possible to analyze the effect some students have on the whole class; i.e. that specific students' distributions represent different interaction degrees. As expected, the number of possible configurations a classroom can adopt makes this problem a difficult one; it is not practical or feasible to test all the alternatives. Hence we propose a way to solve the problem unsing a non-standard combinatory approach, i.e. we use genetic algorithms.

This article is structured as follows; the first section is made up of the present introduction; the second section describes the problem; the third section is devoted to genetic algorithms, while in section four we present our proposal. Section five is devoted to testing and results and we end with the conclusions in section six.

2   The Problem

The problem we are dealing with consists in proposing a mechanism that can
help to distribute students in a classroom, based on their potential bullying capabilities. A set of tests allows to detect some features that can be useful in
classifying students in terms of relationships; as they are seen by other classmates. In other words, based on a set of tests, we can establish which are the
specific students that have a possible negative impact on the others, and find a
distribution, in a classroom, for minimizing this menace.

To deal with this problem we have considered real cases, taking into account
several courses in a school. After taking tests, and asking to teachers in an
independent way, results are very coincidental, in the sense that students identify
through these tests the conflictive classmates, based on their own perception; and
teachers identify the same groups, based on their experience.

Tests are the tool that allows to collect input data, and once we have the
data, we are able to create a sociogram. A sociogram is a set of indexes and
graphics obtained from procedures developed for analyzing intra group relationships; aimed to measure and describe the structures of relationships underlying
in small groups. It is a kind of snapshot of a group from a particular perspective
in a specific time. It is interesting because it allows to compare different snapshots through time, given that every human group has a dynamic behavior. The
procedure for obtaining a sociogram is very simple; it doesn't require special
material, tests can be taken simultaneously to all group members and it takes
about 15 minutes to be completed. This technique consists of asking the students
with which particular classmates the would like to associate in a specific activity,
or, on the other side, with which of them they prefer not to meet. The idea is to
have a graphical representation of groups relationships, to visualize key points
in a graph.

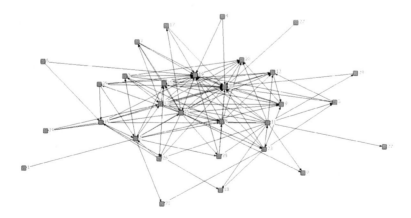

Fig. 1. A typical graph showing relationships among students

A typical graph is shown in Figure 1. Students are identified through numbers and, in this particular graph, a student having a high number of in-degree connections is a student that is visualized as a menace by other classmates.

The graph generated in this way is the basis for creating a particular configuration in the classroom. In doing so, we are going to use a particular classroom layout, but it can easily be extended for different situations. In this work, we use genetic algorithms for generating different groups distribution, and then we choose the one that, according to a specific criteria, represent the less conflictive configuration. After a research in existing literature, pointing to different ways in which this problem is faced, we have found no references involving the use of genetic algorithms as part of the solution to this challenge.

3   Genetic Algorithms

Genetic algorithms can be described as a set of algorithms taking inspiration from the principles of natural evolution. They have been used successfully on hard problems where other optimization methods fail or are trapped in suboptimal solutions. These suboptimal solutions are frequently found in problems in which to explore the complete space of solutions is not possible, i.e., when the number of alternatives to explore makes the exhaustive process not possible.

The structure of a genetic algorithm consists of a simple iterative procedure on a population of genetically different individuals. The phenotypes are evaluated according to a predefined fitness function, the genotypes of the best individuals are copied several times and modified by genetic operators, and the newly obtained genotypes are inserted in the population in place of the old ones. This procedure is continued until a solution is found [3].

Genetic operators

Different genetic operators were considered for this work. These genetic operators are briefly described below:

- Selection. Selection is accomplished by using the roulette wheel mechanism [3]. It means that individuals with a best fitness value will have a higher probability to be chosen as parents. In other words, those classroom configurations that are not a good representation of a less conflictive one, are less likely selected.
- Cross-over. Cross-over is used for exchanging genetic material. This allows part of the genetic information belonging to one individual, to be combined with part of the genetic information of a different individual. It allows us to increase genetic variety, in order to search for better solutions. In other words, if we have two individuals, in this case, two different classroom configurations, we exchange the genetic material in such a way that part of the first configuration is combined with part of the second configuration. The role of the crossover operator is to recombine information from two parent

solutions into what we hope are even better offspring solutions. The problem is to design a crossover operator that combines characteristics of both parents while producing a valid solution [5]. The classic crossover operator takes a first part from one of the parents and the second part of the new chromosome from the other one. The problem with this approach is that for two parents, student x can appear in the first part of the first parent and in the second part of the second parent, resulting in that the same student, take the case of student x, will appear in an offspring twice. It means that the same student appears in a specific classroom in two different positions while, as a consequence of this, some students are lost and do not appear in the classroom layout. To overcome this problem, OX operator is used. This operator builds offspring by choosing a subsequence of one parent chromosome while preserving the relative order of *students* from the other parent [4].

- Mutation. By using this genetic operator, a slight variation is introduced into the population so that a new genetic material is created. In this work, mutation is accomplished by randomly exchanging, with a low probability, two particular students in a given configuration.

4   The Proposal

This work attempts to find a mechanism, based on genetic algorithms, to supply a group students distribution in a classroom in such a way that negative influences are minimized. Different group distributions are going to be generated and evaluated, the best of them is selected as the solution. A key element in this process is the function used to evaluate different configurations as explained in the following paragraphs.

The graph obtained to represent interactions in a class, is the fundamental data we are going to deal with. The graphs have been generated through tests and validated by teachers that know the way in which students interact in the classroom.

It is necessary to find a representation for the elements that we consider should be present in the model. The basic component in this model is, as expected, the set of students that belong to a class.

To represent students in a particular course, we fill entries in a matrix that reflects, approximately, the classroom distribution. See figure 2. Values in this matrix represent the potential influence that a particular student has. To illustrate this, we could say that this value corresponds to the in-degree the node has in the graph.

The second element to take into account is a specific classroom configuration, as shown in figure 3. This is a particular layout having five columns and six rows. Obviously, different layouts are possible, for example by including aisles into a particular configuration. In this specific configuration we assign arbitrary weights to different positions, each value representing the strength of the particular position in terms of influence on the neighborhood.

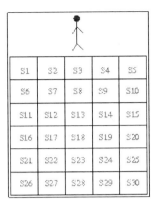

S1	S2	S3	S4	S5
S6	S7	S8	S9	S10
S11	S12	S13	S14	S15
S16	S17	S18	S19	S20
S21	S22	S23	S24	S25
S26	S27	S28	S29	S30

Fig. 2. An arbitrary students distribution

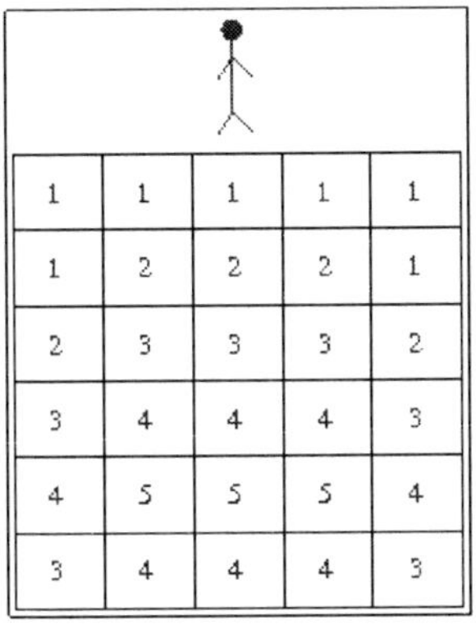

1	1	1	1	1
1	2	2	2	1
2	3	3	3	2
3	4	4	4	3
4	5	5	5	4
3	4	4	4	3

Fig. 3. An arbitrary classroom layout

When thinking on genetic individuals, we propose that one individual is represented by a new matrix, in which each entry represents the product obtained by multiplying the values that reflects de degree of influence a student has, by the value associated to the relative importance of a particular location in the classroom.

To get a chromosome, in genetic terms, we proceed in two steps. First, the chromosome is selected to be a particular classroom configuration.

In this work, we suggest that configurations that minimize the interaction for the complete class are better that those configurations in which the influence of interaction is high, due to the fact that we represent in the matrix their potentially negative behavior.

The next step is to define the fitness, i.e., the measure that will reflect the aptitude an specific individual (chromosome) has.

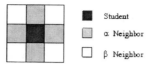

Fig. 4. Neighborhood that considers direct and diagonal neighbors of a location

We have decided to propose a specific mechanism to evaluate each group distribution, that considers the degree of influence a student presents, according to the tests, lets denote it *influence degree* (ID) and the neighborhood surrounding that specific student. In doing so, for each student there is value that is computed as follows: first, the influence degree for each student is the number of incident arcs that the node representing the student has, in the graph. Second, the relevance of the neighborhood for the student is computed by adding a value α if the neighbors are *direct* neighbors, and a value β in case of *diagonal* neighbors. Direct neighbors and diagonal neighbors are illustrated in figure 4. If a neighbor doesn't exists, the corresponding α and β are replaced by zero. The third element to be considered is the relative importance of each specific location.

So, for a specific student the final influence value, lets denote it FIV, is computed taking into account the influence degree, ID, the neighborhood, and the weight (W) of the location, as follows:

$$\text{FIV} = \text{ID} * \text{W}(\alpha_1 + \ldots + \alpha_n + \beta_1 + \ldots + \beta_m), \text{ with } m, n \leq 4$$

A particular group distribution will have a value, the fitness for that chromosome, that reflects the summation of students for the complete class. As it reflects the (negative) influence as a general value, the lowest value for the complete configuration is to be considered as the best group distribution.

At this point, we have a specific value given for every student on every position in the classroom. That is the basic chromosome we are dealing with. Now we need to reflect the most important issue in this set of relationships; the influence of each critical point on their neighborhood.

5   Preliminary Results

An arbitrary class has been selected for testing, Table 1 illustrates the way in which the influence degree is associated to every student in the class. It is obtained from the graph.

Table 1. Influence degree (ID) for each student

S1	S2	S3	S4	S5	S6	S7	S8	S9	...	...	S20	S21	S22	S23	S24	S25	S26	S27	S28	S29	S30
1	12	1	2	1	7	2	5	1	...	...	1	1	3	1	11	3	3	1	8	2	1

Table 2 illustrates the relative importance a particular location has in the classroom configuration. L_i refers to a particular location.

Table 2. Weights associated to a particular location in the classroom

L1	L2	L3	L4	L5	L6	L7	L8	L9	...	...	L20	L21	L22	L23	L24	L25	L26	L27	L28	L29	L30
5	8	8	8	5	8	12	12	12	...	...	8	8	12	12	12	8	5	8	8	8	5

Table 3 summarizes values considered for testing.

Table 3. Parameters used for testing

Item	Value(s)
Population	100, 500, 1000
Number of generations	100, 500, 1000, 5000
Crossover	80%
Mutation	0%, 2%, 5%

Values for α and β are arbitrary and can be modified depending on particular constraints. Examples of results we show in this section have considered $\alpha = 2$ and $\beta = 1$, on the students distribution shown in figure 2.

Starting from an initial arbitrary value, that corresponds to a specific students distribution in a classroom, the genetic algorithm generates different students distributions, after a finite number of generations. The best solution, according to our criteria, is the one in which critical students are located in positions in which their influence on the rest of classmates is diminished. Typically, those critical students are located in the corners of the first row, i.e. the row that is closer to the classical teacher location. At the same time, students identified by their classmates as non aggressive, appear in positions with higher influence values.

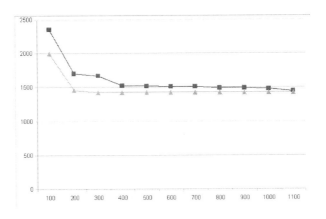

Fig. 5. Fitness evolution for 1000 individuals

Evolution of fitness shows a convergence to a stable value after 1000 generations, as shown in figure 5. This figure shows that mutation is useful in the sense that the stable value is reached in early stages of evolution. The upper line represents the evolution without mutation, while the second line (closer to the x axis) represents the evolution when using mutation.

6   Conclusions

Genetic algorithms have shown to be an interesting approach for solving combinatorial problems, and obtained results in this particular problem confirm this issue.

Different classroom configurations produce different fitness for the chromosomes that represent solutions. Specific values for α and β may impact in different ways the results; so it is clear that a tunning stage is necessary for each specific group of students, and for each different classroom configuration.

This proposal is currently under testing and evaluation taking into account different actors. Further steps will include penalization of undesirable individuals; e.g, as shown in the graph in figure 1, given two students sharing a neighborhood relationship, and given they are neighbors in the resulting configuration; the chromosome will be penalized adding an arbitrary value to its fitness. Nevertheless, teachers that interact with students think that the proposed distribution is probably a better solution for reducing the effect of an undesirable influence that specific students can have on the class. The cases presented to the teachers considered the existence of two aisles, as in the real school. In these real situations, the fitness value is, as expected, lower that the value obtained from the block distribution we used to explain the core idea in this work.

Nevertheless, bullying behaviors in children are influenced by family issues, problems in neighborhoods, and school size. Prevention strategies should also focus on these risk factors.

References

1. Ani, Z.C., Yasin, A., Husin, M.Z., Hamid, Z.A.: A Method for Group Formation Using Genetic Algorithm. International Journal on Computer Science and Engineering 02(09), 3060–3064 (2010)
2. Farrington, D., Baldry, A., Kyvsgaard, B., Ttofi, M.: Effectiveness of Programs to Prevent School Bullying. Institute of Criminology, Sidgwick Avenue, Cambridge CB3 9DT, UK (2008)
3. Floreano, D., Mattiussi, C.: Bio-Inspired Artificial Intelligence. Theories, Methods, and Technologies. The MIT Press, Cambridge (2008)
4. Michalewicz, Z., Fogel, D.B.: How to Solve It: Modern Heuristics. Springer, Heidelberg (2000)
5. Poon, P.W., Carter, J.N.: Genetic algorithm crossover operators for ordering applications. Computer Ops. Res., 135–147 (1995)
6. Vreeman, R.C., Carroll, A.E.: A systematic review of school-based interventions to prevent bullying. Archives of Pediatric and Adolescent Medicine 161, 78–88 (2007)
7. Wolke, D.: Bullying: Facts and Processes. University of Warwick institutional repository (2010), http://www.wdms.org/publications.htm

A Speaker Recognition System Based on an Auditory Model and Neural Nets: Performance at Different Levels of Sound Pressure and of Gaussian White Noise

Ernesto A. Martínez–Rams[1] and Vicente Garcerán–Hernández[2]

[1] Universidad de Oriente, Avenida de la América s/n, Santiago de Cuba, Cuba
eamr@fie.uo.edu.cu
[2] Universidad Politécnica de Cartagena, Antiguo Cuartel de Antiguones
(Campus de la Muralla), Cartagena 30202, Murcia, España
vicente.garceran@upct.es

Abstract. This paper performs the assessment of an auditory model based on a human nonlinear cochlear filter-bank and on Neural Nets. The efficiency of this system in speaker recognition tasks has been tested at different levels of voice pressure and different levels of noise. The auditory model yields five psychophysical parameters with which a neural network is trained. We used a number of Spanish words from the 'Ahumada' database as uttered by native male speakers.

1 Introduction

The performance of an auditory model can be evaluated through Speaker Recognition techniques, studying how each model affects recognition level. Traditional automatic speaker recognizers are based either on measurements of individual characteristics of the speaker's voice [4,5], on short-time analysis procedures [6,2,3], on methods derivative of the voice production model, or on long-term features such as prosody. Feature parameters extracted from a speech wave are compared with the stored templates or models for each registered speaker. In speaker modeling, Gaussian Mixture Models (GMM) and Hidden Markov Models (HMM) are widely accepted.

The performance of recognizers often degrades dramatically with noise, with different talking styles, with different microphones, etc. because extracted features are distorted, causing mismatched likelihood calculations. However, human cochlear models [7] that mimic some aspects of the human cochlea and psychoacoustic behavior have been proposed to lessen the impact of such problem. In general, these models incorporate 'spectral analysis', 'automatic gain control', 'neural adaptation', 'rectification' and 'saturation effects', and have shown superior results for speech recognition [8,9,10,11,12,13,14,15] and speaker recognition [16,17,18]. In [19] we presented a new method based on the model of perception, applied to a DRNL [1] voice processor for speaker recognition applications using neural nets. The goal of our current research has been to investigate the voice processor's performance at different levels of voice pressure and noise.

J.M. Ferrández et al. (Eds.): IWINAC 2011, Part II, LNCS 6687, pp. 157–166, 2011.

2   Data and Methodology

2.1   Speech Data

The vocabulary used in the experiments comes from the 'Ahumada' database
[20], designed and collected for speaker recognition tasks in Spanish. Although
this database contains both records from high quality microphones and from
telephone lines, we only used the former. In order to compute our feature sets,
656 utterances of the digit "uno" (which means "one" in Spanish) uttered by
eleven speakers, were used as input signals, all of them with a duration longer
than 200 ms. Table 1 shows an analysis of speaker utterances.

Table 1. Speaker utterances

Speaker	L1	L2	L3	L4	L5	L6	L7	L8	L9	L10	L11	Total
Training	41	29	40	43	44	40	43	30	35	22		367
Validation	13	9	12	13	14	13	13	10	12	6		115
Test	13	9	12	13	14	13	13	10	12	6		115
Impostor											59	59
TOTAL	67	47	64	69	72	66	69	50	59	34	59	656

We then added Gaussian white noise with different values of signal to noise
ratio, obtaining five sets of noisy signals with 5, 10, 20 and 30 dB SNR and one
set without noise.

2.2   Feature Extractions

Speech Feature (i.e. parameter) extraction is based on the auditory model shown
in Fig. 1, [1,21,22,19]. The auditory model's input signal is scaled at three sound
pressure levels: 30, 60 and 80 dB SPL. The output consists of 30 channels distributed between 100 and 7,000 Hz in the logarithmic scale.

A frequency analysis of these channels returned that frequency components
below 250 Hz can be significant for speaker modeling. For this reason, the envelope extraction method is preferred, [19]. Fig. 2 depicts the Envelope Component
Extractor for one auditory model channel.

Loudness parameter LA is derived from loudness component 'el' and has been
calculated considering the time necessary for the lowpass filter output (F_5) to
achieve stable state, which is equal to $3 \cdot \tau_5$.Roughness parameters RA and RF
(amplitude and frequency) were derived from roughness component 'er'. The
square of the FFT module was then calculated. Roughness amplitude parameter
RA is the maximum spectrum value in the $F_4 - F_3$ range, while roughness
frequency parameter RF is its frequency. Similarly, virtual tone parameters VTA
and VTF (amplitude and frequency) are derived from virtual tone component
'ev'. The square of the FFT module was then calculated. Virtual tone amplitude
parameter VTA is the maximum spectrum value in the F_2-250 Hz range, whereas
virtual tone frequency parameter VTF is its frequency.

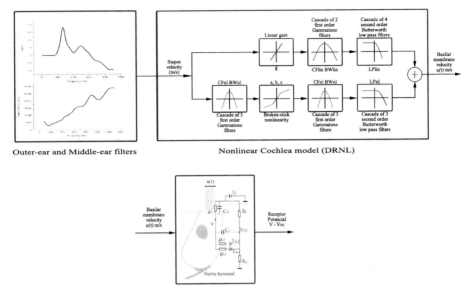

Outer-ear and Middle-ear filters   Nonlinear Cochlea model (DRNL)

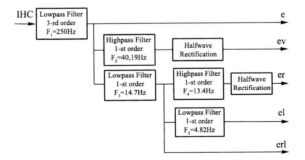

Inner Hair Cell model

Fig. 1. The auditory model: an outer-ear filter that adapts the headphone to eardrum response; a middle-ear filter to obtain stapes velocity; a nonlinear cochlea model that obtains basilar membrane velocity and an inner hair cell model that carries out the receptor potential

Fig. 2. Envelope Component Extractor (ECE). The Inner Hair Cell (IHC) signals generated by the auditory model are taken as ECE input. Signal 'er' is the roughness component; signal 'el' is the loudness component and signal 'ev' is the virtual tone component. Other 'e' and 'erl' signals are available but were not used in this research.

In pattern classification with Neural Nets, the larger an input vector, the larger its effect on the weight vector. Thus, if an input vector is much larger than the others, the smaller ones must be represented many times so as to produce a perceivable effect. A possible solution would be to normalize each input vector in the process path, and there are several possible methods to achieve it. In this paper three normalization methods have been analyzed [19]:

Method A. Each IHC signal channel is normalized in relation to its global maximum.

Method B. The selected amplitude parameters are normalized with their respective global maximum.

Method C. Each amplitude component is normalized to an rms value equal to one.

3   Neural Nets

Figure 3 shows the single neuronal network architecture used to evaluate each parameter (LA, VA and VTA) in speaker recognition tasks, whereas Figure 4 depicts the compound network architecture used to evaluate the three parameters at the same time. Note that output layers have 10 neurons, one for each speaker (L1 to L10). For both architectures a Feedforward-Backpropagation-Multilayer is used. For hidden layer neurons we used the tansig transfer function. The neuron of the last layer uses the linear transfer function. To train the single neural network, we used the Levenberg-Marquardt algorithm, and the so-called early stopping method for improving generalization. The training data are used to compute the gradient and to update network weights and biases. The error from validating data is monitored during the training process. The layer's weights and biases in the compound network are initialized according to the Nguyen-Widrow initialization algorithm. This algorithm chooses values in order to distribute the active region of each neuron in the layer approximately evenly across the layer's input space. Values contain a degree of randomness, so they are not the same each time this function is called.

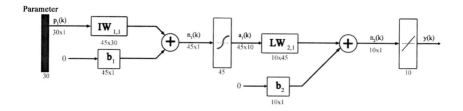

Fig. 3. Single Neural Network Architecture. The input layer has 30 neurons, one for each auditory channel parameter (LA, VA or VTA). The hidden layer has 45 neurons, and the output layer has 10 neurons.

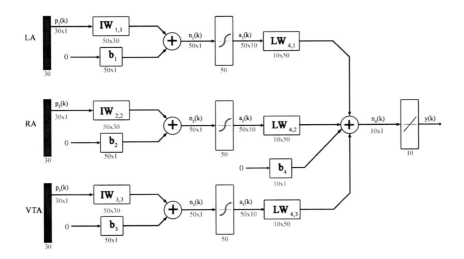

Fig. 4. Compound Neural Network Architecture. Three input layers, one for each type of parameter. Three hidden layers with 50 neurons each, connected to their respective inputs, and a common output layer, with 10 neurons.

In the simulation phase, the parameters obtained from a locution by a specific speaker (as network input) produce numerical values in the output layer. One of these 10 neurons reaches a value greater than the rest of neurons in that layer. If this neuron identifies a specific speaker, it is considered a success. In the impostor L11 speaker test, a threshold is previously defined for the rest of speakers. If one neuron of the output layer reaches a value above the threshold, it is treated as an error.

4   Results

This section presents the results for each of the three normalization methods. These tests serve a double purpose: to analyze the behavior of the auditory model using signals with different sound pressure and noise levels, at the same time determining which of the three proposed normalization methods presents better recognition ratios.

4.1   Analysis of the Behavior of the Auditory Model Using Signals with Different Sound Pressure Levels

Figure 5 shows the success ratio in speaker recognition when the neural net from Fig. 3 is trained with parameters LA, RA and VTA, at different sound pressure levels and without noise. Parameter LA (Loudness Amplitude) presents better results in the three normalization methods. However, Method C is the worst.

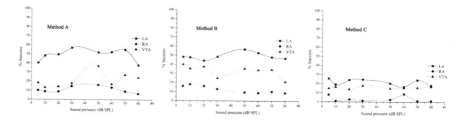

Fig. 5. Success ratio in speaker recognition, using validation + test utterances. The network is trained with parameters LA, RA and VTA at different sound pressure levels and without noise.

4.2   Analysis of the Behavior of the Auditory Model with Noisy Signals

Figure 3 shows a neural network typology used for speaker recognition, in which input parameters are combined. The training phase stops when the mean square error (MSE) is less than 0.01 or when 200 epochs are reached. There are differences between the three methods of normalization: while Method C reaches MSE=0.01 when epoch=21, Method A and B reach 200 epochs with MSE=0.04. Nevertheless, testing the network with the same training parameters, a 100% of success is obtained.

In order to draw a comparison among the three parameter normalization methods used for neuronal network operation, Figure 6 shows the success outcome in speaker recognition for each method and the different sound pressure levels (SPL) and noise used. This figure shows that the best recognition results are obtained when parameters are normalized according to method B and the sound pressure level of the signal is equal to 60 dB SPL and 30 dB SNR, for the validation set (69.6%); and also when the sound pressure level of the signal is equal to 60 dB SPL and the signal has no noise, for the test set (67.8%). In general, 80% of success measurements decrease to 80 dB SPL with methods B and C. A point of inflexion at 60 dB SPL appears for most SNR values. This seems to be due to the behavior of roughness parameter RA.

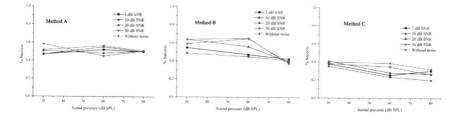

Fig. 6. Comparative results of success for validation + test, at different noise levels (5, 10, 20 and 30 dB SNR and without noise) and different sound pressure levels (30, 60 and 80 dB SPL) for the three methods (A, B y C). Epochs=200.

4.3   Analysis of the Evolution of Recognition Performance as Function of Noise

Finally, we studied the evolution of recognition performance based on noise levels, but in a different way than before. The network already trained using a signal without noise, is now simulated using the same signal but with different values of added noise. The level of sound pressure is the same in both signals. This analysis was carried out for each of the three normalization methods (A, B and C). The results are summarized in Figures 7, 8 and 9. In the three analyses the network is trained with locutions without noise. Fig. 7 shows the speaker recognition level when the network is simulated with the same locutions they are used in training. Fig. 8 shows the speaker recognition level when the network is simulated with the validation and test locutions. And finally, Fig. 9 shows the speaker recognition level when the network is simulated with the impostor's locutions.

If the network is trained using signals without noise and simulated using signals with noise from the same training set, it can be inferred from our analysis in figure 7 that the success rate in the three methods grows alongside the value of SNR. At 30 dB SPL and with method C, this increase is almost a linear function; at 60 and 80 dB SPL (in all three methods) the increase is almost a

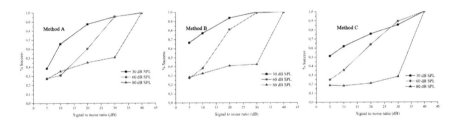

Fig. 7. Success rate in a network trained using locutions without noise and simulated using locutions with different noise values. Results are shown for the three methods and three sound pressure levels. All locutions are training.

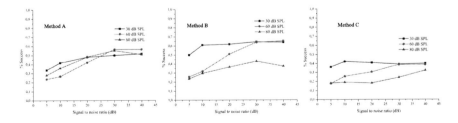

Fig. 8. Success rate in network trained with locutions without noise and simulated using locutions with different noise levels. Results are shown for the three methods and three sound pressure levels. All locutions are validation+test.

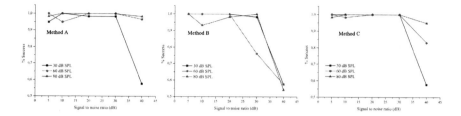

Fig. 9. Success rate of network trained using locutions without noise and simulated using locutions with different noise values. Results are shown for the three methods and three sound pressure levels. All locutions are impostor. Error is shown.

linear function too, interval-wise. The best results are obtained at 30 dB with method B. If the sum of simulation results from validation and test locutions are considered at 30 dB SPL and using method C (Fig. 8, success dispersion remains within a smaller range of values (between 36.09% and 40.00%) than with method B (between 50.00% and 63.92%). With regard to the test of the impostor signal, Fig. 9, it is possible to conclude that better results are obtained when signals are not degraded with noise, and the smallest errors take place at 30, 60 and 80 dB SPL in method B, and only at 30 dB SPL in both metods A and C. In general terms, method B at 30 dB SPL contributes to better indices of recognition for all training, validation and test signals.

5   Conclusions an Future Developments

The results obtained for speaker recognition are highly dependent both on the parameters and on the auditory models used (outer-ear, middle-ear and an inner hair cell). The two first models are responsible for the delay in the output of the IHC signal and -possibly- of the auditory system response to low frequency. It would be possible to develop a set of models for each part of the auditory system, and thus be able to compare their behavior. For example, to compare the performance of the auditory system modeled using the DRNL filter with respect to others (linear filters bank, gamma-tone filters, etc.) holding the remaining models of the ear. On the other hand, so as to improve system performance, the source of signals -the 'Ahumada' database- should be evaluated in more depth. Our proposal would be to make the previous study of locutions and then "select" those more suitable.

Acknowledgements

This project was funded by the Spanish Ministry of Education and Science (Profit ref. CIT-3900-2005-4).

References

1. Lopez-Poveda, E.A., Meddis, R.: A human nonlinear cochlear filterbank. J. Acoust. Soc. Am. 110(6), 3107–3118 (2001)
2. Atal, B.S., Hanauer, S.L.: Speech analysis and synthesis by linear prediction of the speech wave. Journal of The American Acoustics Society 50, 637–655 (1971)
3. Merkel, J.D., Gray, A.H.: Linear prediction of speech. Springer, Heidelberg (1976)
4. Furui, S.: Cepstral analysis techniques for automatic speaker verification. IEEE Transaction on Acoustics, Speech and Signal Processing 27, 254–277 (1981)
5. Mermelstein, P.: Distance measures for speech recognition, psychological and instrumental. In: Chen, C.H. (ed.) Pattern Recognition and Artificial Intelligence, pp. 374–388. Academic, New York (1976)
6. Fant, G.: Acoustic Theory of Speech Production. Mouton, The Hague (1970)
7. von Békésy, G.: Experiments in Hearing. McGraw-Hill, New York (1960); (reprinted in 1989)
8. Anderson, T.R.: A comparison of auditory models for speaker independent phoneme recognition. In: Proceedings of the 1993 International Conference on Acoustics, Speech and Signal Processing, vol. 2, pp. 231–234 (1993)
9. Anderson, T.R.: Speaker independent phoneme recognition with an auditory model and a neural network: a comparison with traditional techniques. In: Proceedings of the Acoustics, Speech, and Signal Processing, pp. 149–152 (1991)
10. Anderson, T.R.: Auditory models with Kohonen SOFM and LVQ for speaker Independent Phoneme Recognition. In: IEEE International Conference on Neural Networks, vol. 7, pp. 4466–4469 (1994)
11. Jankowski Jr., C.R., Lippmann, R.P.: Comparison of auditory models for robust speech recognition. In: Proceedings of the Workshop on Speech and Natural Language, pp. 453–454 (1992)
12. Kasper, K., Reininger, H., Wolf, D.: Exploiting the potential of auditory preprocessing for robust speech recognition by locally recurrent neural networks. In: Proc. Int. Conf. Acoustics, Speech and Signal Processing (ICASSP), pp. 1223–1226 (1997)
13. Kim, D.S., Lee, S.Y., Hil, R.M.: Auditory processing of speech signals for robust speech recognition in real-world noisy environments. IEEE Transactions on Speech and Audio Processing, 55–69 (1999)
14. Koizumi, T., Mori, M., Taniguchi, S.: Speech recognition based on a model of human auditory system. In: 4th International Conference on Spoken Language Processing, pp. 937–940 (1996)
15. Hunt, M.J., Lefébvre, C.: Speaker dependent and independent speech recognition experiments with an auditory model. In: International Conference on Acoustics, Speech, and Signal Processing, pp. 215–218 (1988)
16. Colombi, J.M., Anderson, T.R., Rogers, S.K., Ruck, D.W., Warhola, G.T.: Auditory model representation and comparison for speaker recognition. In: IEEE International Conference on Neural Networks, pp. 1914–1919 (1993)
17. Colombi, J.M.: Cepstral and Auditory Model Features for Speaker Recognition. Master's thesis (1992)
18. Shao, Y., Wang, D.: Robust speaker identification using auditory features and computational auditory scene analysis. In: International Conference on Acoustics, Speech, and Signal Processing, pp. 1589–1592 (2008)

19. Martínez–Rams, E., Garcerán–Hernández, V.: Assessment of a speaker recognition system based on an auditory model and neural nets. In: Mira, J., Ferrández, J.M., Álvarez, J.R., de la Paz, F., Toledo, F.J. (eds.) IWINAC 2009. LNCS, vol. 5602, pp. 488–498. Springer, Heidelberg (2009)
20. Ortega-Garcia, J., González-Rodriguez, J., Marrero-Aguiar, V., et al.: Ahumada: A large speech corpus in Spanish for speaker identification and verification. Speech Communication 31(2-3), 255–264 (2004)
21. Shamma, S.A., Chadwich, R.S., Wilbur, W.J., Morrish, K.A., Rinzel, J.: A biophysical model of cochlear processing: intensity dependence of pure tone responses. J. Acoust. Soc. Am. 80(1), 133–145 (1986)
22. Lopez Poveda, E.A., Eustaquio-Martín, A.: A biophysical model of the Inner Hair Cell: The contribution of potassium currents to peripherical auditory compression. Journal of the Association for Research in Otolaryngology JARO 7, 218–235 (2006)

Automatic Detection of Hypernasality in Children

S. Murillo Rendón[1], J.R. Orozco Arroyave[2], J.F. Vargas Bonilla[2],
J.D. Arias Londoño[3], and C.G. Castellanos Domínguez[1]

[1] Universidad Nacional de Colombia, s. Manizales
[2] Universidad de Antioquia, Medellín
[3] Universidad Antonio Nariño, Bogotá

Abstract. Automatic hypernasality detection in children with Cleft Lip and Palate is made considering five Spanish vowels. Characterization is performed by means of some acoustic and noise features, building a representation space with high dimensionality. Most relevant features are selected using Principal Components Analisis and linear correlation in order to enable clinical interpretation of results and achieving spaces with lower dimensions per vowel. Using a Linear-Bayes classifier, success rates between 80% and 90% are reached, beating success rates achived in similiar studies recently reported.

Keywords: Cleft Lip and Palate, Hypernasality, Principal Components Analysis, speech.

1   Introduction

The Cleft Lip and Palate (CLP) is one of the most frequent congenital malformation around the world [1], and there exists several speech pathologies associated to it, such as: hypernasality, hyponasality, glottal stop, among others. Hypernasality is the most common pathology present in CLP patients.

The speech with inappropriate nasal resonance is mainly produced by deficient control of articulatory muscles asociated to soft palate and velopharyngeal cavity [12]. In order to develop complete control over such muscles in the shortest possible time, constant phonoaudiologic therapy is required. To achieve good results, the phonoaudiologist must have a trained ear to judge about the existence of hypernasality in the child's voice. The health system on different countries changes constantly the assigned therapist for each child; considering that every specialist has different ear capacity and different perceptual criteria, it is very important the development of automatic system for the detection of hypernasality in CLP patients to be used as computer-aided medical diagnosis tool. Accuracy of the computarized system will depend on the discriminancy power of the measured features; thus characterization is a fundamental step in the process of classification between healthy and hypernasal voices. 51 features are considered here to achieve a complete acoustical analysis of children's voices.

J.M. Ferrández et al. (Eds.): IWINAC 2011, Part II, LNCS 6687, pp. 167–174, 2011.

Detection of hypernasality requires a lot of features and evaluate all of them increases the complexity of the problem and its computational cost. It is important to find the features with most discriminant capacity, and to eliminate those that provide less information to the pathology identification task; this procedure is done by means of a Principal Components Analysis (PCA) transformation and a elimination of features which present linear correlation with other. For classification process a linear-Bayes classifier is trained and validated with unknown records.

Similar works, performed multidimensional analysis of hypernasality in children's voice [4], [5], considering intonation features and 24 Mel Frequency Cepstral Coefficients (MFCC). In these studies no features selection is made and any acoustic feature is implemented, difficulting clinical analysis of the characteristics that most information provide to the hypernasality detection process. In the same works, success rates rounding 71.8% are reached in vowels, and up to 75.8% for words.

In this work, success rates between 80% and 90% are reached for spanish vowels, and the proposed methodology considers acoustic and noise features as well as eleven MFCC. After characterization, automatic feature selection is made allowing further clinical analysis for the pathology.

The rest of the paper is organized as follows. In section 2, a description of the followed methodology is presented, section 3 provides details about experimental framework describing the database used and some experimental details about the followed methodology for classification and error estimation. The section 4 contains graphics and tables with success classification rates. Finally, in section 5, some conclusions are provided.

2   Methodology

In the following, details about characterization, features selection and classification processes are provided.

Characterization. In general terms, the features considered in this paper, correspond to acoustic and noise analysis of hypernasal voice, in addition to the cepstral analysis by means of MFCC.

First, perturbations of amplitude and time in the fundamental period (Pitch) of voice signals are measured taking voice windows of $20ms$; on each time interval the variation of the Pitch was calculated as Jitter, and the variation of the maximum amplitude of the Pitch was considered as Shimmer. Considering the craniofacial malformation that suffers most of CLP patients and its implications on their vocal tract, with these two features, is possible to find general problems with their vocal folds movement as improper or incomplete closure of the vocal folds. Jitter indicates variations on the frequency vibration in vocal folds due to the lack control of vocal fold muscles; shimmer represents a reduction of glottic resistance and mass lesions in the vocal folds [6].

After time and amplitude perturbation analysis, some noise measurements were considered to quantify the presence of noise in the voice of CLP patients and considering its influence in hypernasal voices [7]. Harmonics to Noise Ratio (HNR) is implemented according to the procedure in [8]. Considering again $20ms$ windows, the complete voice signal $x(t)$ is divided and averaged by the number of time intervals n resulting the harmonic component of the signal denoted by $x_A(t)$, thus the energy of harmonic and noise components of the signal are defined as in 1.

$$H = n \int_0^T x_A^2(t)\, dt, \qquad N = \sum_{i=1}^n \int_0^{T_i} |x_i(t) - x_A(t)|^2 dt \qquad (1)$$

where T is the duration of each time window, $x_i(t)$ is the voice signal in the $i - th$ time interval and $x_i(t) - x_A(t)$ is the noise wave in each window. The relation H/N is the HNR.

In order to estimate noise content on different spectral components, HNR can be calculated in Cepstral domain. According to [9], voice signal is windowed and is transformed to calculate its Cepstral version, considering the definition of Cepstral tranformation given in equation 2:

$$C(t) = \int_{-\infty}^{\infty} \log\left(|S(f)|\, e^{j2\pi ft}\right) df \qquad (2)$$

Where $S(f)$ is the spectrum of the signal. The summed energy of the noise in the signal is subtracted from the energy of the original harmonic spectrum to provide Harmonics to Noise Ratio in Cepstral domain (CHNR).

Normalized Noise Energy (NNE) is also considered and calculated following the algorithm presented in [10] summarized in expression 3.

$$NNE = 10 \log \left(\frac{\frac{1}{L} \sum_{k=N_L}^{N_H} \sum_{i=1}^{L} \left|\hat{W}(k)\right|^2}{\frac{1}{L} \sum_{k=N_L}^{N_H} \sum_{i=1}^{L} |X_i(k)|^2} \right) \qquad (3)$$

Where $\hat{W}(k)$ and $X_i(k)$ are spectrums of the additive noise $w(n)$ and $i - th$ window's voice, respectively. $N_L = [N f_L T]$ and $N_H = [N f_H T]$, L is the number of frames, T is the sample period, f_L and f_H are respectively, lower and upper frequencies in the band where noise energy is calculated.

Additional to mentioned problems, velopharyngeal insufficiency or incompetence suffered by CLP patients, leads them to the need for compensatory movements in vocal tract, producing glottal stops [11] and general problems with glottal articulation movement [12] Glottal to Noise Excitation Ratio (GNE), can be used to measure possible glottal problems by means of the relation between the amount of vocal excitation due to vocal folds vibration versus the amount of excitation for turbulent noise on the vocal tract. This feature is implemented according to the procedure in [13].

Finally, besides acoustic and noise features described above, eleven MFCC are calculated considering its wide usage on the framework of voice and speech recognition.

Each feature is calculated using temporal windowing; due to this mean value, standard deviation and variance of each feature vector were calculated achieving a static characterization of voices.

Automatic Features Selection. For the automatic features selection, a linear transformation of the initial representation space is made using Principal Components Analysis (PCA) to find an optimal representation of the orginal data in a lower dimension space.

In PCA transformation the main objective is to find maximum variance of the projections of the initial feature space in other with lower dimension; this problem can be solved finding the maximum proper value associated to the proper vector in the transformed space variance matrix. Following this, the relevance of the features can be sorted beginning with higher ones.

After the relevance analysis, features with high linear correlation are eliminated condering that they are providing the same information for the detection of the pathology. Thus final features are sorted according to its relevance, finding an optimal subset of features that best describes the phenomena [14].

Classification. Classification stage takes place using a Linear-Bayes classifier based on the covariance of the features looking to divide the space in 2 regions, minimizing error rates by means of Mahalanobis distance optimization, defined in the equation 4.

$$r = \sqrt{(X - \mu)^T \Sigma^{-1} (X - \mu)} \tag{4}$$

Where X is the features vector, μ is it mean value, Σ^{-1} is de inverse of the covariance feature matrix, and $(X - \mu)^T$ is the transpose of $(X - \mu)$.

3   Experimental Framework

Database. The database is composed by 110 healthy voices and 156 recordings from patients with CLP which were labeled as hypernasal according to the criteria of a phonoaudiology specialist. The age of the people from both groups, healthy and hypernasal was in the range of 5 − 15. Registers are formed with utterances of five spanish vowels.

The audio records were captured in low noise conditions, with professional wiring and using a frequency sample of 44.100Hz and 16 quantization bits. All registers were automatically segmented to isolate each vowel phonation considering both its energy level and its zero crossing rate.

Error Classification Estimation. For the classification error calculation the database is splited into two subsets, training with 70% and validation with the remaining 30%. When classifier is trained, validation subset is used to measure classifcation errors.

This strategy is implemented seeking to find realistic success rates. 30% validation group is chosen to simulate new children coming into clinical consult.

4   Results

Initially, considered features are indexed with numbers in order to facilitate its asociation in further analysis as is shown in table 1.

Table 1. Indexes allocation for features

FEATURES	Jitter	Shimmer	HNR	CHNR	NNE	GNE	11MFCC
Means Index	1	2	3	4	5	6	7 a 17
Standard Deviation Index	18	19	20	21	22	23	24 a 34
Variance Index	35	36	37	38	39	40	41 a 51

The Fig.1 is built adding features to the experiment according to the relevance calculated in the automatic feature selection process. This figure shows the increasing success rates classifying normal and hypernasal voices as more features are considered for each vowel.

In the same figure, features are added to the system according to the order presented in table 3, the points from which acoustic and noise features begin to be consider by the system are shown. Note that for all cases, when acoustic and noise features appear, the classification rates begin to grow.

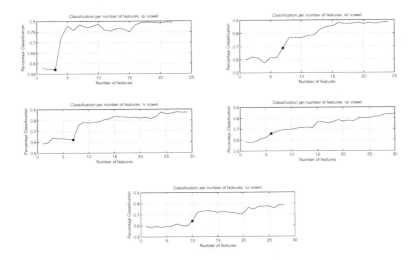

Fig. 1. Increase in success rates per vowel as more features are considered

In table 2 success rates before and after consider acoustic and noise features are presented. In general, an increase of 20% is reached by the automatic system when consider cepstral coeffcients, acoustic and noise features.

Table 2. Indexes of selected features after PCA transformation and correlation analysis

Vowels	Success Rates Before Consider Acoustic Features	Success Rates After Consider Acoustic Features
/a/	57,06%	79,57%
/e/	68,89%	88,82%
/i/	62,08%	87,49%
/o/	65,95%	84,10%
/u/	64,13%	78,86%

Table 3 compiles feature selection results per vowel. Representation spaces are reduced from dimensionalty 51 to 22 for vowel /a/, 24 for /e/, 29 for /i/, 30 for /o/ and 28 for /u/. These results imply that original representation spaces are reduced between 42% and 67%. It is important to note that dimensionalty reduction avoids overfitting problems in the classification stage and also implies lower complexity of the system.

Table 3. Indexes of selected features after PCA transformation and correlation analysis

Vowels	Feature Index														
/a/	24	29	31	20	3	4	8	15	13	10	16	14	6	23	21
	12	22	7	1	11	17	9								
/e/	24	32	30	31	29	7	13	4	2	8	17	5	11	12	23
	3	10	22	1	16	21	9	15	14						
/i/	25	27	28	24	31	30	34	4	2	1	26	33	20	3	5
	32	7	11	23	14	8	10	9	16	21	17	13	15	12	
/o/	24	28	33	32	31	25	4	30	29	2	27	19	1	34	3
	6	7	21	14	11	5	23	12	10	8	9	15	17	16	13
/u/	32	29	25	28	33	27	24	26	31	7	3	2	4	23	5
	21	9	16	6	11	12	17	10	13	8	15	14	1		

5   Conclusions

Using a simple linear-Bayes classifier, the success rates are superior in relation with other works in the state of the art which have addressed the same problem. This is due to the use of acoustic and noise features that provides a special information related to the physiological characteristics of the hypernasality.

Results showed in table 3 indicate that mainly using standard deviation of MFCCs, is possible to achieve acceptable success rates; however, according to learning curves showed in figure 1 and the results in table 2, to achieve higher success rates is necessary to consider means and standard deviations of acoustic features such as shimmer and jitter and noise measures such as HNR, CHNR, NNE and GNE.

In table 3, information of variance of considered features is not relevant. This can be because variance and standard deviation are two statistics measures highly correlated, indicating that proposed methodology effectively detect linear relations between features.

Considering the implemented classifier, is possible to infere that using a more complex one better results will be obtained.

Acknowledgments. This work was granted by ARTICA, financed by COL-CIENCIAS and Ministerio de TIC in Colombia. Project N°1115-470-22055 and was accomplished in association with the Clínica Noel in Medellín, Colombia.

References

[1] Congenital malformations worldwide. International Clearinghouse for Birth Defects Monitoring Systems. Amsterdam, Holland. Tech. Rep. (1991)

[2] Castellanos, G., Daza, G., Sanchez, L., Castrillón, O., Suarez, J.: Acoustic speech 177 analysis for hypernasality detection in children. In: 28th Annual International Con. 178 ference of the IEEE Engineering in Medicine and Biology Society, pp. 5507–5510 (2006)

[3] Cairns, D., Hansen, J., Riski, J.: A noninvasive technique for detecting hypernasal 181 speech using a nonlinear operator. IEEE Transactions on Biomedical Engineering 43(1), 35–45 (1996)

[4] Maier, A., Hönig, F., Hacker, C., Shuster, M., Nöth, E.: Automatic Evaluation of Characteristic Speech Disorders in Children with Cleft Lip and Palate. In: Proc. of 11th Int. Conf. on Spoken Language Processing, Brisbane, Australia, pp. 1757–1760 (2008)

[5] Maier, A., Shuster, M., Haderlein, T., Nöth, E.: Hypernasality in Speech of Children with Cleft Lip and Palate: Automatic Evaluation. In: 8th Int. Seminar on Speech Production, Strasbourg, France, pp. 277–280 (2008)

[6] Wertzner, H.F., Schreiber, S., Amaro, L.: Analysis of fundamental frequency, jitter, shimmer and vocal intensity in children with phonological disorders. Rev. Bras. Otorrinolaringol. 71(5), 582–588 (2005)

[7] Setsuko, I.: Effects of Breathy Voice Source on Ratings of Hypernasality. The Cleft Palate-Craniofacial Journal 42(6), 641–648 (2005)

[8] Yumoto, E., Gould, W.J.: Harmonics-to-noise ratio as an index of the degree of hoarseness. Journal of Acoustic Society of America 71(6), 1544–1550 (1982)

[9] Krom, G.: Cepstrum based technique for determining a harmonics-to-noise ratio in speech signals. Journal of Speech, Language and Hearing Research 36(2), 254–266 (1993)

[10] Manfredi, C.: Adaptive Noise Energy Estimation in Pathological Speech Signals. IEEE Transactions on Biomedical Engineering 47(11) (2000)

[11] Golding, K.K.: Therapy Techniques for Cleft Palate Speech and Related Disorders. Singular Thomson Learning (2001)

[12] Henningsson, G.E., Isberg, A.M.: Velopharyngeal movement patterns in patients alternating between oral and glottal articulation: a clinical and cineradiographical study citation. The Cleft Palate-Craniofacial Journal 23(1), 1–9 (1986)

[13] Michaelis, D., Gramss, T., Strube, H.W.: Glottal to Noise Excitation Ratio - a new measure for describing pathological voices. Acustica/Acta 83 (1997)

[14] Daza-Santacoloma, G.D., Arias-Londoño, J.D., Godino-Llorente, J.I., Sáenz-Lechón, N., Osma-Ruiz, V., Castellanos, G.: Dynamic feature extraction: An application to voice pathology detection. Intelligent Automation and Soft Computing 15(4), 665–680 (2009)

Characterization of Focal Seizures in Scalp Electroencephalograms Based on Energy of Signal and Time-Frequency Analysis

Alexander Cerquera[1], Laura V. Guío[2], Elías Buitrago[3],
Rafael M. Gutiérrez[3], and Carlos Medina[2]

[1] Faculty for Electronic and Biomedical Engineering, Complex Systems Group,
Antonio Nariño University. Bogotá, Colombia
`alexander.cerquera@uan.edu.co`
[2] Central League Against Epilepsy, Bogotá, Colombia
`victoriaguilam@gmail.com medinamalo@epilepsia.org`
[3] Complex Systems Group, Antonio Nariño University, Bogotá, Colombia
`elias.buitrago@uan.edu.co, director.sistemas.complejos@uan.edu.co`

Abstract. This work presents a method for characterization of focal seizures from scalp digital electroencephalograms (EEG) obtained in the Central League Against Epilepsy in Bogotá, which were acquired from patients between 13 and 53 years old. This characterization was performed in segments of 500 ms with presence of focal seizures that had been initially identified and labeled by a specialist during their visual examination. After selection of the segments and channels for analysis, the energy of the signals were calculated with the idea that the energy of focal seizures could be larger than the one of their side waves in the segments. This procedure produced peaks of energy corresponding to the seizures and, some times, to noise and artifacts. In order to identify the peaks of energy of the seizures an analysis with continuous wavelet transform was performed. It was found that the mother wavelet 'bior2.2' allowed more easily the identification of such seizures from the seventh scale of the analysis. The method allowed the identification of the 65% of the seizures labeled by the specialist.

1 Introduction

Epilepsy is one of the most known neurological disorders characterized by recurrent unprovoked seizures that arise as consequence of sudden and excessive electrical discharges in a group of brain cells. According to information from the World Health Organization (2011), around 50 million people worldwide suffer epilepsy and nearly 90% of these persons are located in developing countries.

One of the most useful tools employed by specialists for diagnosis of epilepsy is the scalp electroencephalogram (EEG), which is a graphical register of the electrical activity on the cerebral cortex. Generally, specialists examine by visual inspection the morphological aspects of the EEG signals to detect waves corresponding to epileptic seizures. This detection is relatively easy to perform

J.M. Ferrández et al. (Eds.): IWINAC 2011, Part II, LNCS 6687, pp. 175–184, 2011.

in generalized or distributed epilepsies, since their seizures are characterized by waves of high amplitudes and short duration present simultaneously in all the channels of an EEG register of these epilepsies. However, other kind of seizures, commonly known as focal or partial, are difficult to detect during the visual examination. This situation is found when the epileptic discharges are located in different parts of the cerebral cortex. Among other reasons, the difficulty to detect them is due to the low amplitudes that characterize them frequently, being sometimes masked by noise or any other artifacts present in the register. Unlike the seizures in the generalized epilepsies, the focal seizures do not arise in all the channels of the EEG register (Niedermeyer and Lopes da Silva (2005)). Therefore, their success detection is highly dependent on the training and experience of the specialist who tries to identify them by visual inspection.

One approach for the automatic detection of focal seizures was presented by Van Hoey and colleagues (2000), employing a principal component analysis (PCA) decomposition and a single dipole source to analyse the spatial properties of the EEG signals. This method was combined by Van Hese and collaborators (2008) with the analysis of temporal properties of the signals to improve the results of detection. The combination of both methods in this work showed an enhancement in the detection of focal seizures avoiding activity that does not match epileptiform activity. Thus, the results exhibited a mean sensitivity and selectivity of 92% and 77%, respectively. Other studies with a similar aim were presented by Vanrumste et al. (2004a and 2004b). In these works, the authors investigated the nature of apparent non-epileptiform seizures from the same brain area as epileptiform activity in eight EEG registers of pediatric patients with focal epilepsy. The method employed was based on detection of seizures with a single underlying source having a detection of seizures with a single underlying source having a dipolar potential distribution. The results were compared with the seizures labeled by an specialist and it was found that in five out of the eight patients a substantial number of other detections arose from the same area as the labeled seizures. These observations suggested that the morphology of a high proportion of these other detections did not resemble typical epileptiform activity. However, these waves are likely to be related to the underlying epileptogenic process.

Another work published by Henriksen and collaborators (2010) presented an adaptation of methods for analysis of intracranial electroencephalography for the characterization of scalp electroencephalography. In this study, the authors represented the EEG data employing wavelet transform and classified with a support vector machine. In this way, they reported an improvement in the performance by inclusion in the analysis of high frequency containing lower levels in the wavelet transform, whose results showed a sensitivity of 96.4%. Although the results of these researches and other related ones have seemed to be promising, it has been difficult so far the development of automatic systems for detection of seizures as reliable as interpretations done by specialists, especially due to the large number of false positive detections. In addition, each study in this topic has been carried out with the own electroencephalogram databases of the researchers (Halford 2009).

The Central League Against Epilepsy (LICCE from its name in Spanish *Liga Central Contra la Epilepsia*) in Bogotá, Colombia, is an institute dedicated to the treatment of patients with epilepsy and other neurological disorders. In this institute, one of the activities carried out by physicians taking their specialization in neurology is the analysis and interpretation of EEG signals. These activities include the detection of focal seizures by visual inspection. As mentioned above, this is a very difficult procedure that requires a high level of experience and training. Therefore, the purpose of this work is the application of techniques based on temporal and time-frequency analysis to characterize these seizures, being the basis that leads in future to a system that guide physicians about the possible time points in an EEG register with presence of focal seizures.

The study has been accomplished in framework of the project "Characterization of electroencephalograms for detection and classification of epileptic patterns employing measures of complexity and nonlinear synchronization", which is currently financed by the Central League Against Epilepsy and the Vice-Rectorate for Science, Technology and Innovation of the Antonio Nariño University.

2   Analysis of the Data

2.1   Acquisition of the EEG Registers and Organization of the Database

The EEG registers were acquired from four patients between 13 and 53 years old diagnosed with focal epilepsy in the LICCE, employing an equipment for digital acquisition of electroencephalograms Grass-Telefactor Comet with an amplifier system AS40. This equipment was designed for electroencephalographic and polysomnographic monitoring. The registers were recorded during 30 minutes with a sample rate of 200 Hz and stored in .edf format. An intermittent photic stimulation was applied during the whole time of registration and the electrodes were placed in bipolar longitudinal montage form (Jasper (1958)), i.e. surface electrodes distributed according to the international system 10–20. This montage allowed the registration of 18 channels identified as: Fp1–F7, F7–T3, T3–T5, T5–O1, Fp1–F3, F3–C3, C3–P3, P3–O1, FZ–CZ, CZ–PZ, Fp2–F8, F8–T4, T4–T6, T6–O2, Fp2–F4, F4–C4, C4–P4 and P4–O2. After the acquisition an specialist of the LICCE performed a visual examination of the registers using the software Grass-Telefactor EEG TWin 3.8 to look for focal seizures. In this way, the registers were labeled by the specialist in their time stamps with presence of focal seizures. After this step, the data were converted from .edf to .mat format in order to make possible their processing and analysis with Matlab. This conversion was done using the software EDFbrowser 1.32.

2.2   Preprocessing of the Data

It was considered that the initial sample rate of 200 Hz utilized in the acquisition of the registers could be too low for analysis purposes, especially because

the low resolution that might come up in the time-frequency analysis as will
be explained in Section 2.4. Thus, the sample rate was increased with a factor of 5/1, obtaining a new sample rate of 1000 Hz. Subsequently, the time
stamps of the seizures labeled by the specialist were located and extracted in
segments of 500 ms, trying to place them in the middle of their segment. After
this procedure the data were normalized. Since the data in each channel contained negative values, the normalization was performed in two steps as indicated
in Equation 1 (Teknomo (2006))::

$$x_a\,[n] = x\,[n] - (\min\,\{x\,[n]\} + 0.001)$$

$$x_b\,[n] = \frac{x_a[n]}{\max\{x_a[n]\}},$$

$$(1)$$

where $x\,[n]$ is the original set of data from each channel and $x_b\,[n]$ indicates the
normalized values of $x\,[n]$.

As mentioned in Section 1, when focal seizures arise they do not appear in all
the channels of the EEG register but in some of them. Therefore, in this work
only the channels in the segments of 500 ms with the presence of focal seizures
were selected for analysis. Figure 1(a) shows one of these segments with their
respective channels and with the labeled seizures fenced by dashed line circles.
After selection of segments and channels, the EEG data were filtered with a high-
pass zero-phase Butterworth filter of two poles, employing a cut-off frequency
of 0.5 Hz. The aim of this procedure was to remove, as much as possible, the
fluctuation of low frequency in the basal line. Figure 2 shows the results of the
high-pass filtering.

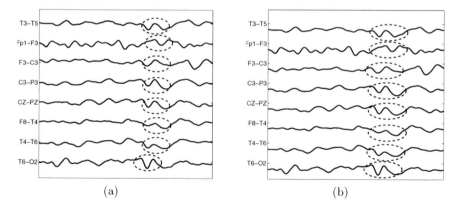

(a)   (b)

Fig. 1. (a) Original segment of 500 ms with presence of focal seizures labeled by a
specialist of the LICCE. These seizures are fenced inside the dashed circles and the
channels are indicated on the left side. (b) Signals shown in Figure 1(a) after processing
with the high-pass zero-phase filter.

2.3   Calculation of the Energy from the Signals

It can be observed from the graphics of Figure 1 that seizures may be characterized by an increase of their energies in comparison with their lateral waves. For this reason, the energy values of the filtered signals were calculated using sliding windows of 80 ms from the beginning to the end of each selected segment of time. An overlapping of 10% was employed in each iteration. These values were selected taking into account that the focal seizures are waves whose time width is normally between 30 and 70 ms (Fisch (1999)). The calculation of the energy was performed according to Equation 2 (Mitra (2001)):

$$E = \sum_{n=N_1}^{N_2} \left(|x(n)| \right)^2, \tag{2}$$

where N_1 and N_2 are the time limits of the sliding window and $x(n)$ represents the values of every EEG register in each instant of time n. Figure 2(a) shows the energy calculated as mentioned above. It can be observed some peaks of energy on the right middle of the graphic that correspond to the seizures in each channel, which are fenced by two dashed lines. However, it can be noticed other peaks of energy that belong to noise or artifacts located in time points before and after the true seizures. With the aim to emphasize the amplitude of those events in the channels that belong to seizures, as well as to minimize the peaks of energy corresponding to artifacts, each of the elements in the energy vectors were squared. To ensure the removal of peaks of low amplitude that does not correspond to seizures it was established a threshold, so that every value of the squared energies below that threshold become to zero. The results of this last step are shown in Figure 2(b).

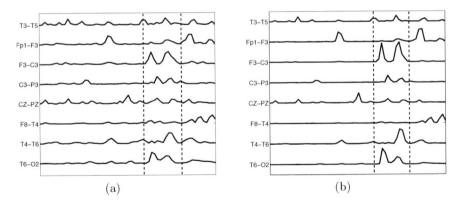

(a)   (b)

Fig. 2. (a) Energy of the segmented channels calculated as indicated in Equation 2. The peaks of energy corresponding to seizures are fenced inside the dashed lines. (b) Squared and thresholded peaks of energy to minimize events corresponding to artifacts and noise.

2.4   Analysis with Continuous Wavelet Transform

In order to detect the peaks corresponding to seizures in the Figure 2, it was proposed an analysis using continuous wavelet transform (CWT) (Mertins (1999)), taking into account that the CWT can detect the peaks of energy like the ones showed in Figure 2. This analysis was done as indicated in Equation 3:

$$CWT_x^{\Psi}(\tau, s) = \frac{1}{\sqrt{|s|}} \int x(t) \Psi^* \left(\frac{t - \tau}{s} \right) dt \tag{3}$$

where s is the scale, $x(t)$ is the signal and $\Psi^* \left(\frac{t-\tau}{s} \right)$ represents the mother wavelet. It was expected that the peaks of energy corresponding to the seizures produce a high degree of correlation with the mother wavelet using a suitable scale s. This is possible if the similarity between the mother wavelet and the peaks or energy of the seizures is as high as possible. In this case, the biorthogonal wavelet 'bior2.2' contained in the Wavelet Toolbox of Matlab was showed to be the most similar. This mother wavelet is shown in Figure 3. The effect of other mother wavelets ('db3', 'bior2.4', 'coif1' and 'mexh') was also observed but the results were not so satisfactory.

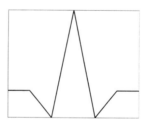

Fig. 3. Biorthogonal wavelet utilized in the analysis, identified as 'bior2.2' in the Wavelet Toolbox of Matlab

The analysis with the CWT was performed with 20 scales in all the channels and segments selected for study, obtaining complex coefficients in each iteration whose magnitude indicates the degree of similarity between the mother wavelet and the time window analysed in the iteration. The results are shown in Figure 4 for two of these channels.

In both plots of Figure 4 it can be observed that from the sixth or seventh scale it is possible the detection of the largest magnitude of these complex coefficients, which correspond to the peaks of energy of the seizures. Therefore, the largest magnitude of these coefficients in the seventh scale were located in all the analysed channels and defined as the position in time of the seizures. Figure 5 shows the time points estimated as detected seizures after application of the method. It can be observed that in three channels (Fp1–F3, CZ–PZ and F8–T4) the analysis estimated the seizures out of the limits labeled previously by the specialist, specially in the channels CZ–PZ and F8–T4. It is possible that

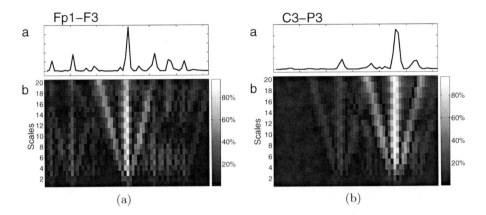

Fig. 4. Results of the analysis with CWT for two of the selected channels (Fp1–F3 and C3–P3). The upper panels (a) show the squared energy of the signals as shown in Figure 2(b). Every patch in the lower panels (b) indicate the magnitude of the complex coefficients obtained from each iteration in the calculation of the CWT; the lighter a patch the larger coefficient of similarity bewteen the mother wavelet and the time window of the iteration in each channel. This calculations were done with 20 scales, such as indicated on the left side of the lower panels.

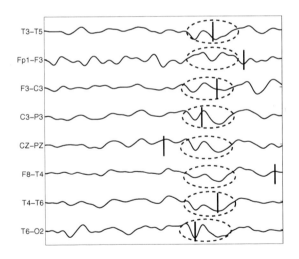

Fig. 5. Results of the detection of focal seizures for the segments and channels selected for the analysis. The dashed circles fence the seizures labeled initially by the specialist of the LICCE, such as shown in Figure 1. The vertical lines indicate the time points indicated as seizures after the analysis with CWT.

the noise present around the seizures entails to a wrong detection of seizures, since the energy of noise and artifacts could be sometimes equal or larger than the energy of the seizures.

The same analysis was done for all the four EEG registers obtaining a correct detection rate of about 65%; 35% of the estimations were considered as false positives.

2.5   Discussion

From the observation of the plots of Figure 1 it is difficult to determine whether the focal seizures might be characterized by sudden changes in frequency and detected by a simple analysis with CWT. For instance, Figure 6 shows the effect of trying to identify the seizure directly from the analysis with CWT on the EEG data in the channel Fp1–F3. In this case, it can be noticed in the row of the twentieth scale two complex coefficient whose magnitudes are sufficiently large and whose time points could be considered as seizures. Therefore, a calculation of the energy of the signal (Section 2.3) resulted to be beneficial for enhancing the peaks of energy corresponding to seizures, although some of them that correspond to artifacts were also enhanced. Nevertheless, this last situation could be partially diminished squaring the values of the energy vectors (see graphics of Figure 2). The calculation of the energy allows the enhancing of the seizures regardless their orientation, which is another resultant benefit of this procedure (the peaks of the seizures can be projected in positive or negative sense).

After the procedures of calculation of energy it seemed to be easier the detection of the seizures with CWT, however, it was necessary a correct selection a priory of a suitable mother wavelet for this purpose. As mentioned in Section 2.4, the mother wavelets 'db3', 'bior2.2', 'bior2.4', 'coif1' and 'mexh' of the Wavelet Toolbox in Matlab were evaluated due to the similarity of their shapes with the energy of the seizures. Nonetheless, the mother wavelet 'bior2.2' (see Figure 3) allowed obtaining the best results. The evaluation of these wavelet mothers for the mentioned purpose are not included in this work due to space limitations. Even so, in the analysis with CWT some artifacts were identified as seizures, which is reflected in a false positives rate of about 35%. Probably, it might be necessary the combination of further methods to improve the capability of detection of true seizures, such as principal component analysis (PCA) decomposition and a single dipole source as proposed in the works of Van Hoey et al. (2000) and Van Hese et al. (2008). Another alternative is the improvement of the preprocessing of the registers with the aim to remove artifacts whose shapes are similar to the ones of the seizures. These activities are planned to be the next steps in this research, in order to perform the analysis over larger segments of time extracted from the EEG registers and improve the detection capability of the method. Likewise, the staff in the LICCE will continue collecting EEG registers with focal seizures. This is also one important task since few epochs with focal seizures are found in a single EEG register. Most of these register contain between three and six epochs in 30 minutes of acquisition.

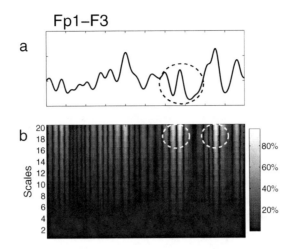

Fig. 6. Attempt to identify the seizure directly from an EEG register. Panel a shows the segment of the channel with the seizure fenced by a dashed circle. In panel b it can be noticed that the identification could be possible with the twentieth scale of the CWT (left white dashed circle). However, there is another artifact afterwerwards that could be wrongly identified as seizure (right white dashed circle). This observation contrasts with the result shown in Figure 4(a) for the same analysed segment.

References

Fisch, B.: Fisch and Spehlman's EEG primer: Basic principles of digital and analog EEG, 3rd edn., pp. 237–240. Elsevier, Amsterdam (1999)

Halford, J.J.: Computerized epileptiform transient detection in the scalp electroencephalogram: Obstacles to progress and the example of computerized ECG interpretation. Clinical Neurophysiology 120, 1909–1915 (2009)

Henriksen, J., Remvig, L.S., Madsen, R.E., Conradsen, I., Kjaer, T.W., Thomsen, C.E., Sorensen, H.B.D.: Automatic seizure detection: going from sEEG to iEEG. In: 32nd Annual International Conference of the IEEE EMBS, Buenos Aires, Argentina, August 31 - September 4, pp. 2431–2434 (2010)

Jasper, H.H.: The ten-twenty electrode system of the International Federation. Electroencephalogr. Clin. Neurophysiol. 10, 371–375 (1958)

Mertins, A.: Signal analysis: wavelets, filter banks, time-frequency transforms and applications, pp. 210–213. John Wiley & Sons Ltd., England (1999)

Mitra, S.K.: Digital Signal Processing: A Computer-Based Approach, 2nd edn., pp. 52–53. McGraw-Hill, New York (2001)

Niedermeyer, E., Lopes da Silva, F.: Electroencephalography: basic principles, clinical applications, and related fields, 5th edn. Lippincott/Williams & Wilkins, Philadelphia, PA (2005)

Teknomo, K.: Tutorial of normalization methods (2006),
    http://people.revoledu.com/kardi/tutorial/Similarity/Normalization.html

Van Hese, P., Vanrumste, B., Hallez, H., Carroll, G.J., Vonck, K., Jones, R.D., Bones, P.J., D'Asseler, Y., Lemahieu, I.: Detection of focal epileptiform events in the EEG by spatio-temporal dipole clustering. Clinical Neurophysiology 119, 1756–1770 (2008)

Van Hoey, G., Vanrumste, B., Van de Walle, R., Boon, P., Lemahieu, I., D'Havé, M., Vonck, K.: Detection and localization of epileptic brain activity using an artificial neural network for dipole source analysis. In: Proceedings of the EUSIPCO 2000 Conference, Tampere, Finland, September 5-8 (2000)

Vanrumste, B., Jones, R.D., Bones, P.J., Carroll, G.J.: Background Activity Originating from Same Area as Epileptiform Events in the EEG of Paediatric Patients with Focal Epilepsy. In: Proceedings of the 26th Annual International Conference of the IEEE EMBS, San Francisco, CA, USA, September 1-5 (2004)

Vanrumste, B., Jones, R.D., Bones, P.J., Carroll, G.J.: Singular Value Decomposition and Dipole Modelling applied to Observe Background Activity Originating from the Same Area as Epileptiform Events in the EEG of Paediatric Patients with Focal Epilepsy. In: Proceedings of the 2nd International Conference on Advances in Biomedical Signal and Information Processing (MEDSIP 2004), Valleta, Malta, September 2004, pp. 92–98 (2004)

Web page of the World Health Organization. Fact sheet N° 999,
    http://www.who.int/mediacentre/factsheets/fs999/en/index.html

An Optimized Framework to Model Vertebrate Retinas

Andrés Olmedo-Payá[1], Antonio Martínez-Álvarez[2], Sergio Cuenca-Asensi[2], Jose M. Ferrández[3], and Eduardo Fernández[1]

[1] Institute of Bioengineering and CIBER BBN, University Miguel Hernandez, Alicante, Spain
e.fernandez@umh.es
[2] Computer Technology Department, University of Alicante, Carretera San Vicente del Raspeig s/n, 03690 Alicante, Spain
amartinez@dtic.ua.es
[3] Department of Electronics and Computer Tecnology, Universidad Politécnica de Cartagena, Spain
jm.ferrandez@upct.es

Abstract. The retina is a very complex neural structure, which contains many different types of neurons interconnected with great precision, enabling sophisticated conditioning and coding of the visual information before it is passed via the optic nerve to higher visual centers. Therefore the retina performs spatial, temporal, and chromatic processing on visual information and converts it into a compact 'digital' format composed of neural impulses. However, how groups of retinal ganglion cells encode a broad range of visual information is still a challenging and unresolved question. The main objective of this work is to design and develop a new functional tool able to describe, simulate and validate custom retina models. The whole system is optimized for visual neuroprosthesis and can be accelerated by using *FPGAs*, *COTS* microprocessors or *GP-GPU* based systems.

Keywords: Artificial retinas, visual neuroprostheses, retina simulation, spiking neurons.

1 Introduction

The retina is essentially a piece of brain tissue that gets direct stimulation from the outside world's lights and images. Visual input to the retina consists on a stream of photons, which can be unequivocally quantified in space and time. The retina performs spatial, temporal, and chromatic processing on visual information and converts it into spike trains. Thus our entire experience of the external visual world derives from the concerted activity of a restricted number of retinal ganglion cells, which have to send their information, via the optic nerve, to higher visual centers. This representation has to be unequivocal and fast, in order to ensure object recognition for any single stimulus presentation within a few hundreds of milliseconds[1]. Therefore, the question of how the information

J.M. Ferrández et al. (Eds.): IWINAC 2011, Part II, LNCS 6687, pp. 185–194, 2011.
© Springer-Verlag Berlin Heidelberg 2011

about the external world is compressed in the retina, and how this compressed representation is encoded in spike trains is an important challenge for visual neuroscience and for the success of any visual prosthesis [2].

Our group is working on the development of a cortical visual neuroprosthesis that can be useful to restore some functional vision to profoundly blind. The first stage of the system have to perform has to extract and enhance the most relevant features of the scene. The goal of developing such a bioinspired retinal encoder is not simply record a high-resolution image, but also to transmit visual information in a meaningful way to the appropriate site(s) in the visual pathways. In order to achieve this goal we have to take into account the coding features of the biological visual system and design constraints related to the number and distribution of the set of working electrodes the visual scene is mapped to [2] [3].

This paper is focused on the design of a retinal-like encoder. In this way, we present a framework to model and test retina models. The framework is conceived as a high–level abstraction tool covering diverse aspects in the development of such devices, allowing the interoperation of every discipline involved in its design. The final objective is to progress towards an uniform functional tool to:

- describe and simulate a custom bioinspired retina model at various levels of abstraction
- perform an *automatic* optimization and acceleration of the processing by means of a sort of supported technologies: *FPGA*, *COTS* microprocessors or *GP-GPU* architectures.
- assist the designer to decide the final target to be used in terms of: real–time capabilities, feasibility, portability (what is the hardware support that best fit the designer constraints?), plasticity (what variables can be defined as custom parameters?), etc . . .
- easily compare synthetics and biological records using *de facto* standards for data sharing (i.e. the *Neural Event Format* [4]) to validate the correctness of the retina model
- perform *in–vivo* stimulation using the generated output

Assuming every requirement exposed above, we can find no tools or environments offering the same as an uniform framework. However there are research projects developing tools in the same direction or carrying out a subset of the previous items such as *Retiner* [5] or *EPIRET3* [6].

Our framework named *RetinaStudio* is designed to model and process bioinspired vision models such as a retina. Bioinspired vision models don't have a very significant utilization in artificial vision due to the fact that they used to have a larger computational costs. Despite this, they suppose a more convenient model to design reliable visual neuroimplants [7]. The bioinspired processing scheme used by *RetinaStudio* is based on the one proposed in the previous citation, and can be inspected in Fig. 1. Succinctly it is composed of a first stage of image capturing (typically taken from one ore more videocameras), a bioinspired spatio–temporal filtering (with primitives such as Difference of Gaussians, Gaussians, Laplacian of Gaussians, derivative, Gabor, Sobel, etc.), a stage to encode

Fig. 1. Main functional blocks of the bioinspired visual processing used by *RetinaStudio*

the raw output using a neuromorphic scheme to generate the spiking information to be finally mapped to a custom microelectrode in the last stage.

The paper is organized as follows: next Section presents the proposed framework to model and test retina models; Section 3 includes some experiments and use cases related with the tool. Finally 4 summarizes some concluding remarks and suggests directions for future work.

2   A High-Level Framework to Model Artificial Retinas

A special problem to be addressed when dealing with different models and tools for the design of visual neuroprosthesis is the fact that there are many variables to be taken into account. These variables usually overlap a number of heterogeneous disciplines such as neuro-science, neuro-engineering, electronics engineering, histology and computer science/engineering. This fact depicts a sort of questions that must be considered to design a reliable bio-model:

- Is the implementation/target fast enough for *in–vivo* stimulation?
- What is the maximum error (worst case) that could be generated using a given technology?
- What is the more convenient data–representation to be used, and how does it affects to the overall system speed?
- Can the model be embedded into an wearable device?
- How many variables from the visual processing system can operate as editable parameters for a rapid tuning of the model?

Our contribution to cover these topics is presented as a high-level abstraction framework to model data-flow based systems, in particular visual neuroprosthesis. *RetinaStudio* allows the specification, testing, simulation, validation and implementation of bio–inspired retina models to be used within visual neuroimplants. It offers an uniform platform that overcomes the drawback generated by the diverse disciplines that take place in the design of the final device. In this way, *RetinaStudio* is an interdisciplinary tool suitable for histologists, neuro/electronics/computer engineers and neuro–surgeons. The interaction of such heterogeneous disciplines and the environment has been carefully designed to be easy and efficient.

Fig. 2 shows a capture of a working session with *RetinaStudio*'s front-end where an easy vision model is been depicted. There are three ways of interaction with *RetinaStudio* to define a custom retina model. First of all, the researcher can describe the retina model and neuroprosthesis device by means of a predefined widgetry based on contextual dialogs and step–by–step druids. Although

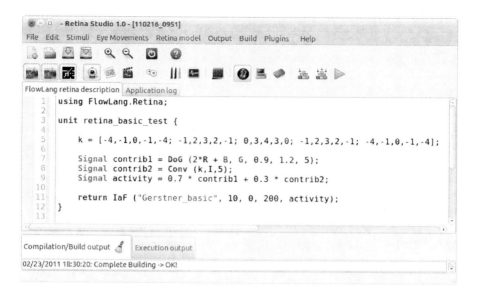

Fig. 2. A screenshot showing a working session with *RetinaStudio*

there are tools working in a similar way, such as *Retiner* [5], *RetinaStudio* in addition, extends its dialog-assisted edition not only to the specification of the retina model, but also to the rest of tasks offered by the framework (processing acceleration on different targets, interface with a neurostimulator [8], etc.). The second way of interaction with the tool is by using its exposed *API* (*Application Programming Interface*). Actually, *RetinaStudio* can also be seen as a vast library to model, simulate, test and accelerate vision models. The last way of interaction with *RetinaStudio* is describing the overall visual system using an *ad-hoc* compilable language named *Flowlang*.

2.1   An Ad-hoc Language to Describe Visual Systems

RetinaStudio includes a support compiler for an *ad-hoc* language for the description of *data–flow* based systems, in particular, bioinspired retina models. *Flowlang*, is intended to provide these items:

- Multitarget and flexible : the same system description can be re–targeted with little or no changes to a *COST* microprocessor, a supported *FPGA* or a *GP-GPU* architecture.
- Easy to learn and easy to read : having a low learning curve.
- *Object Oriented Programming* : with a syntax very similar to *C#*.
- Easy built-in interoperation with custom libraries : the primitives can be easily extended.
- Predefined high–level support libraries covering *data–flow*, image processing and capturing, spiking neurons models, I/O primitives, etc.
- Aggressive optimization of the processing model.

An example of a *data–flow* system generating a very simple Difference of Gaussian *DoG* (a primitive widely used when modeling retinas) is conveniently shown in Fig. 2.

2.2   Built-In Multitarget HW/SW Bioinspired Libraries

Jointly with *Flowlang*, it is presented its build&runtime library support. In this paper, we are focused in the *COTS* microprocessor target. In this sense, *Retina-Studio* offers a rich set of bioinspired accelerated visual primitives for a number of widely used microprocessors architectures such as: *Intel* Core2 Duo/Solo and *Intel Core i3,i5,i7*. For example, we have more than 10 diverse implementation for a 2D–convolution depending of using *SIMD* acceleration (from *MMX* up to *SSE3*), 32 or 64 bits, fixed, floating point or integer arithmetic, etc. Every implementation of the convolution offers the same functionality (calculus of a 2D–convolution) for the designer, but different performance and accuracy, so there is no distinction at at behavior level. We can use directly whatever implementation for the convolution, or ask the compiler to do this for us, depending on the requirements of the application.

2.3   A Multitarget High–Level HW/SW Compiler

After having a visual system described using *Flowlang*, the behavior encapsulated by this representation is then translated by a custom compiler to an optimized version consisting of a set of interconnected processing primitives. Choosing a given microprocessor as the preferred output for our model, the compiler will address our design constraints (i.e. ensure a low error level in the optimization) and will offer us an accelerated library as well as a simulation library to be used out-the-box. The simulation of the overall system (with independence of the technological target to be used) is performed using double precision floating point arithmetic. The final output uses *OpenCV* [9] for capturing and simulation purposes, and also in the cases where a primitive is not defined in *RetinaStudio* runtime library.

3   Results

To show the usefulness of *RetinaStudio* to design, simulate and validate bioinspired retina models, we have depicted, as a case study, a basic bioinspired retina model. The high–level model description to be studied, which has already been shown in Fig. 2, can be mathematically modeled as follows:

$$Model = 0.7\,DoG_{5\times5}(2R+B,G,\sigma_1=0.9,\sigma_2=1.2) + 0.3\,Conv_{5\times5}(I,\kappa) \quad (1)$$
$$= 0.7\cdot(Gauss_{5\times5}(2R+B,0.9) - Gauss_{5\times5}(G,1.2)) + 0.3\cdot(I*\kappa)$$

where

$$\kappa = \begin{pmatrix} -4 & -1 & 0 & -1 & -4 \\ -1 & 2 & 3 & 2 & -1 \\ 0 & 3 & 4 & 3 & 0 \\ -1 & 2 & 3 & 2 & -1 \\ -4 & -1 & 0 & -1 & -4 \end{pmatrix}$$

As can be observed, we are interested in performing a pondered sum of two contributions: a 5×5–*Difference of Gaussians* (*DoG*) between a certain combination of color channels, and a custom edge detection by means of a $2D$–5×5 Laplacian kernel over the intensity channel (I). The *DoG* filter employed, acts as a spatially color–opponent filter, generating a mexican hat–shaped contribution that simulates the behavior of the bipolar cells in the retina. This contribution, together with the edge detection involves an basic approximation as it is shown in this section.

DoG and other bioinspired filters (laplacians, Leakage–Integrate&Fire, ...) are defined in the *Flowlang* runtime library **Flowlang.Retina**. Usual image processing filters such as the convolution, (scalar, vector or matrix) addition, etc. are built-in primitives within the language. To reproduce the spiking events from the retinal ganglion cells, we use a basic leakage–Integrate&Fire spiking model (`IaF` primitive in the code below) from [10].

The resulting model description using *Flowlang* is shown as follows:

```
using FlowLang.Retina;
unit retina_basic_test {
    k = [-4,-1,0,-1,-4; -1,2,3,2,-1; 0,3,4,3,0;
        -1,2,3,2,-1; -4,-1,0,-1,-4];
    Signal contrib1 = DoG (2*R + B, G, 0.9, 1.2, 5);
    Signal contrib2 = Conv (k,I,5);
    Signal activity = 0.7 * contrib1 + 0.3 * contrib2;
    return IaF ("Gerstner_basic", 10, 0, 200, activity);
}
```

As can be noted, *Flowlang* offers an easy way to describe a dataflow computing model using well known data processing and a mathematical primitives. In addition, the model can be automatically edited by means of the supported *RetinaStudio* widgetry.

3.1 Outputs Provided by the *RetinaStudio*

Once the *Flowlang* high–level description of the retina is ready, we can proceed with its simulation inside *RetinaStudio*. In this example, we have selected the *COTS-processor* output target, in particular we are interested in an automatic optimization for the *Intel Core i5* processor. Fig. 3 shows the different results offered by our environment to test and validate the model. The upper-left image shows the input taken from a simple usb videocamera. The upper-right image shows the filtered output from equation 1; the accuracy of the output depends on the level of applied optimization; in this case, *RetinaStudio* has inferred 16–bit fixed–point arithmetic for every convolution primitive (taking advantage of *SSE3* special instructions with a throughput up to 8 instruction per cycle) and 32–bit floating point arithmetic for matrix/image operations (addition, and multiplication by a scalar), using the *Intel64* instruction set. We can obtain this output by returning the variable `activity` in the *Flowlang* model instead of returning the spiking events. The overall accelerated model, automatically compiled using *GNU gcc*, is capable of running in real-time at 50 *frames/s* using a common *Intel Core2 Duo* microprocessor. The lower-left image shows the activity matrix of the filtered input. This operation basically remaps the incoming information to fit the dimensions of a given microelectrode (25×25 in this case). The final image shows the Leakage-Integrate&Fire output from the activity matrix. In particular, this image shows the spiking events in a state of time (t) where not all electrodes stimulated are activated simultaneously; at the next time ($t + \delta t$) other electrodes will be activated. This is because of the Leakage–Integrate & Fire function, that simulates the ganglion cells behavior, only triggers when a threshold is reached.

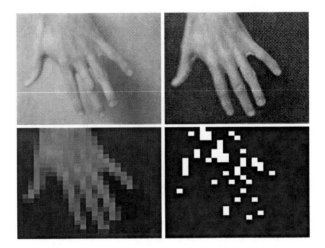

Fig. 3. From the left to right and the top to bottom, the first image corresponds to the original scene, the second image is the input after being processed by a set of colour-opponent filters, the third image show the activity matrix, and the last image correspond to spiking events, for a 25×25 electrode output.

3.2   Comparison with Biological Data

RetinaStudio is able to generate results in *Neural Events File 2* (*NEV2*) and *ASCII* formats compatible with programs as *Neuralc* or *Nev2lkit*, therefore we can compare the biological records with the synthetic records obtained by our tool.

To make the experiment we performed extracellular ganglion cell recordings in isolated superfused rabbit retinas. The retina was isolated from the pigment epithelium and placed with the ganglion cell layer against the hexagonal arranged multi-electrode array (100 electrodes) thus recording their electrical activity. Retinas were superfused with bicarbonate-buffered *AMES* medium (Sigma-Alderich) at 35°C. The electrode signals were amplified, passed through a bandwidth filter (300–2000 Hz) and digitized with a resolution of 16 Bit at a sampling rate 32kHz or 25 kHz respectively using a commercially available acquisition system (Braingate, USA; MultiChannel Systems, Germany). Once retina preparations were placed on the array, recording sessions were started after a 10 minutes period in darkness to allow the specimen to gravitate on the electrodes.

This retina preparations were then visually stimulated using a computer system with a 16-bit BENQ TFT monitor or white LEDs.

We used *Neuralc* to compare records obtained from a rabbit retina with records obtained from *RetinaStudio*, when stimulated it with a periodic flash with full field white light.

An example is shown in a Fig. 4, the top plot shows four stimulus of the records of population of ganglion cells in a real retina. In the bottom plot we simulate only the ON cells. Visually we observed that when the stimulus arrives, the ganglion ON cells are activated in both registers (*RetinaStudio* and biological retina) forming vertical columns of spikes.

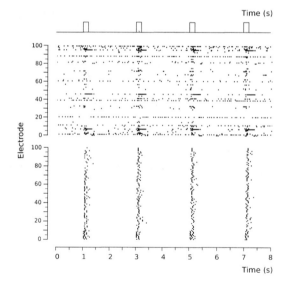

Fig. 4. Full-field flashing stimulation of a retina. The top plot shows a periodic stimuli flash (200ms–ON and 1800ms–OFF), the middle plot shows *in vivo* recording from a rabbit retina, and the bottom plot shows the recording generated by *RetinaStudio*.

In the Fig. 5 is show only two stimulus which shows the distribution of spikes when the stimulus arrive.

As we observed in the two plots, the number of spikes is higher at the beginning of the stimulus and its is decreasing with the time until the stimulus is finished.

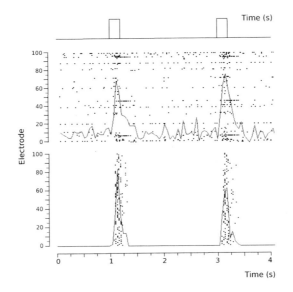

Fig. 5. Spikes distribution with 50ms bin: The top plot shows a periodic stimuli flashes of 200ms–ON and 1800ms–OFF, the middle plot shows a spikes distribution of *in vivo* recording from a rabbit retina, and the bottom plot shows the spikes distribution generated by *RetinaStudio*

4   Conclusions and Future Work

In this paper we have presented a new functional tool to describe, simulate and validate custom retina models. The tool, named *RetinaStudio*, integrates a high-level language for modeling the behavior of vertebrate retinas and can be also used to get a better understanding of visual processing at the biological retinas.

Our main aim is to facilitate the development of retinal models in three different ways: (1) using a predefined widgetry based on contextual dialogs and step–by–step druids, (2) using the *API* or (3) using the Flowlang language in an intuitive, simple and quick way.

To compile the language *RetinaStudio* integrates a retargetable compiler which generates an optimized code using an *API* that implements accelerated functions with *SIMD* instructions (*MMX*, *SEE2/3*). This allows to reduce substantially the computational cost in comparison with other implementations without optimization.

Our preliminary results are very encouraging in terms of performance and speed. In addition, *RetinaStudio* support data formats that are compatible with standard electrophysiological programs such as *Neuralc* or *Nev2lkit*, becoming a useful tool for comparing the results of real electrophysiological experiments on isolated retinas and the output of different retinal models.

As a future work, we will add more modules and features to *RetinaStudio* such as bio-inspired specific code for *FPGAs* and *GPUs*.

Acknowledgment

We would like to thank Markus Bongard for all his help with the electrophysiological recordings. This work has been supported in part by the ONCE (National Organization of the Spanish Blind), by the Research Chair on Retinitis Pigmentosa Bidons Egara and by grant SAF2008-03694 from the Spanish Government.

References

1. Bongard, M., Ferrandez, J.M., Fernandez, E.: The neural concert of vision. Neurocomputing 72, 814–819 (2009)
2. Fernandez, E., Pelayo, F., Romero, S., Bongard, M., Marin, C., Alfaro, A., Merabet, L.: Development of a cortical visual neuroprosthesis for the blind: the relevance of neuroplasticity. J. Neural Eng. 2, R1–R12 (2005)
3. Normann, R.A., Greger, B.A., House, P., Romero, S., Pelayo, F., Fernandez, E.: Toward the development of a cortically based visual neuroprosthesis. J. Neural Eng. 6, 1–8 (2009)
4. NEV 2.0 (Neural Event Format) format specification, http://cyberkineticsinc.com/NEVspc20.pdf
5. Morillas, C.A., Romero, S.F., Martínez, A., Pelayo, F.J., Fernández, E.: A Computational Tool to Test Neuromorphic Encoding Schemes for Visual Neuroprostheses. In: Cabestany, J., Prieto, A.G., Sandoval, F. (eds.) IWANN 2005. LNCS, vol. 3512, pp. 268–316. Springer, Heidelberg (2005)
6. Roessler, G., et al.: Implantation and Explantation of a Wireless Epiretinal Retina Implant Device: Observations during the EPIRET3 Prospective Clinical Trial. Investigative Ophthalmology & Visual Science 50, 3003–3008 (2009)
7. Sousa, L., Tomas, P., Pelayo, F., Martínez, A., Morillas, C.A., Romero, S.: Bioinspired stimulus encode for cortical visual neuroprostheses. In: New Algorithms, Architectures, and Applications for Reconfigurable Computing, ch. 22, pp. 279–290. Springer, Heidelberg (2005)
8. Biomedical Technologies –Stim100, http://biomedical-technologies.com
9. Open Computer vision Library, http://opencvlibrary.sourceforge.net
10. Gerstner, W., Kistler, W.: Spiking neuron models. Cambridge Univeristy Press, Cambridge (2002)

An Expandable Hardware Platform for Implementation of CNN-Based Applications

J. Javier Martínez-Álvarez, F. Javier Garrigós-Guerrero,
F. Javier Toledo-Moreo, and J. Manuel Ferrández-Vicente

Dpto. Electrónica, Tecnología de Computadoras y Proyectos,
Universidad Politécnica de Cartagena, 30202 Cartagena, Spain
jjavier.martinez@upct.es

Abstract. This paper proposes a standalone system for real-time processing of video streams using CNNs. The computing platform is easily expandable and customizable for any application. This is achieved by using a modular approach both for the CNN architecture itself and for its hardware implementation. Several FPGA-based processing modules can be cascaded together with a video acquisition stage and an output interface to a framegrabber for video output storage, all sharing a common communication interface. The pre-verified CNN components, the modular architecture, and the expandable hardware platform provide an excellent workbench for fast and confident developing of CNN applications.

1 Introduction

CNNs are a particular kind of ANNs specifically designed for hardware implementation, usually in embedded and real-time systems, such as image processing applications. CNN were first implemented as dedicated hardware in [1,2]. Since then, several groups have been working in developing CNN-specific circuits [3,4,5,6,7], using analogue and digital approaches, over different VLSI technologies (ASICs and FPGAs). This chips usually include only the basic cells of the CNN, and are just a piece of the platform, requiring the use of external resources for the implementation of a complete processing system [8,10,11].

As most CNN applications are related to image processing, CNN chips must have the maximum performance and size. This implies great challenges for the designer, which has to deal with the high processing speed and scale of integration required by these applications. The great design complexity usually leads to highly customized and application-dependent architectures.

This paper describes a computing system designed to facilitate the development of video processing applications based on CNNs. The proposed platform is composed of multiple FPGA boards, allowing for the emulation of great size and complex CNNs. A new CNN architecture, that we have called Carthagonova architecture, has been specifically designed to support the flexibility, modularity

J.M. Ferrández et al. (Eds.): IWINAC 2011, Part II, LNCS 6687, pp. 195–204, 2011.

and expandability required by a a platform designed to be distributed across several physical modules. Our architecture uses a real-time dataflow processing schema. Its modularity provides an straight design flow, using a hierarchical methodology, where an application is structured in several abstraction levels in order to simplify its design. Several processing boards can be connected using a simple expansion interface, as required for a given application. Every board, called a processing module, includes an FPGA that hosts part of the CNN and the expansion interface. The total computing capacity of the distributed platform is given by the number and size of the CNN implemented. The main features of our approach are:

1. A modular physical architecture, based on an expandable set of processing boards, providing versatility and adaptability to the size of the application.
2. A modular logical architecture of the CNN components, enabling a hierarchical design methodology and a straight forward development cycle of the CNN application.
3. A very efficient hardware implementation of the CNN components, preoptimized and pre-verified for maximum performance over several families of FPGA devices.

2   Platform Structure

The aim is to emulate large CNN by distributing the net through different Processing Modules (PM), in order to increase the versatility and computational performance of the platform. The system supports an Expansion Interface (EI) to allow for cascading of multiple PM and facilitate the incorporation of new modules to the system. PMs and EIs have been designed both to work in a data flow configuration and support a distributed sequential computing architecture. To do this, PMs must support a CNN architecture that is able to first, work in a data flow mode, and second, process data in real time. EIs, on the other hand, have to be able to support and propagate the control signals and data streams with enough versatility. Every PM processes the video data stream received at its input port and provides an output compatible with the input of the next module. Each PM is composed by an FPGA board that hosts a portion of the main CNN and a connector used to implement the EI. Finally, the system incorporates a frame grabber which corrects the latency introduced by the PMs and provides a compatible video output. Figure 1 shows the structure of the hardware platform proposed.

As shown, the digital video source, in progressive mode, is connected in the first position to the system chain, using the input port of EI_1. The video source provides video data to be processed in the PMs and the necessary control signals that are propagated along the EIs. These are mainly synchronisms used to initiate the cascade of processing PMs and by the frame grabber. While the video stream is been processed and propagated along the chain, with a latency determined by each PM, the modules are activated and involved in processing.

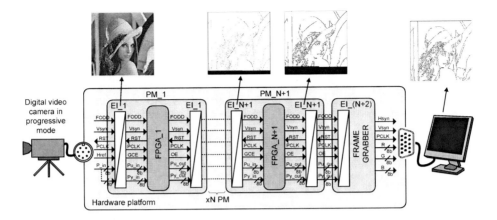

Fig. 1. Hardware platform

Since the latency of the modules produces a mismatch between video stream and frame synchronization, the frame grabber is used to correct the mismatch. The frame grabber is connected to the system using the same expansion interface and provides a compatible video output.

3   Proccesing Module and Expansion Interface

The expansion interface provides the modules (either PMs or frame grabber) with the video data and control signals necessary to perform a sequential processing. On the other hand, the modules connected to the interface, supply the processed video data and control signals to be used in further stages of the chain. Figure 2 shows the input/output ports defined for the EI and its connection to the FPGA. The PM module shows in this example a processing consisting in several CNNs connected in series. In general, the complexity of the internal structure of each PM depends essentially on the size and computing capacity of CNN required by the application, and on the resources available in the FPGA, which are both independent of the interface.

In order for the system to support sequential computing, the EI has been designed using a simple scheme based on pairs of ports: one input and one output, among which a direct connection can be established. This feature allows PM modules to be connected and disconnected from the system quickly taking into account only two considerations: the first one is that PMs must be plugged and unplugged when the system is powered off and, second, that pairs of ports on an unused connector must be bypassed so that information can reach the following modules. This feature allows over-sizing the number of connectors in the system and using only those which are necessary according to each application. Once the PMs have been connected and the system powered, the FPGAs can be independently configured with the required CNN structure. The video signal provided by the interface is supplied to the modules through two input ports:

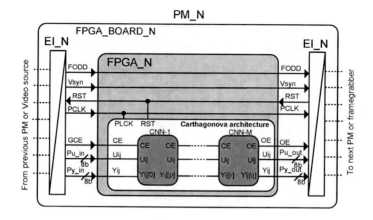

Fig. 2. Processing modules and expansion interface

Pu_in, which spread the video stream to the chain from the main entrance of the system, and *Py_in*, which forwards the video data that has been processed by a given PM, to the following. Within each PM, the two video streams are combined and processed by the CNN. The result is returned to the interface through the output port *Py_out* to be used in the next module. The output port, *Pu_out*, propagates the system's video input (*Pu_in*) to the next module, without any special processing but the necessary delay stages to keep the system synchronized.

Other ports used by the EI are: PCLK (Pixel Clock), RST (Reset), GCE (General Cell Enable), OE (Output Enable), Start and FODD. The CGE port is used to enable the modules connected to the interface, while OE is an output port that is activated when a PM provides valid results on its ports, *Pu_out* and *Py_out*. To maintain information consistency and synchronization, OE output must be connected to the GCE input of the next module. Signals Start and FODD are generated by the main video source and are propagated through the system chain to be used by the frame grabber. These signals are used to enable the frame grabber and to check whether the video stream is being generated in progressive or interlaced mode, respectively.

4   Processing Approach

In general, two basic schemes can be considered for the implementation of CNN on FPGAs: traditional architectures for spatial processing, with multiple inputs receiving data in parallel, and architectures for temporal processing, with only one input receiving data serially.

Architectures for spatial CNN processing, such as the ACE16K [5] and the Q-eye [6], use a *Rows × Cols* cell structure for the CNN, as shown in Figure 3-a. These architectures lead to implementations that maximize the spatial parallelism of the network and achieve the highest speed and computing power.

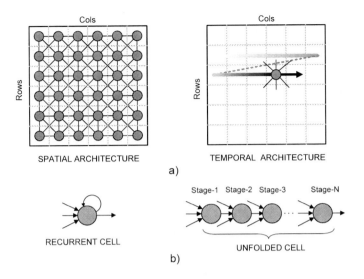

Fig. 3. a) Spatial and temporal approaches. b) Iterative and unfolded cells.

However, architectures for spatial processing require such a number of resources in terms of processing elements, memory, I/O pins, etc. that only CNNs with a few cells can be implemented on off-the-shelf FPGA devices, being necessary to turn to more complex hardware platforms if larger CNNs are required.

When the CNN is too large, architectures for temporal processing may be seen as an alternative solution. As depicted in Figure 3-a, these architectures usually have only one processing unit that, multiplexed in time, emulates the processing of the whole CNN; the processing unit computes the data sequentially, acting like a cell that goes all over the input array, from left to right and from top to bottom. With this model, input data are processed in data streaming in the same order that are generated by a progressive camera. The temporal processing approach keeps the hardware resource consumption to a minimum, reducing also the I/O port number, which simplifies both the connection of the FPGA with other external components, and the design of multi-FPGA systems. On the negative side, these advantages are at the expenses of a reduction of the processing speed, with respect to the spatial approach.

However, processing performance can be increased by exploiting parallelism at the cell level: instead of executing N iterations on the same cell, the cell operation can be unfolded and executed on N stages connected in cascade. With this approach, a cell comprised of N stages could process data N times faster at the expenses of consuming N times more area, roughly. Figure 3-b depicts the simplified scheme of both types of cells: iterative and unfolded.

The PMs and the EIs of the proposed platform have been designed to use CNN architectures for temporal processing, which compute data streaming using unfolded cells.

5   The Carthagonova Architecture

The Carthagonova architecture has been developed to be implemented in an FPGA and perform a sequential emulation of the discrete approach to the CNN model. As its previous version Carthago [7,12], the Carthagonova architecture combines the advantages of temporal processing and parallelism at cell level in order to efficiently deal with complex CNN algorithms without the need of off-chip memories. Further, the definition of a simple I/O interface makes it easy the expansion of the system, and so, to adapt the hardware to an enormous variety of applications. The Carthagonova architecture is a discrete version of the CNN model obtained from the Euler method, whose behavior is determined by equations 1 and 2.

$$X_{ij}[n] = \sum_{k,l \in Nr(ij)} A_{kl}[n-1]Y_{kl}[n-1] \; + \sum_{k,l \in Nr(ij)} B_{kl}[n-1]U_{kl} \; + \; I_{ij} \; , \qquad (1)$$

$$Y_{ij}[n] = \frac{1}{2}\left(|X_{ij}[n]+1| - |X_{ij}[n]-1|\right) \qquad (2)$$

where I, U, Y and X denote input bias, input data, output data and state variable of each cell, respectively. Nr(ij) is a neighborhood function of radius r for cell (i,j), with i and j denoting the position of the cell in the network and k and l the position of the neighboring cells relative to the cell in consideration. B is the non-linear weights template for the inputs and A is the corresponding non-linear template for the outputs of the neighboring cells. Nonlinearity means that templates can change over time and over space.

Unfolding a cell into stages allows to take advantage of parallelism and so to achieve faster processing. The data to be processed are shared out into the stages of the cell, making it possible to use small buffers for local storage instead of the external memories required with an iterative approach. With these local memories not only the design is simpler, but also faster, since the data access time is lower than with external memories. At the same time, the parallel processing of the data at stage level implies that it is possible to emulate much more complex and dynamic behaviors through time-varying values of the templates in each stage.

Carthagonova based CNNs are developed following a hierarchical scheme that systematizes in a fast and simple way the design of complex systems. As shown in Figure 4, a CNN can lie on several FPGA devices and it can be developed at the highest level from independent sub-networks which synthesize the functions and sub-tasks of the main network. Moving down in the hierarchy, each one of these sub-networks can be split into layers whose functionality comes determined by the internal organization of their respective cells, i.e., the layer morphology, and by the configuration of each stage of every cell. This hierarchical and modular design process, based on the customizable instantiation of components at different levels with the stage of a cell being the basic processor at the bottom of the hierarchy, helps to define and configure complex CNN structures.

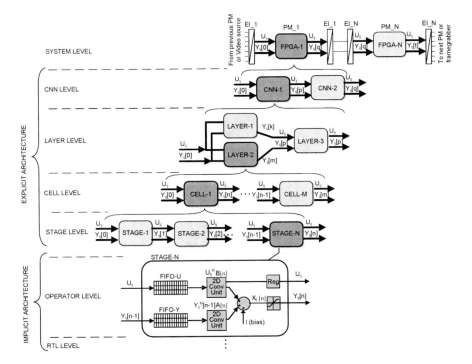

Fig. 4. Hierarchical and recursive high level description of systems based on Carthagonova architecture

As shown in Figure 4, the basic stage of the Carthagonova architecture has a simple input/output interface made up of two input ports, Uij and Yij[n-1], and two output ports, Yij[n] and Uij. The Uij input is propagated to the Uij output, what is intended for transferring the value at the input system along the cascade of stages and so to facilitate the design of CNN structures where the input data U are processed in successive layers and iterations.

The internal architecture of the stage has two local buffers where data rows are stored, two bi-dimensional convolution units, a three-input adder and a comparison circuit that implements the standard activation function. The key component is the convolution unit: an efficient implementation, wherein the inherent parallelism of FPGA internal structure is exploited, provides it with the maximum performance, at the minimum cost. The architecture has been enhanced to support space and time variant templates. It has been also optimized to make the most efficient use of the internal resources available on Virtex4, Virtex5 and Virtex6 FPGA families from Xilinx Inc., without requiring the use of external memory.

5.1 The Optimized Carthagonova RTL-Microarchitecture

The Carthagonova architecture is an enhanced version of the Carthago architecture. A new organization of the internal structure of the cell stages and a higher

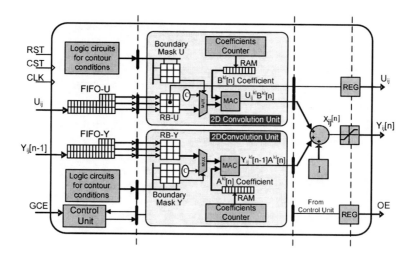

Fig. 5. RTL description of the Carthagonova stage microarchitecture, exemplified for a stage with a 3x3 neighborhood radius and only one multiplier in the convolution units

degree of parallelism lead to a higher processing speed and a more efficient use of FPGA specific resources.

Figure 5 shows, exemplified with a neighborhood radius equals to 1, the detailed RTL microarchitecture of a basic Carthagonova stage: the local memories FIFO-U and FIFO-Y load the incoming data from the input port, Uij, and the values of Yij at previous iterations from the input port Yij[n-1], respectively; two bidimensional convolution units working in parallel, a three-input adder and a comparison circuit emulating the standard activation function carry out the operations involved in the computation of the stage output, Yij[n]. The pixel where the convolution is spatially centered is presented at the Uij output, what is intended for transferring its value along a cascade of stages.

The FIFOs are configured as single-input multiple-output memories: the inputs to the FIFOs are connected to the inputs of the stage; the outputs from the FIFOs, so many as the value $(2r + 1)$ (being r the radius of neighborhood), feed in parallel the register banks RB-U and RB-Y of the convolution units. Each register bank has $(2r + 1)^2$ registers and it is configured as a multidimensional shift register where the data involved in a computation of the convolution are first stored in and then right-shifted. For data loading, the bank is managed as R different sets of R registers in cascade and by hence it offers $(2r + 1)$ inputs, which makes it possible the parallel load of data from the FIFO. The bank output is connected to the multiplication unit through multiplexing logic that manages the values of the data to be used in the computations on the array edges: the data input from the register banks or the boundary values (equal to zero under Dirichlet conditions). This multiplexing is controlled by the boundary mask, a $(2r + 1)^2$-bit register whose b-th bit is used to control the selection of the b-th register in the register bank as an input to the MAC. The value of the boundary mask is determined at each new computation of the convolution centered on any piece of data, Uij, by a logic circuit that gives a value to the preset signal of each bistable in the register according to the position (i,j) of the data in the input array. The other input to the multiplication unit is the coefficients template. The coefficients templates Akl and Bkl are read from a synchronous RAM. The memories contents are given by the templates coefficients, but their depths and widths depend on the coefficients resolution and on the configuration of the convolution unit. The first step on the computation of the convolution centered in Uij consist in reading data from the FIFOs and loading it into the RB-U and RB-Y register banks; in parallel, boundary conditions for the (i,j) position are evaluated and the boundary masks are loaded with the suitable values. Next, the MAC unit carries out the multiplication between the templates coefficients and the data. The results $UijklBkl$ and $Yijkl[n-1]Akl$ of the two convolution units, plus the bias I, are added together. This value is the input to the comparison circuit that implements the activation function, whose output will be the response of the Yij stage at iteration n.

In order to maximize the clock frequency, the cell stage has been pipelined in three substages controlled by the main clock signal, CLK. As shown in Figure 5, the additional pipelining registers are placed at the input and at the output of the convolution units, preceding the stage output registers. This increments the stage latency in 2 units, for a total of $r \times (cols + 1) + 5CLK$ cycles.

6 Conclusions

In this paper, we have proposed a new CNN architecture suitable for hardware emulation of large CNNs on reconfigurable devices. Our architecture has the benefits of been modular and expandable, permitting the development of applications using complex CNNs, including non linear templates, time-variant coefficients, multi-layer structure, etc.

We have also presented an implementation platform, based on FPGA processing modules that share a common interface. To be efficient, the platform was designed to inherit the modularity of logical CNN architecture, been also modular, distributed and expandable, so that it can be configured to accommodate CNNs of very different size and structure. The proposed module architecture can be set up as an standalone or embedded system.

The proposed CNN architecture and implementation platform represent a valuable set of tools for the development of CNN based applications, providing a reliable design methodology that uses a hierarchical approach, based on pre-designed, pre-verified and optimized hardware components.

Our next work will be addressed to the development of a software workbench to assist the designer in the tasks of CNN definition and automatic synthesis of the corresponding hardware architecture, using the predefined modules as processing primitives.

Acknowledgements

This work has been partially supported by the Fundacin Séneca de la Región de Murcia through the research projects 08801/PI/08 and 15419/PI/10, and by the Spanish Government through project TIN2008-06893-C03.

References

1. Harrer, H., Nossek, J., Stelzl, R.: An analog implementation of discrete-time cellular neural networks. IEEE Transactions on Neural Networks 3, 466–476 (1992)
2. Harrer, H., Nossek, J., Roska, T., Chua, L.: A current-mode DTCNN universal chip. In: IEEE International Symposium on Circuits and Systems, vol. 4, pp. 135–138 (1993)
3. Paasio, A., Dawidziuk, A., Porra, V.: A QCIF Resolution Binary I/O CNN-UM Chip. J. VLSI Signal Processing Systems 23, 281–290 (1999)
4. Malki, S., Spaanenburg, L.: CNN Image Processing on a Xilinx Virtex-II 6000. In: Proceedings ECCTD 2003, Krakow, pp. 261–264 (2003)
5. Rodriguez-Vazquez, A., Linan-Cembrano, G., Carranza, L., Roca-Moreno, E., Carmona-Galan, R., Jimenez-Garrido, F., Dominguez-Castro, R., EMeana, S.: ACE16k: the third generation of mixed-signal SIMD-CNN ACE chips toward VSoCs. IEEE Transactions on Circuits and Systems I 51(5), 851–863 (2004)
6. AnaFocus Ltd. (2004), http://www.anafocus.com eye-RIS v1.0/v2.0 Datasheet
7. Martínez, J.J., Garrigós, F.J., Toledo, F.J., Ferrández, J.M.: High performance implementation of an FPGA-based sequential DT-CNN. In: Mira, J., Álvarez, J.R. (eds.) IWINAC 2007. LNCS, vol. 4528, pp. 1–9. Springer, Heidelberg (2007)
8. Voroshazi, Z., Kiss, A., Nagy, Z., Szolgay, P.: Implementation of embedded emulated-digital CNN-UM global analogic programming unit on FPGA and its application. International Journal of Circuit Theory and Applications 36, 589–603 (2008)
9. Nagy, Z., Szolgay, P.: Configurable multilayer CNN-UM emulator on FPGA. IEEE Trans. on Circuits and Systems I 50(6), 774–778 (2003)
10. Laiho, M., Poikonen, J., Virta, P., Paasio, A.: A 64x64 cell mixed-mode array processor prototyping system. In: International Workshop on Cellular Neural Networks and Their Applications, CNNA 2008, pp. 14–16 (2008)
11. Fujita, T., Okamura, T., Nakanishi, M., Ogura, T.: CAM2-universal machine: A DTCNN implementation for real-time image processing. In: International Workshop on Cellular Neural Networks and Their Applications, CNNA 2008, pp. 219–223 (2008)
12. Martínez, J.J., Toledo, F.J., Fernández, E., Ferrández, J.M.: A retinomorphic architecture based on discrete-time cellular neural networks using reconfigurable computing. Neurocomputing 71(4-6), 766–775 (2008)

Classification of Welding Defects in Radiographic Images Using an Adaptive-Network-Based Fuzzy System

Rafael Vilar[1], Juan Zapata[2], and Ramón Ruiz[2]

[1] Departamento de Estructuras y Construcción,
Rafael.Vilar@upct.es
[2] Departamento de Electrónica, Tecnología de Computadores y Proyectos,
Universidad Politécnica de Cartagena, Cartagena 30202, Spain

Abstract. In this paper, we describe an automatic system of radiographic inspection of welding. An important stage in the construction of this system is the classification of defects. In this stage, an adaptive-network-based fuzzy inference system (ANFIS) for weld defect classification was used. The results was compared with the aim to know the features that allow the best classification. The correlation coefficients were determined obtaining a minimum value of 0.84. The accuracy or the proportion of the total number of predictions that were correct was determined obtaining a value of 82.6%.

1 Introduction

In the last five decades, Non Destructive Testing (NDT) methods have gone from being a simple laboratory curiosity to an essential tool in industry. With the considerable increase in competition among industries, the quality control of equipment and materials has become a basic requisite to remain competitive in national and international markets. Although it is one of the oldest techniques of non-destructive inspection, radiography is still accepted as essential for the control of welded joints in many industries such as the nuclear, naval, chemical or aeronautical. For the correct interpretation of the representative mark of a heterogeneity, a knowledge of welded joint features and of the potential heterogeneities and types of defect which can be detected using radiographic welded joint inspection is necessary. Limitations to correlating the heterogeneity and the defect are imposed by the nature of the defect (discontinuities and impurities), morphology (spherical, cylindrical or plain shape), position (superficial or internal location), orientation and size. Therefore, the radiographic welded joint interpretation is a complex problem requiring expert knowledge. In our view, a system of automatic inspection of radiographic images of welded joints should have the following stages: digitalisation of the films, image pre-processing seeking mainly the attenuation/elimination of noise; contrast improvement and discriminate feature enhancement, a multi-level segmentation of the scene to isolate the areas of interest (the weld region must be isolated from the rest the elements that

J.M. Ferrández et al. (Eds.): IWINAC 2011, Part II, LNCS 6687, pp. 205–214, 2011.

compose the joint), the detection of heterogeneities and, finally, classification in terms of individual and global features through tools of pattern recognition. Over the last 30 years, there has been a large amount of research attempting to develop an automatic (or semiautomatic) system for the detection and classification of weld defects in continuous welds examined by radiography [1,2]. Today, the stage corresponding to the classification of patterns has been one of the most studied in terms of research [3,4,5,6,7,8,9,10,11,12].

The aim of this approach is to present the methodology used and the results obtained by an ANFIS to estimate the classification accuracy of the main classes of weld defect. Our methodology tries to solve some shortcomings of works carried out in the past, first we try to reduce the number of features used in the input vector using a very few geometrical features (4), second we use different combinations of features with the aim to know the combination which improves the performance of the classification and therefore the most significative combination.

2   Experimental Methodology

Digital image processing techniques are employed to lessen the noise effects and to improve the contrast, so that the principal objects in the image become more apparent than the background. Threshold selection methods, labelled techniques and feature extraction are used to obtain some discriminatory features that can facilitate both the weld region and defects segmentation. Finally, features obtained are input pattern to adaptive-neuro-fuzzy inference system. The modelling of a welding defect classificator with ANFIS was implemented through its training process. Next, the checking process was used for testing the generalisation capability of the fuzzy inference system. Fig. 1 shows the major stages of our welding defect detection system.

Radiographic films can be digitised by several systems. The most common way of digitisation is through scanner. In this present study, an UMAX scanner was used, model: Mirage II (maximum optical density: 3.3; maximum resolution for films: 2000 dpi) to scan the IIW (International Institute of Welding) films [13]. The spatial resolution used in the study was 500 dpi (dots per inch), totalling an average image size of 2900 pixels (horizontal length) × 950 pixels (vertical length), which resulted in an average pixel size of 50 μm. Such resolution was adopted for the possibility of detecting and measuring defects of hundredths of millimetres, which, in practical terms of radiographic inspection, is much higher than the usual cases.

2.1   Image Preprocessing

After digitising the films, it is common practice to adopt a preprocessing stage for the images with the specific purpose of reducing/eliminating noise and improving contrast. Radiographic films usually have noise and deficient contrast due to intrinsic factors involved in the inspection technique, such as non-uniform illumination and the limited range of intensities of the image capture device. Noise

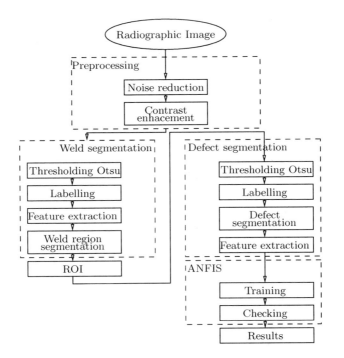

Fig. 1. Procedure for the automatic welding defect detection system

in scanned radiographic images is usually characterised as randomly spread pixels, with intensity values that are different from their neighbouring pixels. Two preprocessing steps were carried out in this work: in the first step, for reducing/eliminating noise an adaptive 7-by-7 Wiener filter [14] and 3-by-3 Gaussian low-pass filter were applied, while for adjusting the image intensity values to a specified range to contrast stretch, contrast enhancement was applied, mapping the values in intensity input image to new values in the output image, so that values between that representing the bottom 1% (0.01) and the top 1% (0.99) of the range are mapped to values between [0 1]. Values below 0.01 and above 0.99 are clipped. Finally, the image was divided into five bands of 640 × 480 pixels located on the horizontal axis for processing of the following stages. In this way, we have an image divided into more computationally user friendly sizes.

2.2 Weld Region Segmentation

In this stage, the weld region for each band must be isolated from the rest of the elements that form this joint. In this way, the scene of interest is framed for later analysis aimed at detecting possible defects. The process is developed in three phases. The goal of the first phase is to find an optimal overall threshold that can be used to convert the gray scale image into a binary image, separating an object's pixels from the background pixels. For this, we uses Otsu's method [15], a threshold selection method from gray-level histograms, which chooses the threshold to minimise the intraclass variance of the thresholded black and white pixels. In a second phase, the connected components in the binary image are labelled. The implement algorithm uses the general procedure proposed by Haralick and Shapiro [16]. The algorithm returns a matrix, the same size as the input image, containing labels for the 8-connected objects in the input image. The pixels labelled 0 are the background, the pixels labelled 1 represent one object, the pixels labelled 2 represent a second object, and so on. To conclude, in a third phase, as a criterion to select between labelled objects, the maximum area is established. In this way, we identify the weld region from among all the objects of the image. Finally, as preparation of the following stage, the smallest rectangle containing the weld region (bounding box) is calculated.

2.3   Defect Candidates Segmentation

This stage takes as input the bounding box image obtained in the previous stage and produces, as output, an image consisting of only potential defects. The process is developed in three phases. In the first phase, the bounding box image obtained in the previous stage is binarised. For this, we use Otsus's method to choose the optimum threshold. The second phase uses a binary image, where nonzero pixels belong to an object and 0-pixels constitute the background. The algorithm traces the exterior boundary of objects, as well as boundaries of holes inside these objects. It also descends into the outermost objects (parents) and traces their children (objects completely enclosed by the parents). Using the algorithm on the radiographic image obtained in the previous step (bounding box) it is easy to deduce that defects are objects included within a weld region, i.e. objects included in the objects labelled as 1. In this way, an image formed only of defect candidates is obtained. The results and output of applying this algorithm are shown in Fig. 2.

2.4   Feature Extraction

The next stage is the feature extraction in terms of individual and overall characteristics of the heterogeneities. The output of this stage is a description of each defect candidate in the image. This represents a great reduction in image information from the original input image and ensures that the subsequent classification of defect type and cataloguing of the degree of acceptance are efficient. In the present work, features describing the shape, size, location and intensity information of defect candidates were extracted. This features are: area, centroid (X and Y coordinates), major axis, minor axis, eccentricity, orientation, Euler number, equivalent diameter, solidity, extent and position. In this stage, the procedure generates an input vector (12 components) for each defect candidate and human experts in weld defects produce an associated target vector.

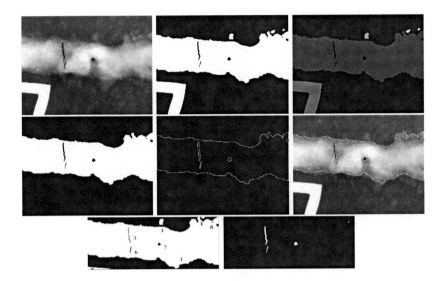

Fig. 2. Results and outputs of applying algorithm to radiographic image. From left to right and up to bottom. Original image. Binary image. Labelled image. Segmented weld region. Weld region boundary. Boundary added to original image. Objects 5, 6, 7 and 8 are included in object 1 (welding joint object). Objects included in weld region (defect candidates).

2.5   ANFIS: Adaptive Neuro-Fuzzy Inference System

The basic structure of the classic fuzzy inference system is a model that maps input characteristics to input membership functions, input membership function to rules, rules to a set of output characteristics, output characteristics to output membership functions, and the output membership function to a single-valued output or a decision associated with the output. All this process is developed using only fixed membership functions that were chosen arbitrarily. This fuzzy inference system is applied to only modelling systems whose rule structure is essentially predetermined by the user's interpretation of the characteristics of the variables in the model.

In some modelling situations, it is very difficult to discern what the membership functions should look like simply from looking at data. Rather than choosing the parameters associated with a given membership function arbitrarily, these parameters could be chosen so as to tailor the membership functions to the input/output data in order to account for these types of variations in the data values. In such cases, the neuro-adaptive learning techniques can help. The shape of the membership functions depends on parameters, and changing these parameters change the shape of the membership function. Instead of just looking at the data to choose the membership function parameters, membership function parameters are chosen automatically.

The neuro-adaptive learning method works similarly to that of neural networks. Neuro-adaptive learning techniques provide a method for the fuzzy modelling procedure to learn information about a data set. Then, Fuzzy Logic computes the membership function parameters that best allow the associated fuzzy inference system to track the given input/output data.

The most frequently investigated ANFIS architecture is the first-order Sugeno model, due to its efficiency and transparency. A representative ANFIS architecture with two inputs (x and y), one output (f), and four rules is illustrated in Fig. 3, which consists of five layers: Adaptive Fuzzification, Fuzzy Rule, Firing Strength Normalisation, Adaptive Implication, and Output.

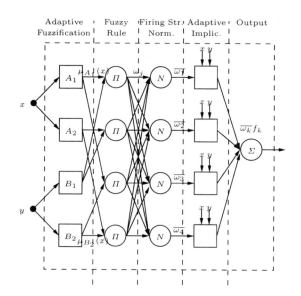

Fig. 3. ANFIS architecture with two inputs, one output, and four rules

To use the ANFIS as a defect classifier system, a schematic illustration is given in Fig. 4. The comprehensive classification process of the weld defects was automated through five independent ANFIS, one for each class of defect, non defect, slag inclusion, porosity, transversal crack and longitudinal crack. All systems were first-order Sugeno-type with 4 inputs, 1 output and 3 bell membership functions per input. The inputs to ANFIS represented some geometric features to determine the type of defect for that vector, while its output was oriented to be 1 or -1 corresponding or not to the defect class.

3   Results and Discussions

In this work, five independent ANFIS were developed to automate the process of classification in five types of defect, non defect, slag inclusion, porosity, longitudinal crack and transversal crack. The training input-output data is a structure

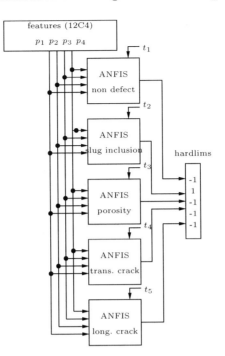

Fig. 4. Classifier system based on ANFIS architecture

whose first component is the four-dimensional input p, and whose second component is the one-dimensional output t. Where p is some of 12 features chosen 4, and t is the output for each class of defect corresponding to a particular ANFIS. There are 375 input-output data values. The system uses 300 data values for the ANFIS training (these become the training data set), while the others are used as checking data for validating the identified fuzzy model. This division of data values results in two data structures, training data and checking data. Training data y checking data are shown in parallel to the five ANFIS systems.

To start the training, it is need a FIS structure that specifies the structure and initial parameters of the FIS for learning. Our FIS structures provide two generalised bell membership functions on each of the four inputs, eight altogether for each FIS structure. The generated FIS structure contains 16 fuzzy rules with 104 parameters. The ratio between data and parameters is about three (300/104). There are two main methods that ANFIS learning employs for updating membership function parameters: back-propagation for all parameters (a steepest descent method), and a hybrid method (our case) consisting of back-propagation for the parameters associated with the input membership functions, and least squares estimation for the parameters associated with the output membership functions. As a result, the training error decreases, at least locally, throughout the learning process. Therefore, the more the initial membership functions resemble the optimal ones, the easier it will be for the model parameter training to converge.

The checking data was used for testing the generalisation capability of the fuzzy inference system at each epoch. The checking data had the same format as that of the training data, and its elements were distinct from those of the training data. This data set is used to cross-validate the fuzzy inference model. This cross-validation requires applying the checking data to the model and then seeing how well the model responds to this data. The checking data is applied to the model at each training epoch. The model parameters that correspond to the minimum checking error was elected. The use of the minimum checking data error epoch to set the membership function parameters assumes, that the checking data is similar enough to the training data that the checking data error decreases as the training begins, and that the checking data increases at some point in the training after the data overfitting occurs. After training, the final FIS is the one associated with the minimum checking error.

All possible combinations without repetition of 12 features chosen 4 was used to know the features that allow the best classification. In this way, the best four features can be used to classify the heterogeneities. In Fig 5 is shown the correlation coefficients for each input combination and for each defect class. The best combination for input vector was obtained by means of the best mean in the correlation coefficients for the five class: eccentricity, orientation, equivalent diameter and solidity with next correlation coefficients for each class, non defect 0.87, slag inclusion 0.84, porosity 0.87, transversal crack 1 and longitudinal crack 0.96.

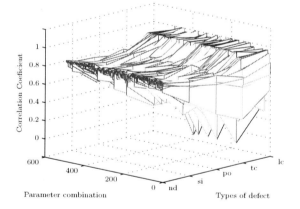

Fig. 5. Correlation coefficients for each defect class and for each combination of input vector of characteristics

In Table 1, we show the confusion matrix of our classifier based on ANFIS. With this table we can to see if the system is confusing two classes (i.e. commonly mislabelling one as another). But in this case, our ANFIS is not confusing any class because the class with less samples is perfectly classified. A table of confusion can also discern the number of true negatives, false positives, false negatives, and true positives. True positives are the value of the intersection of rows and columns represented with the same label (i.e., for the non defect class is 118), falses positives are the rest of values of that row (i.e., for the non defect class is $12 + 6 + 0 + 1 = 19$), falses negatives are the rest of values of that column (i.e., for the non defect class is $10 + 9 + 1 + 2 = 22$), true negatives are the rest of values of the confusion table (i.e., for the non defect class is 216). The accuracy (AC) is the proportion of the total number of predictions that were correct. It is determined using the equation: $AC = \dfrac{TP + TN}{TP + FP + FN + TN}$.

Table 1. Confusion matrix contains information about actual and predicted classifications done by our classification system based on ANFIS

		Actual				Accuracy
		No Defect	Slag Incl.	Poros.	T. Crack	L. Crack	
Predicted	No defect	118	12	6	0	1	0.89
	Slag Incl.	10	100	10	0	0	0.87
	Poros.	9	10	71	0	0	0.96
	T. Crack	1	1	0	8	0	0.99
	L. Crack	2	3	0	0	13	0.98
	Total	140	126	87	8	14	0.82

4 Conclusions

The developed work is devoted to solving one of the stages, maybe the most delicate, of a system of automatic weld defect recognition: the automatic recognition of the boundaries of the weld regions and classification of the defects. The main conclusions and contributions to this end are listed next: this paper presents a set of techniques dedicated to implementing a system of automatic inspection of radiographic images of welded joints: digitalisation of the films, image pre-processing directed mainly at the attenuation/elimination of noise, contrast improvement and discriminate feature enhancement facing the interpretation, multi-level segmentation of the scene to isolate the areas of interest (weld region), heterogeneity detection and classification in terms of individual and overall features by means of an adaptive neuro-fuzzy inference system.

The aim of this paper is to obtain the best performance of un classifier of weld defects based on ANFIS. With this purpose, an ANFIS was used to classify welding defects with different input vectors representing four features. This paper presents and analyses the aspects in the design and implementation of a new methodology for the automatic recognition of weld region and classification of weld defects. After a test phase and updating for the specific proposed technique, an valuation of the relative benefits is presented. From the validation process developed with 375 heterogeneities covering five types of defect were extracted form 86 radiographs of the collection of the IIW, it can be concluded that the proposed technique is capable of achieving excellent results when the input feature vector: eccentricity, orientation, equivalent diameter and solidity is presented as input combination.

Acknowlegment

This work was supported by the Spanish Ministry of Ciencia e Innovaciön (MICINN) under grant TIN2009-14372-C0302 and for the Fundaciön Sëneca under grant 15303/PI/10.

References

1. Silva, R.R., Mery, D.: State-of-the-art of weld seam inspection using X-ray testing: partI-image processing. Materials Evaluation 9(65), 643–647 (2007)
2. Silva, R.R., Mery, D.: State-of-the-art of weld seam inspection using X-ray testing: part II-pattern recognition. Materials Evaluation 9(65), 833–838 (2007)
3. Da Silva, R.R., Caloba, L.P., Siqueira, M.H., Rebello, J.M.: Pattern recognition of weld defects detected by radiographic test. NDT& E International 37(6), 461–470 (2004)
4. Liao, T.: Fuzzy reasoning based automatic inspection of radiographic welds: weld recognition. Journal of Intelligent Manufacturing 15(1), 69–85 (2004)
5. Liao, T.: Improving the accuracy of computer-aided radiographic weld inspection by feature selection. NDT & E International 42(4), 229–239 (2009)
6. Shafeek, H., Gadelmawla, E., Abdel-Shafy, A., Elewa, I.: Automatic inspection of gas pipeline welding defects using an expert vision system. NDT & E International 37(4), 301–317 (2004)
7. Lim, T., Ratnam, M., Khalid, M.: Automatic classification of weld defects using simulated data and an mlp neural network. Insight: Non-Destructive Testing and Condition Monitoring 49(3), 154–159 (2007)
8. Mery, D., Berti, M.: Automatic detection of welding defects using texture features. In: International Symposium on Computed Tomography and Image Processing for Industrial Radiology, Berlin (2003)
9. Mirapeix, J., García-Allende, P.B., Cobo, A., Conde, O.M., Loópez, J.M.: Real-time arc-welding defect detection and classification with principal component analysis and artificial neural networks. NDT & E International 40, 315–323 (2007)
10. Wang, G., Liao, T.: Automatic identification of different types of welding defects in radiographic images. NDT & E International 35, 519–528 (2002)
11. Vilar, R., Zapata, J., Ruiz, R.: Classification of welding defects in radiographic images using an ANN with modified performance function. In: Mira, J., Ferrández, J.M., Álvarez, J.R., de la Paz, F., Toledo, F.J. (eds.) IWINAC 2009. LNCS, vol. 5602, pp. 284–293. Springer, Heidelberg (2009)
12. Vilar, R., Zapata, J., Ruiz, R.: An automatic system of classification of weld defects in radiographic images. NDT & E International 42(5), 467–476 (2009)
13. http://www.umax.comS
14. Lim, J.: Two-dimensional signal and image processing, pp. 536–540. Prentice-Hall, Englewood Cliffs (1990)
15. Otsu, N.: A threshold selection meted from gray-level histograms. IEEE Transactions on Systems, Man and Cybernetics 9(1), 62–66 (1979)
16. Haralick, R., Shapiro, L.: Computer and robot vision, vol. 1, pp. 28–48. Addison Wesley, NY (1992)

Reinforcement Learning Techniques for the Control of WasteWater Treatment Plants

Felix Hernandez-del-Olmo and Elena Gaudioso

Artificial Intelligence Department,
E.T.S.I. Informatica, UNED
{felix,elena}@dia.uned.es

Abstract. Since water pollution is one of the most serious environmental problems today, control of wastewater treatment plants (WWTPs) is a crucial issue nowadays and stricter standards for the operation of WWTPs have been imposed by authorities. One of the main problems in the automation of the control of Wastewater Treatment Plants (WWTPs) appears when the control system does not respond as it should because of changes on influent load or flow. Thus, it is desirable the development of autonomous systems that learn from interaction with a WWTP and that can operate taking into account changing environmental circumstances. In this paper we present an intelligent agent using reinforcement learning for the oxygen control in the N-Ammonia removal process in the well known Benchmark Simulation Model no.1 (BSM1). The aim of the approach presented in this paper is to minimize the operation cost changing the set-points of the control system autonomously.

1 Introduction

Since water pollution is one of the most serious environmental problems today, control of wastewater treatment plants (WWTPs) is a crucial issue nowadays and stricter standards for the operation of WWTPs have been imposed by authorities [15].

WWTPs should be controlled so as to minimize the plant operating costs (OC) while the effluent standards are maintained [14]. Traditionally, in this context, methods of control based on PID controllers [13] have been used. Nevertheless, these control systems in WWTPs do not always respond as they should when the quality of the influent changes in load or flow. In these cases, setpoints in the control loops should be modified in order to lead back the process to the normal state and, thus, to avoid the evolution towards states that diminish the quality of the effluent [13]. In this context, the traditional PID controllers are able neither to predict this problematic situation nor to lead back the process towards optimal conditions. Therefore, it is an *indispensable* requisite to have a more intelligent control to modify the setpoints of the PID controllers. Currently, this role (of control or intelligent agent) is mainly played by plant operators.

In the next section, we review some Artificial Intelligence Systems that have appeared over the years to help plant operators in their decisions and we briefly

J.M. Ferrández et al. (Eds.): IWINAC 2011, Part II, LNCS 6687, pp. 215–222, 2011.

describe the basis of reinforcement learning techniques and their suitability for control tasks in WWTP. Section 3 describes the study carried out with an autonomous and adaptive agent. To conclude, in Section 4 we present the conclusions obtained from this experiment.

2   Artificial Intelligence in the Control of WWTP

2.1   Review of Related Work

Artificial Intelligence (AI) has been widely applied to assist plant operators in the control of Wastewater treatment plants (WWTPs). Of special relevance is the development of specialized Expert Systems (ES) in the control of WWTPs [12]. Usually, however, because of its helper nature, these systems are usually known as Decision Support Systems [8]. According to [8], some of the AI methods, often used in the development of Decision Support Systems in environmental domains in past years are: Rule-Based Reasoning, Planning, Case-Based Reasoning or Model-Based Reasoning.

Focusing on the area of WWTPs and among the various AI techniques used in the development of intelligent control systems, are especially relevant fuzzy logic [10] and artificial neural networks [9]. For example, fuzzy logic has been applied to supervise a pilot-scale WWTP [4] or to diagnose acidification states in a WWTP [5]. Moreover, artificial neural networks have been applied, for example, to provide better predictions of nitrogen contents in treated effluents [6] or to predict plant performance [2].

Regardless of the underlying AI techniques used to build an expert system, the assistant nature implies that the ES acts indirectly on the plant through the plant operator. In other words, ESs are not free to act directly on the WWTP actuators, avoiding the direct interaction with the plant and, lastly, avoiding the autonomous learning of the ES with its own environment. Thus, the designing and building of the traditional ESs requires *interviews* to experts in order to extract the knowledge in which these systems are based. The main disadvantage of this approach is that the expert knowledge does not evolve once extracted and placed into the ES. Some proposals [4] have started to overcome these difficulties by providing the expert system with the ability to both (i) upgrade its knowledge base, in a somewhat automatic way by using an automatic data acquisition system, and (ii) send the relevant orders to the actuator system of the plant. However, among other things, in this case the knowledge base never starts/emerges from the scratch.

In line with the creation of more adaptive decision support systems, it has been recently proposed the use of fuzzy neural networks to implement an adaptive software sensor for an WWTP [10,9]. Fuzzy neural networks are hybrid systems that combine the theories of fuzzy logic and neural networks so as to make an effective use of the easy interpretability of the fuzzy logic as well as the superior learning ability and adaptive capability of neural networks.

In this paper, we are looking for a technique that: (i) provides a direct interaction with the environment, (ii) allows to act in this environment without the need of an initial model and finally, (iii) allows to react to changing environmental conditions. Specifically, we have built an agent whose learning capability is achieved through *model-free* reinforcement learning techniques.

2.2   Reinforcement Learning

Reinforcement learning (RL) algorithms are based on an agent's interaction with its environment. The environment is defined as any external condition that can not be changed directly by the agent [16], they can only be changed through the agent's actions.

The usual way the environment is modeled in RL is by means of Markov Decision Processes (MDP) [16]. Here, the MDP environment is modelled as (i) a space of states S, (ii) a space of actions $A(s)$ that can be done over this environment, given that the environemnt is in state s, and (iii) a set of transition probabilities from one state s to another state s' once the agent has executed action a over this environment $P(s'|s, a)$ besides (iv) the expected reward to be obtained from this environment $E\{r|s', a, s\}$ when changing from state s to state s' having executed action a.

Once the agent has got this model of the environment, it can resolve the optimal policy $\pi(s, a)$ by several methods, for instance dynamic programming [3]. However, if the model of the environment is not provided to the agent, it still can learn this model by means of the so called *model-free* RL methods [11]. Now, with these model-free RL methods, the agent must interact with its environment so as to get, step by step, the model of the environment as well as the optimal policy to act upon it.

More specifically, the agent interacts with its environment making decisions according to its observations, via perception and action. At each step t, the agent observes the current state of the environment s_t and chooses an action to execute, a_t. This action causes a transition between states and the environment provides a new state s_{t+1} and a reward r_{t+1} to the agent. The ultimate goal of the agent is to choose those actions that tend to increase its *return*: the long-term sum of the future reward values r_t. This return, in a continuous environment, is usually set as $R_t = \sum_{t'=t}^{\infty} \gamma^{(t'-t)} r_t$, where $0 < \gamma < 1$ stands for a kind of *Optimization Horizon* (OH, as we will see later). In other words, the higher γ (up to 1), the further the future time considered into the return R_t. Therefore, a model-free reinforcement learning agent learns (i) the model of its environment and (ii) how to best behave on it by systematic trial and error.

Reinforcement learning has already been successfully applied to some related domains. For instance, in [17] it is presented a reinforcement learning approach for pH control on a laboratory process plant. The goal of this approach is to maintain the pH levels stabilized within a band. In this paper, we present an experience with a model-free reinforcement learning agent that learns to choose the dissolved oxygen setpoint guided by the operation cost of the N-ammonia removal process.

3   Autonomous Adaptation of the Oxygen Setpoint in the N-Ammonia Removal Process

N-ammonia removal is a widely known process in the wastewater community, mostly because around 50% [13, chap. 3] of the energy spent in the plant is due to this process. Therefore, many different strategies have been proposed over the years [13, chap. 5].

In this paper, we present a detailed study based on an autonomous adaptive system that learns to choose the dissolved oxygen setpoint guided by the operation cost of the N-ammonia removal process. It is adaptive because it learns it by its own direct interaction with the plant. It is autonomous because it learns it by itself with little human intervention.

The agent proposed for the intelligent control of this WWTP is integrated as another device of the plant. It may be seen in Figure 1.

In the WWTP, the blower of the tank 5 is controlled, as usual, by means of a PI controller. The feedback loop of this PI is closed by an error signal which

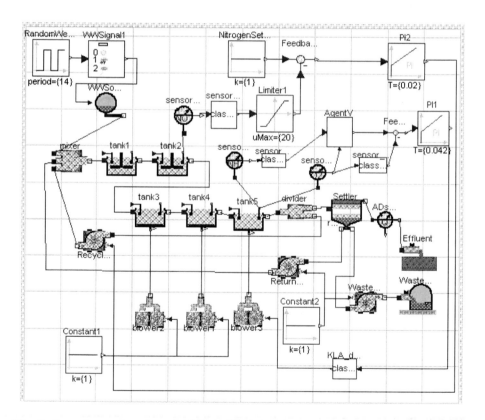

Fig. 1. Experimental setting: the BSM1 WWTP schema and the Agent (AgentV) that controls the DO setpont in the N-ammonia removal process

consists of the difference between the dissolved oxygen (DO) level (of the tank 5) and the DO setpoint.

The agent has two inputs: the measures of NH_4 and O_2 obtained from two sensors placed at the tank 5. The agent also has a single output: the DO setpoint.

The agent acts on the plant by changing this DO setpoint. In fact, the agent must choose among $1.5mg/l$, $1.7mg/l$ or $2.0mg/l$ DO setpoints every 15 minutes.

The agent's goal is to lower the operation costs (OC) as much as possible. In other words, to lower the energy costs while keeping the effluent fines as low as possible. In more specific terms, our model-free reinforcement learning agent's goal is to lower its *return* (see section 2.2), this objective can be set as the minimization of an equation like (eq 1). In this equation, the parameter γ defines the time interval to be considered in the agent's return, also called Optimization Horizon (OH, see section 2.2). In a real setting, this parammeter is to be configured by the plant operator.

$$R_t = \sum_{t'=t}^{\infty} \gamma^{(t'-t)} OC(t') \tag{1}$$

We will show the validity of the agent in the well-established environment COST/IWA Benchmark Simulation Model number 1 (BSM1) [7,1]. In addition, a known operating cost [14,15] has been chosen.

This operation cost OC is calculated as follows:

$$OC(t) = \gamma_1(AE(t) + ME(t) + PE(t)) + \gamma_2 SP(t) + EF(t) \tag{2}$$

where AE, ME and PE stand for aeration, mixing and pumping energy respectively; SP stands for sludge production and EF, for effluent fines. Weights γ_1 and γ_2 are set in proportion to the weights in the operating cost index (OCI) in the benchmark. Based on the experience from the WWTPs operation they were chosen to be 1 and 5 respectively. In this case, following the work [15], a rough estimate of average electricity price in EU ($0.1€/kWh$) is also taken into account and thus γ_1 is set to $0.1€/kWh$ and γ_2 was set to $0.5€/kg$.

Effluent fines EF are paid in proportion to the discharge of pollution into receiving waters. In this study, only nitrogen is considered: effluent ammonia ($S_{nh,eff}$) and total nitrogen ($S_{TN,eff}$) (see i. e. [14]). A formal description of this $EF_X(t)$ is the following:

$$EF_X(t) = Q_{eff}(t) \times \begin{cases} \Delta\alpha_X X_{eff}(t) & \text{if } X_{eff}(t) \leq X_{limit} \\ \Delta\alpha_X X_{limit} + \beta_{0,X} + & \text{if } X_{eff}(t) > X_{limit} \\ +\Delta\beta_X(X_{eff}(t) - X_{limit}) & \end{cases} \tag{3}$$

where $X_{eff}(t)$ and $Q_{eff}(t)$ are the effluent concentration and flow rate respectively and X stands for ammonia S_{nh} or total nitrogen S_{TN}.

Finally, in the following study, OC stands for the monthly mobile average of $OC(t)$.

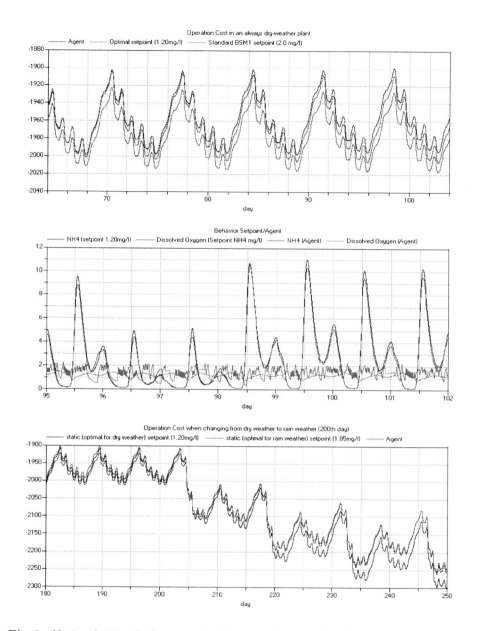

Fig. 2. Notice that in all three graphs the operation cost is represented as a negative number in order to be maximized (no minimized). The top figure presents the comparison of the operational cost for the plant on dry weather conditions with three different settings. The middle figure presents the comparison of the agent's behavior against the optimal setpoint for the dry weather conditions. The bottom figure shows the change of the operational costs (static setpoints and agent) when changing from dry weather to rain weather conditions.

To focus the attention on the autonomous adaptation of the agent, let's see first the agent's evolution in a BSM1 setting that repeats every 14 days the dry-weather influent. This experiment produced the OC showed in Figure 2. The agent's OC is compared with the OC produced by the optimal (static) setpoint for the dry-weather condition (1.20 mg/l).

It can be noticed the quickly agent's adaptation that just in a few days achieved a performance similar to the one achieved by the optimal setpoint. Moreover, it outperformed the optimal setpoint during the first year of learning (notice that the agent started with no knowledge at all). The agent's behavior in the process compared with that of the optimal setpoint is shown in Figure 2.

Second, let's see the agent's adaptation to a different situation while it is operating with the knowledge acquired on a previous one. In this case, we made the plant change from a constant dry-weather condition to a constant rain-weather condition the 200th day (the agent started the experiment with the knowledge acquired in the first experiment). This experiment produced the OC showed in Figure 2. The agent's OC is compared with the OC produced by the optimal (static) setpoint for the dry-weather condition (1.20 mg/l) and for the rain-weather condition (1.85 mg/l). Among other things, notice the quickly adaptation to the new situation thanks to the knowledge acquired during the first experiment.

4   Conclusions

The results presented here show an *autonomous* and an *adaptive* approach for the N-amonnia removal process.

Our goal were not to outperform the latest state-of-the-art control strategy. On the contrary, we wanted to ilustrate how this approach works in different scenarios. We showed that the control strategy performed by our agent *emerged* in a different way for each scenario without having to count on a plant operator or engineer for this task.

This approach seems even worthier when we focus on small wastewater treatment systems. Notice that we invest just *once* in a single agent. Afterwards, this agent will adapt to each location (country/city) *by itself*. Moreover, by means of this approach, the agent will also *supervise* and *modify* the plant (i.e. setpoints) in case of environmental changes. Thus, we get for free an *autonomous* agent that tries to optimize the processes of the plant 24h/day without human intervention.

As a conclusion, this approach could be considered as a solution for plants that cannot afford the important tuning costs that traditional ones have. In addition, this approach allows to adapt quickly to environmental changes. Therefore, it could be appropriate for plants located in places with very unstable environmental conditions.

References

1. Alex, J., Benedetti, L., Copp, J., Gernaey, K.V., Jeppsson, U., Nopens, I., Pons, M.N., Rosen, C., Steyer, J.P., Vanrolleghem, P.: Benchmark Simulation Model no. 1 (BSM1). In: IWA Taskgroup on Benchmarking of Control Strategies for WWTP, p. 61 (April 2008)
2. Belanche, L., Valdes, J.J., Comas, J., Roda, I.R., Poch, M.: Prediction of the bulking phenomenon in wastewater treatment plants. Artificial Intelligence in Engineering 14(4), 307–317 (2000)
3. Bertsekas, D.P.: Dynamic Programming and Optimal Control, 3rd edn., vol. 2. Athenas Scientific (2007)
4. Carrasco, E.F., Rodriguez, J., Punyal, A., Roca, E., Lema, J.M.: Rule-based diagnosis and supervision of a pilot-scale wastewater treatment plant using fuzzy logic techniques. Expert Systems with Applications 22(1), 11–20 (2002)
5. Carrasco, E.F., Rodriguez, J., Punyal, A., Roca, E., Lema, J.M.: Diagnosis of acidification states in an anaerobic wastewater treatment plant using a fuzzy-based expert system. Control Engineering Practice 12(1), 59–64 (2004)
6. Chen, J.C., Chang, N.B., Shieh, W.K.: Assessing wastewater reclamation potential by neural network model. Engineering Applications of Artificial Intelligence 16(2), 149–157 (2003)
7. Copp, J.: The COST Simulation Benchmark: Description and Simulator Manual, p. 154, Office for Official Publications of the European Community, Luxembourg (2002); ISBN: 92-894-1658-0
8. Cortes, U., Sanchez-Marre, M., Ceccaroni, L., R-Roda, I., Poch, M.: Artificial intelligence and environmental decision support systems. Applied Intelligence 13(1), 77–91 (2000)
9. Huang, M., Wan, J., Ma, Y., Li, W., Sun, X., Wan, Y.: A fast predicting neural fuzzy model for on-line estimation of nutrient dynamics in an anoxic/oxic process. Bioresource Technology 101(6), 1642–1651 (2010)
10. Huang, M., Wan, J., Yongwen, M., Yan, W., Weijiang, L., Xiaofei, S.: Control rules of aeration in a submerged biofilm wastewater treatment process using fuzzy neural networks. Expert Systems with Applications 36(7), 10428–10437 (2009)
11. Kaelbling, L.P., Littman, M.L., Moore, A.W.: Reinforcement learning: A survey. Journal of Artificial Intelligence Research 4, 237–285 (1996)
12. Liao, S.: Expert system methodologies and applications: a decade review from 1995 to 2004. Expert Systems with Appications 28, 93–103 (2005)
13. Olsson, G., Nielsen, M., Yuan, Z., Lynggaard-Jensen, A., Steyer, J.P.: Instrumentation, Control and Automation in Wastewater Systems. IWA Publishing (2005)
14. Samuelsson, P., Halvarsson, B., Carlsson, B.: Cost-efficient operation of a denitrifying activated sludge process. Water Research 41, 2325–2332 (2007)
15. Stare, A., Vrecko, D., Hvala, N., Strmcnik, S.: Comparison of control strategies for nitrogen removal in activated sludge process in terms of operating costs: A simulation study. Water Research 41, 2004–2014 (2007)
16. Sutton, R., Barto, A.: Reinforcement Learning: An Introduction. MIT Press, Cambridge (1998)
17. Syafiie, S., Tadeo, F., Martinez, E.: Model-free learning control of neutralization processes using reinforcement learning. Engineering Applications of Artificial Intelligence 20(6), 767–782 (2007)

Genetic Programming for Prediction of Water Flow and Transport of Solids in a Basin

Juan R. Rabuñal[1,2], Jerónimo Puertas[2], Daniel Rivero[1],
Ignacio Fraga[2], Luis Cea[2], and Marta Garrido[2]

[1] Department of Information and Communication Technologies,
University of Coruña, Facultad de Informática,
Campus de Elviña, A Coruña, 15071, Spain
[2] Centre of Technological Innovations in Construction and Civil Engineering
(CITEEC), University of Coruña, Campus de Elviña, A Coruña, Spain
{juanra,jpuertas,drivero,camifc01,lcea,mgarrido}@udc.es

Abstract. One of the applications of Data Mining is the extraction of knowledge from time series [1][2]. The Evolutionary Computation (EC) and the Artificial Neural Networks (ANNs) have proved to be suitable in Data Mining for handling this type of series [3] [4]. This paper presents the use of Genetic Programming (GP) for the prediction of time series in the field of Civil Engineering where the predictive structure does not follow the classic paradigms. In this specific case, the GP technique is applied to two phenomenon that models the process where, for a specific area, the fallen rain concentrates and flows on the surface, and later from the water flows is predicted the solids transport. In this article it is shown the Genetic Programming technique use for the water flows and the solids transport prediction. It is achieved good results both in the water flow prediction as in the solids transport prediction.

Keywords: Evolutionary Computation, Genetic Programming, Civil Engineering, Hydrology.

1 Introduction

In many fields of Civil Engineering the most common way of designing models and structures is application of the laws of physics; experimentation, trial resulting data for their study and validation of the results are the following steps. However, the AI advances have had strong influence on different areas of Civil Engineering, as engineers can use these techniques in several ways and in very different problems [3] [4] [5].

The present work shows the application of Genetic Programming to a specific field of Civil Engineering, the Hydrology, and more specifically to the modelling of water flow, generated after a rain event, in a given basin. This article is the continuation and improvement of the results that were obtained in [6] and [7]. Once the water flow prediction has been obtained, the following step is the prediction of the solids transport and its quantity through the river basin. These

J.M. Ferrández et al. (Eds.): IWINAC 2011, Part II, LNCS 6687, pp. 223–232, 2011.

predictions are very important in the field of Civil Engineering because they serve to predict flooding and the times that exist since a spilling is produced until a fixed concentration reaches to a downstream point (for example, a town with inhabitants).

2   State of Art

One of the most important processes in Hydrology is the so-named "Rainfall-Runoff transformation process" [8], meaning the process in which the rain fallen over an area concentrates and runoff-flows over the surface. As the importance of this process lies basically in it being the cause of overflows, the fact of having a model for the rainfall-runoff transformation process enables forecasting the increase of water levels due to rain episodes. Another important process in Hydrology is the transport modelling of sediments or pollutants. Once the water flow and the pollutant quantity that was spilled into it have been known, it is necessary to know how this waste will evolve downstream in the following temporary moments. This prediction is of vital importance for the vigilance of the water resources that the town inhabitants that are affected will have.

In the field of Civil Engineering there are currently several methods based on mathematical equations for modelling the rainfall-runoff process; some of them are the Hydraulic Equations and the Unit Hydrographs. The Hydraulic Equations try to model, by means of physical equations, the hydraulic processes occurred in a given basin that has been figured as a group of "channels" that represent the roofs, the streets and their interconnections, etc. These models are based on Saint-Vennant equations [9] and they are used quite frequently; however, they are complex to use and also require a wide knowledge about the basin to be modelled. The Unit Hydrographs are based on a common representation of how rain fall-flow is generated. They are based on experimental model, being the SCS Unitary Hydrogram the most used one.

Actually, there is not any method that allows obtaining with reliable results the solids quantity that is transported through the river water flow in case of predicting the pollutant transport when a spilling is produced. As well as in the water flow prediction from the rain, it is needed to generate a time series that produce the solids transport prediction in the following temporary moments from a spilling quantity and the existent water flow.

3   Genetic Programming

The Evolutionary Computation techniques are search algorithms supported by Darwin's Theory of Evolution (1859) according the Principle of Natural Selection [10]. Genetic Programming (GP) [11] [12] is one of these techniques and provides solutions to problems through induction of programs and algorithms to solve them. The result may be a program, an algorithm, or a mathematical expression. The encoding of the solutions is represented by a structure called "tree", which encodes a mathematical formula. This so-called tree structure represents in the "nodes" the operators (arithmetic, trigonometric functions, etc.) and in the "leaves" the constants and variables. One of the great advantages of Genetic Programming is the ability to provide results as mathematical equations. This is a key issue for the civil engineer, because they need to understand and analyze the results of the predictions. This is the main cause of civil engineers do not like the way to get the results of an Artificial Neural Network (ANN) (as a black box structure), which only produces results [4] [13] [14] [15]. Also, using the mathematical expression can be seen the degree of complexity of it and can modify their behavior by altering the terms of the equation. Also, from the viewpoint of hydrology, we can get information on the number of previous values that are needed to predict future values, which in terms of hydrology can be understood as the time of concentration of the basin. This information is essential for experts in civil engineering.

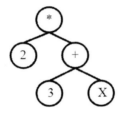

Fig. 1. Tree for the mathematical expression 2*(3+X)

4   Problem Description

In the Centre of Technological Innovations in Construction and Civil Engineering of A Coruña (CITEEC) there is a physique model (to scale) of a basin to be able to study the behavior of the rain fallen transformation and its water flow that was generated as well as the solids transport in the water flow.

This model to scale simulates a basin where rain events can be produced and the water flow that is generated is measured. It can be also added a solid quantity in suspension to the model and it can be studied its transport in the water flow that has been generated. With the obtained data, and using the GP, we can obtain a mathematical expression that predicts the behavior. In the figure 2 it can be observed a photograph of such simulator.

4.1   Water Flow Generated by Rain

It has been made several tests on this simulator and it has been obtained time series of rain and water flow. As it can be observed in figure 3, when a rain event is produced, later it is included a water flow that is proportional to such rain. This water flow time series that has been generated is the main objective of the prediction.

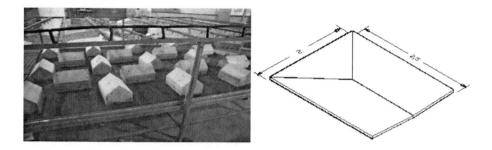

Fig. 2. Rain simulator on an urban or rural basin of the CITEEC. In the photograph it is observed the sprinklers that simulate the rain at the top of the structure and the Basin with a structure of houses with grass at the bottom.

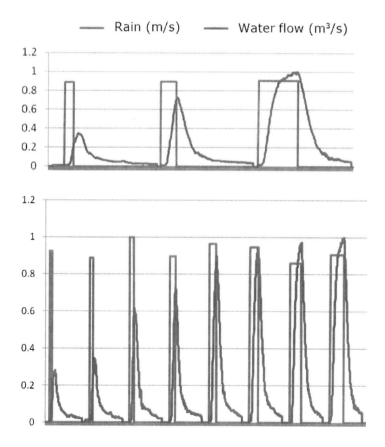

Fig. 3. Training (above) and test (below) data. Time series input data (rain) and Time series output data (water flow).

4.2   Solids Transport

In this artificial model it has been made several solids transport tests. It has been used some material based on sand for the study of sediment entrainment by rainwater. The sediment that has been used has been spread out across the simulator, modifying the concentration in the different tests.

In figure 4 is observed the pollutogram (relation among the water flow and the suspended solids) that were obtained for the most representative test that were made in the artificial basin. The characteristics of the experiment are in table 1.

Table 1. Parameters of the experiment

Parameter	Value
Flow input:	0.082 l/s
Duration:	92.15 seconds
Rainfall:	235 mm/h
Sediment concentration:	0.02 g/cm^2
Sediment type:	White sand

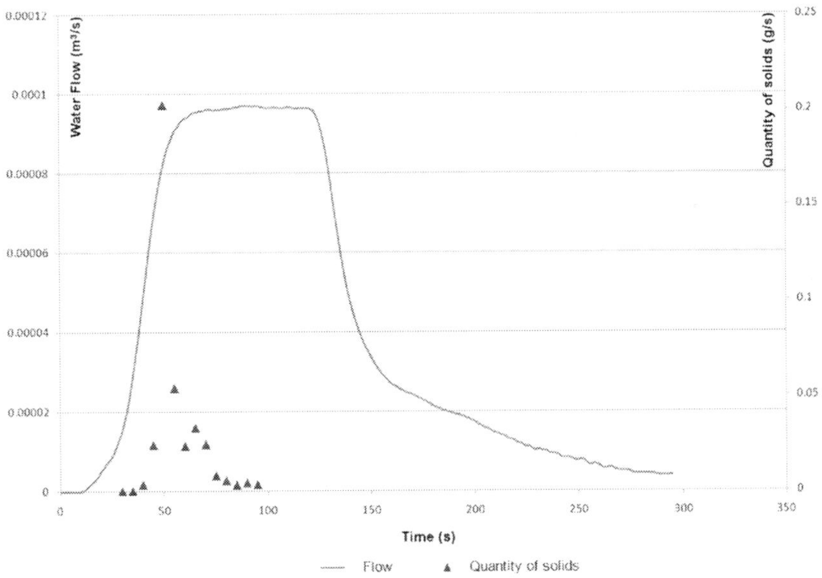

Fig. 4. Training data. Time series input data (water flow) and output data (quantity of solids).

5   Water Flow Prediction

The hydrology characteristics of the artificial basin represents a concentrate time
of 7 seconds, then it is necessary the 7 previous values of the rain intensity values
(one per second). The GP algorithm parameters used are in table 2.

With a low parsimony the algorithm can obtain the best fitness although
the expression may be complicated. The expression obtained (1) makes use of
the rainfall intensity value at the actual time instant (t) and the 6 previous

Table 2. GP parameter settings

GP Parameter	Value
Number of Individuals:	1000
Crossover rate:	90%
Mutation rate:	5%
Selection algorithm:	tournament
Parsimony level:	0.0001
Variables:	P(t), P(t-1), P(t-2), P(t-3), P(t-4), P(t-5), P(t-6)
Constants:	10 random values between -3 and 3
Operators:	+, -, *

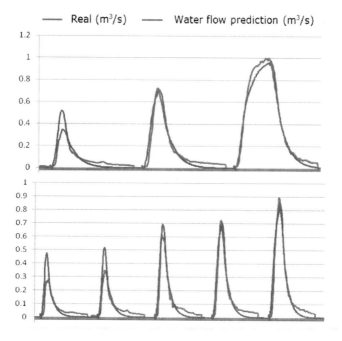

Fig. 5. Comparison of the mathematical expression predictions (training and test data)

values "P(t) to P(t-6)" and the value of water flow predicted by the expression (feedback) in the 4 previous moments "Q(t-1) to Q(t-4)".

$$
\begin{aligned}
\mathbf{Q(t)} =&(0.0251) * P(t) + (-0.083) * P(t-1) + (0.1) * P(t-2) + (-0.068)* \\
& P(t-3) + (0.0651) * P(t-4) + (-0.054) * P(t-5) + (0.0051)* \\
& P(t-6) - (-2.818) * Q(t-1) - (2.662) * Q(t-2) - (-0.784)* \\
& Q(t-3) - (-0.105) * Q(t-4)
\end{aligned}
$$

(1)

Figure 5 is a chart showing the results of the prediction of the mathematical equation and a comparison with the actual flow values obtained in the artificial basin. As can be seen, the flow predictions are very good, allowing sufficient time to obtain flow values that occur in the basin resulting from the rainfall.

6 Solids Transport Prediction

Once the basin water flow prediction has been obtained, it can be modeled the processes that refer to the suspended substances and solids transport in the water flow that are generated. For that modeling, as well as in the previous model, it is made a behavior study in the artificial model that was developed in the CITEEC laboratory.

As in the prediction of water flow from the rain, in this case, we need to predict the transport of solids by the water flow from a number of discharge point in a moment of time and inflow. That is, from the prediction of flow and amount of landfill, the goal is to predict different amounts of solids transported in the following time instants (a concept known in hydrology as pollutogram).

In contrast the prediction of water flow, it is unknown how many previous values of the water flow time series are needed to predict the time series of quantity of solids transported. So there have been several tests using different configurations of time windows for Genetic Programming. It was observed that a time window of 1 produces good results of fitness, however it requires 6 previous values of feedback. The GP parameter settings that produce better results can be seeing in table 1:

The best individual obtained from the GP, converted to mathematical equation (eq. 2) can be seen following:

$$
\begin{aligned}
Y(t) =& \left(\left(\left(\frac{Q(t) - \left(Y(t-2) * Y(t-1)\right)}{Y(t-5)} \right) * \left(\left(Q(t) - \left(Y(t-1) * Y(t-2)\right)\right) - \left(\left(Y(t-2) * Y(t-2)\right) - \left(Y(t-1) * (-0.0119)\right)\right) \right) \right) \right. \\
& \left. - \left(\left(\left(\left(Q(t) - Y(t-5)\right) - \left(Y(t-2) * Y(t-1)\right)\right) - \left(\left(0.8228 * Y(t-2)\right) - \left(Y(t-1) * Y(t-1)\right)\right) \right) * \left(Y(t-3) - \left((-0.0119) - Y(t-1)\right)\right) \right) \right)
\end{aligned}
$$

(2)

The expression obtained (2) makes use of the water flow value at the actual time instant (t) "Q(t)" and the quantity of transported solids in the water predicted by the expression (feedback) in the 6 previous moments "Y(t-1) to Y(t-5)".

Table 3. GP parameter settings

GP Parameter	Value
Number of Individuals:	1000
Crossover rate:	85%
Mutation rate:	10%
Selection algorithm:	tournament
Parsimony level:	0.000001
Variables:	Q(n)
Constants:	10 random values between -1 and 1
Operators:	+, -, *, / (protected division: result = a/b if b≠0 else 1)

As can be seen in the figure 6, the level of prediction is very good, the Medium Square Error (MSE), the fitness of this individual, is $8.38*10^{-18}$. With this equation we can obtain a time series of solids transport from the time series of water flow and a constant amount of solid contribution. This is a significant advancement in hydrology as civil engineers can have mathematical equations to predict the transport of solids in a river, be it rural (a river) or urban (streets, sewer and drainage systems).

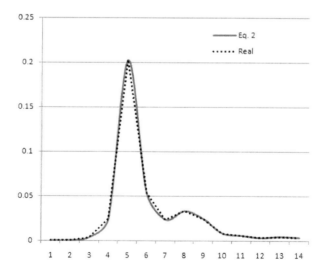

Fig. 6. Comparative of the predictions obtained by the mathematical expression (eq. 2)

7   Conclusions

The results achieved by GP when modelling the runoff flow of an artificial basin by using a mathematical expression shows that they have a satisfactory performance.

Once the water flow prediction is achieved, and it is known the quantity of spilled pollutant in such water flow, it is also possible to model and predict the pollutant transport through the basin. This prediction is very important for civil engineers, because it allows predicting the response times that exist in a Basin in case of possible spilling of a dangerous pollutant in the river and determine the risk conditions for a population.

The modelling of the solids transport process through a mathematical expression obtained using Genetic Programming produces very good results. This means a new advance in the field of Civil Engineering because, until now there was no model or system that allow obtaining this type of results in terms of a mathematical equation and in a satisfactory way.

The following step in the solids transport modelling is being able to obtain a unique mathematical equation that produce satisfactory results in case of different spilling quantities. It is necessary for that to incorporate a new input to the Genetic Programming training algorithm to be representative of the spilling quantity. To be able to obtain a mathematical expression that allows predicting in case of different possible quantities of solids, it is necessary to make more Laboratory tests in the artificial basin to obtain a significant number of time series for the different quantities of solids. Once the tests have been obtained, the GP can be obtain an expression with 2 variable inputs (one input will be the time series of water flow and the other one will be the quantity value of solid spilled into the water).

Acknowledgments. This work was supported by the Dirección Xeral de Investigación, Desenvolvemento e Innovación (General Directorate of Research, Development and Innovation) de la Xunta de Galicia (Galicia regional government) (Ref. 08MDS003CT).

References

1. Han, J., Kamber, M.: Data Mining: Concepts and Techniques. Morgan Kaufmann, San Francisco (2006)
2. Tan, P., Steinbach, M., Kumar, V.: Introduction to Data Mining. Addison-Wesley, Reading (2006)
3. Arciszewski, T., De Jong, K.A.: Evolutionary computation in civil engineering: research frontiers. Civil and structural engineering computing, 161–184 (2001)
4. Flood, I.: Neural Networks in Civil Engineering. Civil and Structural Engineering Computing, 185–209 (2001)
5. Govindaraju, R.S., Rao, A.R.: Artificial Neural Networks in Hydrology. Water Science and Technology Library, vol. 36. Kluwer Academic Publishers, Dordrecht (2000)

6. Miguélez, M., Puertas, J., Rabuñal, J.R.: Artificial neural networks in urban runoff forecast. In: Cabestany, J., Sandoval, F., Prieto, A., Corchado, J.M. (eds.) IWANN 2009. LNCS, vol. 5517, pp. 1192–1199. Springer, Heidelberg (2009)
7. Freire, A., Aguiar, V., Rabual, J.R., Garrido, M.: Genetic Algorithm based on Differential Evolution with variable length. Runoff prediction on an artificial basin. In: International Conference on Evolutionary Computation, ICEC (2010)
8. Viessmann, W., Lewis, G.L., Knapp, J.W.: Introduction to Hydrology. Harper Collins, New York (1989)
9. Huber, W.C., Dickinson, R.E.: Storm Water Management Model, user's manual, version 4. U.S. Envir. Protection Agency, Athens, Ga (1992)
10. Darwin, C.: On the origin of species by means of natural selection or the preservation of favoured races in the struggle for life. Cambridge University Press, Cambridge (1859)
11. Koza, J.R.: Genetic Programming: On the Programming of Computers by Means of Natural Selection. MIT Press, Cambridge (1992)
12. Koza, J.R., Bennet, F.H., Andre, D., Keane, M.: Genetic Programming III. Darwinian invention and problem solving. Morgan Kaufman Publishers, San Francisco (1999)
13. Garrote, L., Molina, M., Blasco, G.: Application of bayesian networks to Real-Time flood risk estimation. Geophysical Research Abstracts 5, 131–171 (2003)
14. Lingireddy, S., Brion, G.: Artificial Neural Networks in Water Supply Engineering. Editorial American Society of Civil Engineers (2005)
15. Wu Jy, S., Han, J., Annambhotla, S., Bryant, S.: Artificial Neural Networks for Forecasting Watershed Runoff and Stream Flows. Journal of Hydrologic Engineering 10(3), 216–222 (2005)

Comparing Elastic Alignment Algorithms for the Off-Line Signature Verification Problem

J.F. Vélez[1], A. Sánchez[1], A.B. Moreno[1], and L. Morillo-Velarde[2]

[1] Departamento de Ciencias de la Computación
Universidad Rey Juan Carlos
28933 Móstoles (Madrid), Spain
jose.velez@urjc.es, angel.sanchez@urjc.es, belen.moreno@urjc.es
[2] Investigación y Programas, S.A.
28016 Madrid, Spain
laura@ipsa.es

Abstract. This paper systematically compares two elastic graph matching methods for off-line signature verification: shape-memory snakes and parallel segment matching, respectively. As in many practical applications (i.e. those related to bank environments), the number of sample signatures to train the system must be very reduced, we selected these two methods which hold that property. Both methods also share some other similarities since they use graph models to perform the verification task and require a registration pre-processing. Experimental results on the same database and using the same evaluation metrics have shown that the shape-memory snakes clearly outperformed to the parallel segment matching approach on the same signature dataset (9% EER compared to 24% EER, respectively).

Keywords: off-line signature verification, snakes, graph matching, elastic alignment, pattern recognition, fuzzy sets.

1 Introduction

Handwritten signature verification is a well-established biometric modality which provides secure means for authentication and authorization in legal documents. The potential applications of this biometry has fostered the research in efficient automatic solutions for the signature recognition and verification problems [8][6]. In the signature recognition (or identification) problem, a given signature is looked up in the database to establish the signer's identity. The signature verification problem is concerned with determining if a particular signature is genuine or if it is a forgery. The techniques for solving both problems can be classified as on-line and off-line [6][8]. In the on-line techniques, data are obtained using an electronic tablet or other device while the subject is signing. In the off-line techniques, the images of signatures are scanned after signing on a piece of a paper and no dynamic information from the act of signing is available.

The literature of off-line signature verification is too vast and the first automatic methods were presented more than thirty years ago [8]. Many proposed

J.M. Ferrández et al. (Eds.): IWINAC 2011, Part II, LNCS 6687, pp. 233–242, 2011.

techniques are based on extraction from the signature image of different types of features (global, local, statistical, geometric, etc) [4]. In some few cases, the signature is considered a holistic pattern and it is directly used (or by reducing its resolution) by the classification algorithm [5]. The problem has been considered by many authors under controlled conditions with promising results (see for example [1]). However, we are far away from an automatic verification system which performs this task under practical conditions with the same effectiveness as a person barely trained for solving the considered problem.

In many types of biometric applications which make use of handwritten signature verification, an important problem is the lack of multiple training signatures from each subject. Although it is not impossible to ask for one client who opens a bank account to sign around ten times to capture his/her intra-personal signing variability, it becomes unconvenient to ask the client for signing more than twice. This practical problem has been solved by some authors by using synthetically generated signatures from one original sample for training their algorithms [10] [14]. These new signatures usually combine some small variations in scale, rotations, differences in the lenght of the strokes or the addition of structural noise on the original signature. Another solution is to devise signature verification methods which make use of a very reduced number of original signatures, as it is the case of the two compared approaches in this paper.

Many graph-based matching methods have been proposed for pattern description and recognition in the last thirty years in the field of Structural Pattern Recognition [3]. One technique which have applied to Computer Vision problems is Elastic Graph Matching (EGM) where a graph model structure is superimposed on the image to adjust, and iteratively this structure is deformed (according to some constraints) to best match the image features. The compared methods in this paper are variants of the EGM framework.

In this paper, we compare in a systematic way two kinds of elastic alignment algorithms which consider the lack of training samples in off-line signature verification: shape-memory snakes [12][13] and parallel segment matching [15]. Both methods have reported competitive results under the above constraint and also share some common useful features such as: using a some kind of graph model to represent signatures, requiring from an initial registration tranformation (i.e. a global adjustment between the model and test signature) and making use of an iterative energy-minimization process. In order to produce a just right evaluation of both methods, they have to be tested not only on the same data sets but also using the same performance metrics for them and adjusting their parameters in a similar form. Moreover, the information related with the evaluation should be detailed sufficiently to allow other researchers to reproduce the evaluation [11].

The work is organized as follows. Section 2 summarizes the two compared verification methods: first, the two shape-memory snake approaches (crisp and fuzzy, respectively) and next the parallel segment alignment algorithm. Section 3 outlines the signature database used for comparing both methods and the experiments carried out. Finally, Section 4 concludes the paper.

2   Considered Elastic Alignment Algorithms

Compared elastic alignment methods for signature verification are outlined.
Next, the common and diferenced aspects of both approaches are highlighted.

2.1   Shape-Memory Snakes

Snakes [7] are energy-minimizing splines based on the analysis of the movement of
a closed or open parametric contour S over an image to which it tries iteratively
to adjust through an energy minimization procedure. As presented in eq. (1)
the snake's energy consists of two terms: the internal or shape energy which is
related to various restrictions of elasticity and flexibility imposed to the curve
and the external or image energy component determined by the influence of some
image features (i.e. intensity values of pixels, edges, corners, etc.) which guide
the movements of the snake:

$$E_{snake}(S) = E_{shape}(S) + E_{image}(S) \tag{1}$$

Snakes have been applied to contour-based segmentation problems [7] where the
curve must change drastically its original shape to adapt to a complex contour.
However, when traditional snakes are used for a shape matching application
(as it happens in signature verification), the excessive snake deformation now
becomes a problem. To avoid this effect, the shape-memory snakes [13] "remember" its original geometry. In particular, the relative proportions between the
adjacent snake segments and also the corresponding angles during the iterative
adjustment between a snake model and the test signature being matched.

Crisp and fuzzy shape-memory snakes have were introduced in [12]. In both
models, in order to speed the convergence of the snake, the image energy term
$E_{image}(S)$ makes use of a potential map of the image [2]. Differences between
the two types of snakes are in the $E_{shape}(S)$ term:

$$E_{shape}(S) = E_{angle}(S) + E_{proportions}(S) \tag{2}$$

where in the crisp model the $E_{angle}(S)$ and $E_{proportions}(S)$ terms in eq. (2) are
defined by expressions involving real-valued parameters (the detailed crisp formulation of $E_{shape}(S)$ in [13]), while in the fuzzy model a zero-order rule-based
Takagi-Sugeno fuzzy inference system (FIS) is used to compute the corresponding snake energy. In the fuzzy model, apart from the spatial distance, the "closeness" in orientation between the snake segments and the strokes of the matched
test signature is included as a third dimension [13].

The signature verification algorithm (for both types of shape-memory snakes)
has two stages, as shown by the UML diagram of Fig. 1:

(1) Adjustment stage. It consists on placing the snake P over the image of the test
signature such that the centers-of-gravity (COGs) of both structures are made
coincident. Next, either a crisp or a fuzzy shape-memory snake adjustement can
be applied to adjust P as much as possible to the signature image by defining
the appropriate image and shape snake energy formulations.

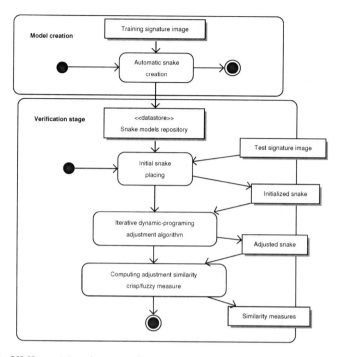

Fig. 1. UML activity diagram describing the shape-memory snake method

(2) Classification stage. After the adjustment stage, the similarity between a test signature and the model snake is determined by a classification technique (i.e. a multilayer perceptron or a first-order Takagi-Sugeno system). This classifier is trained using three extracted discriminative features [13] (refered as coincidence, distance and energy, respectively) that measure the degree of adjustment achieved in the previous stage.

In the crisp model, the adjustment process is directed by eqs. (3), where $v_i(t)$ correspond to the snake points. In the fuzzy model, this process is directed by the more intuitive FIS rules of eqs. (4) which are applied to each snake control point (see [13][12] for details in both cases).

$$E_{angle}(t) = \sum_{i=2}^{N-1} A(v_i(t)) \text{ with } \quad A(v_i(t)) = \begin{cases} \infty & \text{if } \delta > U_{angle} \\ \delta k_{angle} & \text{if } \delta \leq U_{angle} \end{cases}$$
$$E_{prop}(t) = \sum_{i=2}^{N-1} P(v_i(t)) \text{ with } \quad P(v_i(t)) = \begin{cases} \phi k_{prop} & \text{if } \phi \in E(U_{prop}, 1) \\ \infty & \text{in other case} \end{cases} \tag{3}$$

IF (distance is short) $\wedge$ (tangential angle is similar) $\Rightarrow$ (point is near)
IF (distance is large) $\vee$ (tangential angle is quit diferent) $\Rightarrow$ (point is far)
IF (angle changes few) $\wedge$ (prop. changes few) $\wedge$ (point is near) $\Rightarrow$ (energy is low)
IF (angle changes much) $\vee$ (prop. changes much) $\vee$ (point is far) $\Rightarrow$ (energy is high)

$$\tag{4}$$

2.2   Parallel Segment Matching Method

In this method (inspired by a Leung and Suen's paper [9]), to evaluate the similarity between the model and the test signatures, first a global registration between the skeleton of the model signature and the test signature is carried out. This transformation tries to find the global minimum of a defined distance function to make easier the subsequent elastic local alignment between the patterns. After that, both model and test signature skeletons are partitioned into approximately equally-length segments. The sets of segments corresponding to the compared signatures (these can contain different number of elements) are matched between them using the following energy minimization function given by eq. (5) which guide this iterative process:

$$
E_1 = -K_1^2 \sum_{i=1}^{N_I} ln \sum_{j=1}^{N_T} exp\left(- \mid T_j - I_i \mid^2 /2K_1^2\right) f(\theta_{T_j,I_i})
$$
$$
+ \sum_{j=1}^{N_T} \sum_{k=1}^{N_T} w_{jk} \left(d_{T_j,T_k} - d^0_{T_j,T_k}\right)^2
\tag{5}
$$

where N_T and N_I are respectively the number of segments obtained from the model and test signature, T_j and I_i are the position vectors of the midpoints of respective j^{th} segment in the model signature and the ith segment in the test signature, θ_{Tj} and θ_{Ii} are the respective directions of the jth model component and of the ith test component, $\theta_{Tj,Ii}$ is the angle defined by T_j and I_i (restricted within $0°$ and $90°$), $f(\theta_{Tj,Ii}) = max(\cos\theta_{Tj,Ii}, 0, 1)$, $d_{Tj,Tk}$ and $d^0_{Tj,Tk}$ are the respective current and initial value of $\mid T_j - T_k \mid$, parameter w_{jk} is defined by eq. (6), and K_1 and K_2 are parameters representing sizes of Gaussian windows.

$$
w_{jk} = \frac{exp\left(- \mid T_j - T_k \mid^2 /2K_2^2\right)}{\sum_{n=1}^{N_T} exp\left(- \mid T_j - T_n \mid^2 /2K_2^2\right)}
\tag{6}
$$

During the elastic adjustment process, the movement ΔT_j applied to each component segment of the model signature T_j is given by eq. (7):

$$
\Delta T_j = \sum_{i=1}^{N_I} u_{ij}(I_i - T_j) + 2 \sum_{m=1}^{N_T} (w_{mj} + w_{jm}) \left[(T_m - T^0_m) - (T_j - T^0_j)\right]
\tag{7}
$$

where T^0_j is the initial value of T_j and u_{ij} is given by eq. (8):

$$
u_{ij} = \frac{exp\left(- \mid I_i - T_j \mid^2 /2K_1^2\right) f(\theta_{I_i,T_j})}{\sum_{n=1}^{N_T} exp\left(- \mid I_i - T_n \mid^2 /2K_1^2\right) f(\theta_{I_i,T_n})}
\tag{8}
$$

At the end of this iterative elastic adjustment, the mean Euclidean distance between the corresponding matched segments of the two compared signatures is computed. This value is compared to an experimental threshold (which was computed using the three training signatures) to decide whether the test signature is authentic or it is a forgery. The method is outlined by the following UML diagram represented in Fig. 2

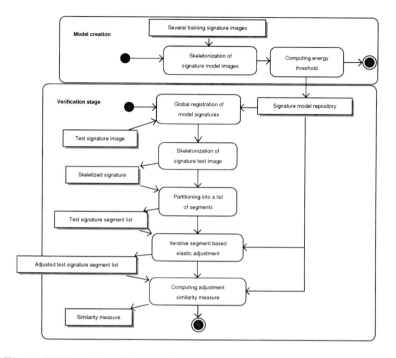

Fig. 2. UML activity diagram for the parallel segment matching method

Table 1. Common and differenced aspects of Elastic Aligment Algorithms

Aspect	Parallel segment matching	Shape memory snakes
1) Few training signatures per class	≥ 3	1
2) Signature skeletonization	Yes (applied to both, model and test signature)	Yes (only applied to the model signature)
3) Global registration pre-processing	Fine registration (shift and rotation)	Rought registration (centers of gravity made coincident)
4) Iterative local energy minimization	As described by eq. (5)	As described by eqs. (2)(3) (crisp approach) and by eq. (4) (fuzzy aproach)
5) Signature division into segments	Equally length segments	Equally length segments
6) Signature segment adjustment	Independent	Dependent
7) Signature verification similarity measure	Segments' mean Square Euclidean distance	Coincidence, distance and energy measures

2.3   Analysis of the Compared Methods

Table 1 summarizes the common and different elements of the compared algorithms for off-line signature verification.

3   Experimental Results

3.1   Signature Database

Experiments were carried out on a database of signatures created by the authors that can be obtained from the following URL: http://gavab.es/recursos.html. It contains six different signatures from a 29 subjects. The set of samples corresponding to each writer were captured in different different years (four signatures the first time and two additional ones after two years) in order to capture the influence of time in intra-personal signatures. One sample signature was used for creating the model (i.e. training stage) and the other five ones for testing in the shape-memory snake method, while three signatures were used for training and the same number for testing purposes in the parallel segment matching method.

3.2   Performed Experiments

We first present a visual comparison of the achieved degree of adjustment on a test signature produced by the compared methods (see Fig. 3). As it can be observed, shape-memory snakes produce a better adjustment than the parallel segment matching algorithm. This adjustment is again improved when using the proposed fuzzy snakes.

As a quantitative performance comparison of the approaches, Fig. 4 presents together the corresponding ROC curves of the signature verification methods. The corresponding ERR values of the curves again demonstrate the advantage of shape-memory snakes (specially, the fuzzy approach) upon parallel matching algorithms on the considered signature database. Table 2, Fig. 5 and Table 3 show the parameter values used in our experiments (for the sake of simplicity, parameters involved in the fuzzy sets [13] for the case of fuzzy shape-memory snakes are displayed as a figure).

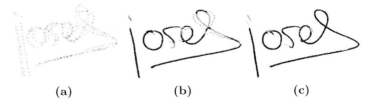

(a)   (b)   (c)

Fig. 3. Adjustment produced by the methods on the same signature: (a) parallel segment matching, (b) crisp shape-memory and (c) fuzzy shape-memory

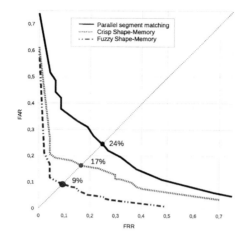

Fig. 4. ROC curves of the compared signature verification methods

Table 2. Crisp snake algorithm parameter values in eq. (3)

Crisp snake parameters	U_{angle}	U_{prop}	k_{angle}	k_{prop}	k_{fc}
Value	5	0.1	0.1	1	10

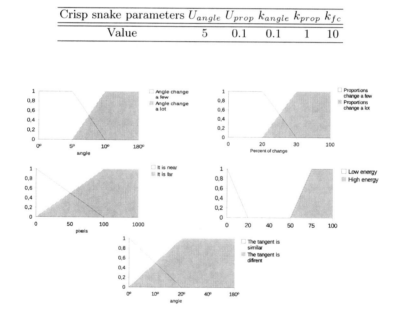

Fig. 5. Visual representation of the fuzzy parameters according to eqs. (4)

Table 3. Parallel segment matching parameters values in eqs. (5)-(8)

Parameters	Initial value	Update value
K_1	2	$K_1 = K_1 - \max(0.4, 0.15K_1)$
K_2	4	$K_2 = K_2 - \max(0.4, 0.1K_2)$

It is important to remark that the 18.6% ERR result reported by You et al [15] is better than the achieved by our comparison (which achieved approximately 24% ERR). These differences could be due to some of the following causes. First, it is possible that the intra-personal signature variability in our database will be higher than the corresponding one used by You et al in their experiments (our database contain samples from the same subjects collected in two different years). Second, the values of some parameters (in particular the constants K_1 and K_2) which appear in the energy formulation which define the elastic alignment method are not detailed in the paper [15]. Third, differently from You et al who used three or more signatures to build the signature model for each writer, we have only used the minimal number of three samples for such purpose.

4   Conclusion

This paper has compared, in a systematic and fair way, using the same database, shape-memory snakes and parallel segment matching algorithms for the off-line signature verification problem. Experimental results using the same evaluation metrics have shown that the shape-memory snakes clearly outperformed to the parallel segment matching method (9% of EER compared to 24% of EER). Future work will consider in this comparative study the use of forgeries (both simple and skilled ones). Another interesting aspect to analyze is the robustness of these methods when comparing signatures captured at different periods of time.

Acknowledgements

This research has been partially supported by the Spanish project TIN2008-06890-C02-02.

References

1. Bajaj, R., Chaudhury, S.: Signature verification using multiple neural classifiers. Pattern Recognition 30, 1–7 (1997)
2. Cohen, L.D., Cohen, I.: Finite element method for active contour models and balloons for 2-D and 3-D images. IEEE Trans. Pattern Analysis and Machine Intelligence 15(11), 1131–1147 (1993)

3. Conte, D., Foggia, P., Sansone, C., Vento, M.: Thirty Years of Graph Matching in Pattern Recognition. Pattern Recognition 36(1), 91–101 (2003); Intl. Journal of Pattern Recognition and Artificial Intelligence 18(3) 265–298 (2004)
4. Fang, B., et al.: Off-line signature verification by the tracking of feature and stroke positions. Pattern Recognition 36(1), 91–101 (2003)
5. Frías-Martínez, E., Sánchez, A., Vélez, J.F.: Support vector machines versus multi-layer perceptrons for efficient off-line signature recognition. Engineering Applications of Artificial Intelligence 19, 693–704 (2006)
6. Impedovo, D., Pirlo, G.: Automatic Signature Verification: The State of the Art. IEEE Trans. Systems, Man, and Cybernetics (Part C) 38(5), 609–635 (2008)
7. Kass, M., Witkin, A., Terzopoulos, D.: Snakes: Active Contour Models. Intl. Journal of Computer Vision 1(4), 321–331 (1988)
8. Leclerc, F., Plamondon, R.: Automatic Signature Verification: The State of the Art. Intl. Journal of Pattern Recognition and Artificial Intelligence 8(3), 643–660 (1994)
9. Leung, C.H., Suen, C.Y.: Matching of Complex Patterns by Energy Minimization. IEEE Trans. Systems, Man and Cybernetics (Part B) 28(5), 712–720 (1998)
10. Mizukami, Y., et al.: An off-line signature verification system using an extracted displacement function. Pattern Recognition Letters 23, 1569–1577 (2002)
11. Phillips, P.J., Martin, A., Wilson, C.L., Przybocki, M.: An Introduction to Evaluating Biometric Systems. IEEE Computer, 56–63 (2000)
12. Vélez, J., Sánchez, A., Fernández, F.: Improved Fuzzy Snakes Applied to Biometric Verification Problems. In: Proc. 9th Intl. Conf. on Intelligent System Design and Applications, pp. 158–163. IEEE Press, Los Alamitos (2009)
13. Vélez, J., Sánchez, A., Moreno, B., Esteban, J.L.: Fuzzy shape-memory snakes for the automatic off-line signature verification problem. Fuzzy Sets and Systems 160(2), 182–197 (2009)
14. Vélez, J.F., Sánchez, A., Moreno, A.B.: Robust off-line signature verification using compression networks and positional cuttings. In: Proc. IEEE 13th Intl. Workshop on Neural Networks for Signal Processing, pp. 627–636. IEEE Press, Los Alamitos (2003)
15. You, X., Fang, B., He, Z., Tang, Y.Y.: Similarity measurement for off-line signature verification. In: Huang, D.-S., Zhang, X.-P., Huang, G.-B. (eds.) ICIC 2005. LNCS, vol. 3644, pp. 272–281. Springer, Heidelberg (2005)

A Fuzzy Cognitive Maps Modeling, Learning and Simulation Framework for Studying Complex System

Maikel León[1,*], Gonzalo Nápoles[1], Ciro Rodriguez[2], María M. García[1], Rafael Bello[1], and Koen Vanhoof[3]

[1] Central University of Las Villas, Santa Clara, Cuba
[2] Cienfuegos University, Cienfuegos, Cuba
[3] Hasselt University, IMOB, Diepenbeek, Belgium

Abstract. This paper presents Fuzzy Cognitive Maps as an approach in modeling the behavior and operation of complex systems; they combine aspects of fuzzy logic, neural networks, semantic networks, expert systems, and nonlinear dynamical systems. They are fuzzy weighted directed graphs with feedback that create models that emulate the behavior of complex decision processes using fuzzy causal relations. First, the description and the methodology that this theory suggests is examined, later some ideas for using this approach in the control process area are discussed. An inspired on particle swarm optimization learning method for this technique is proposed, and then, the implementation of a tool based on Fuzzy Cognitive Maps is described. The application of this theory might contribute to the progress of more intelligent and independent systems. Fuzzy Cognitive Maps have been fruitfully used in decision making and simulation of complex situation and analysis. At the end, a case study about Travel Behavior is analyzed and results are assessed.

Keywords: Fuzzy Cognitive Maps, Particle Swarm Optimization, Travel Behavior, Complex Systems, Decision Making, Simulation.

1   Introduction

Modeling dynamic systems can be hard in a computational sense and many quantitative techniques exist. Well-understood systems may be open to any of the mathematical programming techniques of operations study. First, developing the model usually requires a big deal of effort and specialized knowledge outside the area of interest. Secondly, systems involving important feedback may be nonlinear, in which case a quantitative model may not be possible [1]. In the past years, conventional methods were used to model and control systems but their contribution is limited in the representation, analysis and solution of complex systems.

There is a great demand for the development of autonomous complex systems that can be achieved taking advantage of human like reasoning and description

---

* Corresponding author: mle@uclv.edu.cu

J.M. Ferrández et al. (Eds.): IWINAC 2011, Part II, LNCS 6687, pp. 243–256, 2011.

of systems [2]. Human way of thinking process for any method includes vague descriptions and can have slight variations in relation to time and space; for such situations Fuzzy Cognitive Maps (FCM) seem to be appropriate to deal with. FCM are a combination of some aspects from Fuzzy Logic, Neural Networks and other techniques; combining the heuristic and common sense rules of Fuzzy Logic with the learning heuristics of the Neural Networks. They were introduced by Kosko [3], who enhanced cognitive maps with fuzzy reasoning, that had been previously used in the field of socio-economic and political sciences to analyze social decision-making problems.

The use of FCM for many applications in different scientific fields was proposed, they had been apply to analyze extended graph theoretic behavior, to make decision analysis and cooperate distributed agents, also were used as structures for automating human problem solving skills and as behavioral models of virtual worlds, and in many other fields. That's why, with the elaboration of a framework that allows the design, execution, analyses and simulation of FCM is provided to specialists of diverse knowledge areas of a means for the study, prediction and assessment of situations that characterize diverse real problems.

This work proposes such a computational tool, where is possible to represent knowledge in a graphic and comprehensible way through FCM, to try to study the systems like a whole, settling down how the entities that conform them are affected with others, offering to users, not necessarily specialist in Computer Science, a tool that allows the creation and execution of FCM, including experimentation facilities and as a framework for the study of complex real systems.

The paper is structured as follows. First, there is a section with a background about FCM; after that, it is discussed a procedure where a particle swarm optimization metaheuristic is applied in the learning process of FCM, and a pseudocode of the idea is also presented. Following this, a section with the computational tool is described, showing the main windows to give readers a better comprehension on what is offered. At the end a case study about Travel Behavior is analyzed, with important results of the application of FCM in the problem modeling and an improvement over cognitive structure due to the use of the learning method.

2   Overview about Fuzzy Cognitive Maps

FCM in a graphical illustration seem to be a signed directed graph with feedback, consisting of nodes and weighted arcs (see figure 1). Nodes of the graph place for the concepts that are used to express the behavior of the system and they are connected by signed and weighted arcs representing the causal relationships that exist connecting the concepts [4]. It must be mentioned that the values in the graph are fuzzy, so concepts take values in the range between [0, 1] and the weights of the arcs are in the interval [-1, 1]. The weights of the arcs between concept C_i and concept C_j could be positive ($W_{ij} > 0$) which means that an augment in the value of concept C_i leads to the increase of the value of concept C_j, and a decrease in the value of concept C_i leads to a reduce of the value

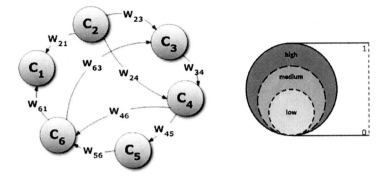

Fig. 1. Simple Fuzzy Cognitive Map. Concept activation level.

of concept C_j. Or there is negative causality ($W_{ij} < 0$) which means that an increase in the value of concept C_i leads to the decrease of the value of concept C_j and vice versa.

Observing this graphical representation, it becomes clear which concept influences other concepts showing the interconnections between concepts and it permits updating in the construction of the graph. Each concept represents a characteristic of the system; in general it stands for events, actions, goals, values, trends of the system that is modeled, etc. Each concept is characterized by a number that represents its value and it results from the renovation of the real value of the system's variable [5].

Beyond the graphical representation of the FCM there is its mathematical model. It consists of a $1 \times n$ state vector A which includes the values of the n concepts and a $n \times n$ weight matrix W which gathers the weights W_{ij} of the interconnections between the n concepts. The value of each concept is influenced by the values of the connected concepts with the appropriate weights and by its previous value. So the value A_i for each concept C_i can be calculated, among other possibilities, by the following rule expressed in (1).

$$A_i = f \left(\sum_{\substack{j=1 \\ j\,is\,not\,i}}^{n} A_j W_{ji} \right) \qquad (1)$$

Where A_i is the activation level of concept C_i, A_j is the activation level of concept C_j and W_{ji} is the weight of the interconnection between C_j and C_i, it is to say, the value of A_i depends of the weighted sum of its input concepts, and f is a threshold or normalization function. So the new state vector A_{new} is computed by multiplying the previous state vector A_{old} by the weight matrix W, see (2). The new vector shows the effect of the change in the value of one concept in the whole FCM [6].

$$A_{new} = f\left(A_{old}W\right) \qquad (2)$$

In order to build an FCM, the knowledge and experience of one expert on the system's operation must be used. The expert determines the concepts that best illustrate the system; a concept can be a feature of the system, a state or a variable or an input or an output of the system; identifying which factors are central for the modeling of the system and representing a concept for each one.

Moreover the expert has observed which elements of the system influence others elements; and for the corresponding concepts the expert determines the negative or positive effect of one concept on the others, with a fuzzy value for each interconnection, since it has been considered that there is a fuzzy degree of causation between concepts.

It is possible to have better results in the drawing of the FCM, if more than one expert is used. In that case, all experts are polled together and they determine the relevant factors and thus the concepts that should be presented in the map. Then, experts are individually asked to express the relationship among concepts; during the assigning of weights, three parameters must be considered: how strongly concepts influence each other, what is the sign of the weight and whether concepts cause [7].

This is one advantage over other approaches like Petri Nets (PN) or Bayesian Networks (BN). PN are a graphical and mathematical modeling tool consisting of places, transitions, and arcs that connect them, that can be used as a visual-communication aid similar to flow charts, block diagrams, and networks. As a mathematical instrument, it is possible to set up state equations, algebraic equations, and other mathematical models governing the performance of systems.

It is well known that the use of PN has as a disadvantage the drawing process by a non-expert in this technique, that's way there is a limited numbers of tools usable for this purpose, and it is not well established how to combine different PN that describe the same system [8].

FCM feedback structure also makes a distinguishing from the earlier forward-only acyclic cognitive maps and from modern Artificial Intelligence expert-system search trees. Such tree structures are not dynamical systems because they lack edge cycles or closed inference loops. Nor are trees closed under combination. Combining several trees does not produce a new tree in general because cycles or loops tend to occur as the number of combined trees increases.

But combining FCM always produces a new FCM. The combined FCM naturally averages the FCM and their corresponding causal descriptions as well as much of their dynamics (example on figure 2).The user can combine any number of weighted FCM, combining into a single averaged FCM by the simple artifice of adding their scaled and augmented (zero-padded) adjacency edge matrices.

The strong law of large numbers then ensures that the sample average of even quantized or rounded-off FCM will converge with probability one to the underlying but often unknown FCM that generates these matrix realizations. So FCM edge-matrix combination improves with sample size. FCM knowledge representation likewise tends to improve as the user combines more FCM from an ever larger pool of domain experts.

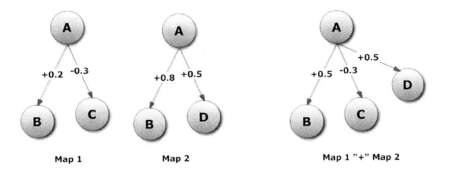

Fig. 2. Combining two FCM

If there will be a collection of individual FCM that must be combined into a collective map (see figure 3) and if there are experts of different credibility, for them, then their proposed maps must be multiplied with a nonnegative "credibility" weight. So the combination of these different FCM will produce an augmented FCM.

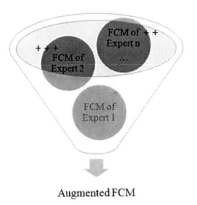

Fig. 3. Combining some FCM into a collective map

As over PN, this is an advantage over BN [9]. BN is a powerful tool for graphically representing the relationships among a set of variables and for dealing with uncertainties in expert systems, but demanding effort caused by specification of the net (structure and parameters) and an expensive algorithm of propagation of probabilities [10]. Also is not evident for a non-expert in this field how to construct a BN, and even more difficult how to compare or combine different BN that describe the same system.

When a FCM has been constructed, it can be used to model and simulate the behavior of the system. Firstly, the FCM should be initialized, the activation level of each of the nodes of the map takes a value based on expert's opinion for the current state and then the concepts are free to interact. This interaction between concepts continues until a fixed equilibrium is reached; a limited cycle is reached or a chaotic behavior is exhibited. So, FCM are a powerful methodology that can be used for modeling systems, avoiding many of the knowledge extraction problems which are usually present in by rule based systems [11].

The threshold or normalization function used over concept value of FCM, serves to decrease unbounded inputs to a severe range. This destroys the possibility of quantitative results, but it gives us a basis for comparing nodes (on or off, active or inactive, etc.). This mapping is a variation of the "fuzzification" process in fuzzy logic, giving us a qualitative model and frees us from strict quantification of edge weights [12].

3   A FCM Learning Process Using PSO Metaheuristic

Problems associated with manual development of FCM encourage researchers to work on automated or semi-automated computational methods for learning FCM structure using historical data. Semiautomated methods still require a relatively limited human intervention, whereas fully automated approaches are able to compute a FCM model solely based on historical data.

Researches on learning FCM models from data have resulted in a number of alternative approaches. One group of methods is aimed at providing a supplement tool that would help experts to develop accurate model based on their knowledge about a modeled system. Algorithms from the other group are oriented toward eliminating human from the entire development process, only historical data are necessary to establish FCM model [13].

Particle Swarm Optimization (PSO) method, which belongs to the class of Swarm Intelligence algorithms, can be used to learn FCM structure based on historical data, consisting of a sequence of state vectors that leads to a desired fixed-point attractor state.

PSO is a population based algorithm, which goal is to perform a search by maintaining and transforming a population of individuals. This method improves the quality of resulting FCM model by minimizing an objective or heuristic function. The function incorporates human knowledge by adequate constraints, which guarantee that relationships within the model will retain the physical meaning defined by experts.

The following idea flow chart (see figure 4) illustrates the application of PSO in the readjusting of the weight matrix, trying to find a better configuration that guaranty a convergence or waited results. PSO is applied straight forwardly using an objective function defined by the user. Each particle of the swarm is a weight matrix, encoded as a vector.

First the concepts and relation are defined, and the construction of FCM is made, and then is possible to make simulations and obtain outputs due to the inference process. If the new values are not adequate, known by the execution of the heuristic function, then it is necessary a learning process (in this case through the use of PSO metaheuristic) having as results new values for the weight matrix.

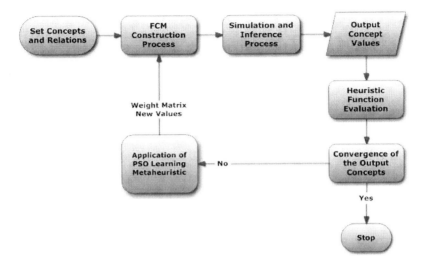

Fig. 4. Application of PSO metaheuristic as a FCM learning method

Generate initial population using W_{ij} as initial approximation
Initial evaluation
Cross over good particles
Mutation of random particles
Initialize the vector X_{pbest} with best solutions found by each particle
Initialize X_{gbest} as the best global found
Initialize $W_{max}= 1.4$, $W_{min}= 0.4$, $c_1 =2.0$, $c_2 =2.0$
For $t=0$ to $N_{generations}$
　$w_k= (W_{max} - W_{min}) * ((N_{cmax} - t) / (N_{cmax} + W_{min}))$
　For each X_i
　　Calculate $V_i(t+1)$ and limit to $[-V_{max}, +V_{max}]$ using w_k
　　Calculate $X_i(t+1)= X_i(t) + V_i(t+1)$
　　Analyze the vector Swarm with $X_i(t+1)$ and Speed with $V_i(t+1)$
　　Evaluate the particle $X_i(t+1)$
　　Analyze the vector X_{pbest} with the best solutions
　　Update X_{gbest} with the best global particle
　endFor
endFor

Fig. 5. Pseudocode of the PSO proposed method

Mixed approaches, like using Genetic Algorithm and Particle Swarm Optimization have been performed so far. Literature reports some results that are really very promising and encouraging further researches and applications in this area.

Using this approach, new zones of the search space are explored in a particular way, through the crossover of good initial particles and the mutation of some others, just to mention two possible approaches. In the following pseudocode illustrated in figure 5, we can appreciate the general philosophy of our proposed method. In this case the genetic algorithm operators are used as initial steps.

4   Tool Based on Fuzzy Cognitive Maps

The scientific literature shows some software products developed with the intention of drawing FCM by non-expert in computer science, as FCM Modeler [14] and FCM Designer [15]. The first one is a rustic incursion, while the second one is a better implementation, but still hard to interact with and it doesn't have experimental facilities. Figure 6 shows the general architecture of our proposing tool to model and simulate FCM, the organization and structuring of the components are presented.

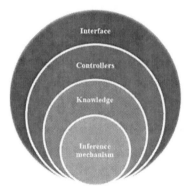

Allows the user-tool interaction through the options to create FCM, definition of parameters and formalization of the information into a knowledge base.

Makes a link between the Interface and the algorithms and data, it is a connectivity layer that guarantees a right manipulation of the information.

Generates the computational representation of the created FCM from an Artificial Intelligence point of view. Processes the input and output data of algorithms in the variables modeling.

Makes the inference process through the mathematical calculus for the prediction of the variable values.

Fig. 6. General architecture of the tool

In figure 7 it is possible to observe the main window of the tool, and a modeled example, in the interface appears some facilities to create concepts, make relations, and define parameters, also to initialize the execution of the inference process, and visualization options for a better understanding of the simulation process. There were defined some facilities and options in the tool, to create, open or save an FCM. Through these amenities a non-expert in Computer Science is able to elaborate FCM describing systems; we had paid attention to these facilities guarantying an usable tool, specifically for simulation purposes.

In figure 8 we can appreciate some important options, where is possible to define the assignment of a delay time in the execution for a better understanding of the running of the FCM in the inference process, also it is possible to define the normalization function that the FCM will use in the running [16]. In simulation

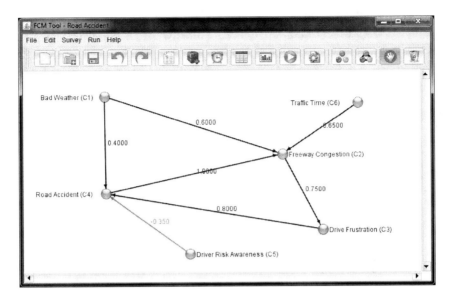

Fig. 7. Main view of FCM Tool

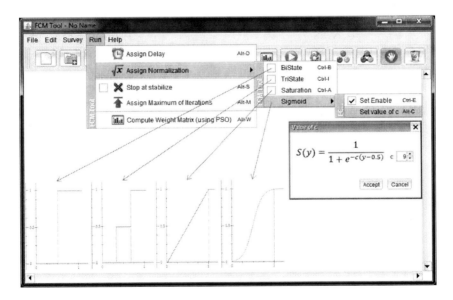

Fig. 8. Run options of the FCM Tool

experiments the user can compare results using these different functions or just can select the appropriate function depending of the problem to model:

- Binary FCM are suitable for highly qualitative problems where only representation of increase or stability of a concept is required.

- Trivalent FCM are suitable for qualitative problems where representation of increase, decrease or stability of a concept is required.

- Sigmoid FCM are suitable for qualitative and quantitative problems where representation of a degree of increase, a degree of decrease or stability of a concept is required and strategic planning scenarios are going to be introduced.

On the other hand, there are also options for experts on the topic, specifically related to the learning process, as it is possible to perceive in figure 9. The necessary definition of parameters is done through this window.

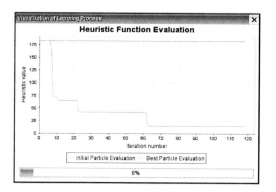

Fig. 9. Parameters definition for PSO method

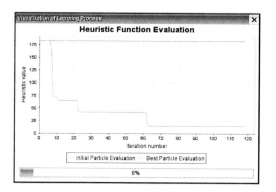

Fig. 10. Heuristic function evaluation in the learning process

In simulation and experiment in general, the visualization consists a fundamental aspect, that´s why it was consevied a panel where the learning process can be observed, figure 10 shows the heuristic function evaluation panel.

It is possible to see how the FCM is updated with a new weight matrix that better satisfy the waited results, in less than 100 iteration, there is a significative convergence.

5   Case Study: Modeling Travel Behavior

In the process of transportation planning, travel demand forecast is one of the most important analysis instruments to evaluate various policy measures aiming at influencing travel supply and demand. In past decades, increasing environmental awareness and the generally accepted policy paradigm of sustainable development made transportation policy measures shift from facilitation to reduction and control.

Objectives of travel demand management measures are to alter travel behavior without necessarily embarking on large-scale infrastructure expansion projects, to encourage better use of available transport resources avoiding the negative consequences of continued unrestrained growth in private mobility [17].

Individual activity travel choices can be considered as actual decision problems, causing the generation of a mental representation or cognitive map of the decision situation and alternative courses of action in the expert's mind. This cognitive map concept is often referred to in theoretical frameworks of travel demand models, especially related to the representation of spatial dimensions, but much features can be taken into account.

However, actual model applications are scarce, mainly due to problems in measuring the construct and putting it into the model's operation. The development of the mental map concept can benefit from the knowledge provided by individual tracking technologies [18]. Researches are focusing on that direction, in order to improve developed models and to produce a better quality of systems.

At an individual level it is important to realize that the relationship between travel decisions and the spatial characteristics of the environment is established through the individual's perception and cognition of space. As an individual observes space, for instance through travel, the information is added to the individual's mental maps (see figure 11).

In the city of Hasselt, Belgium, a study related to travel behavior was made. More than 200 real habitants were asked to specify how they take into account the transport mode they will use for an imaginary shopping activity: situation variables, attributes, benefits; and starting from that data, a FCM structure per person was constructed.

At the same time, virtual scenarios were presented, and the personal decisions were stored. Figure 12 shows the acting of the initial modeled FCM, for example, only the 24% of the maps were able to predict 100% scenarios. A FCM learning method, based on the PSO metaheuristic was applied, having the stored scenarios as training data, and the results show that after the learning process, 77% of maps were able to predict 100% of scenarios.

Fig. 11. Abstraction levels of mind related to Travel Behavior

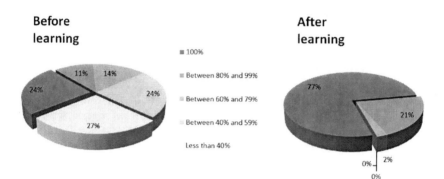

Fig. 12. Improvement made to data

It is considered a significant improvement over the maps, having structures able to simulate how people think when visiting the city center, specifically the transport mode they will use (car, bus or bike), offering policy makers, a tool to play with, in order to test new policies, and to know in advance the possible resounding in the society (buses cost, parking cost, bike incentive, etc.).

6   Conclusions

It has been examined Fuzzy Cognitive Maps as a theory used to model the behavior of complex systems, where is extremely difficult to describe the entire system by a precise mathematical model. Consequently, it is more attractive and practical to represent it in a graphical way showing the causal relationships between concepts. Since this symbolic method of modeling and control of a system is easily adaptable and relies on human expert experience and knowledge, it can be considered intelligent.

A learning algorithm for determining a better weight matrix for the through-put of FCM was presented. It is an unsupervised weight adaptation methodology that have been introduced to fine-tune FCM causal links and accompanied with the good knowledge of a given system or process can contribute towards the establishment of FCM as a robust technique. Experimental results based on simulations of the process system verify the effectiveness, validity and advantageous behavior of the proposed algorithm.

The area of FCM learning is very promising because the FCM obtained are directly interpretable by a human and are an useful tool to extract information from data about the relations between the concepts or variables inside a domain.

The development of a tool based on FCM for the modeling of complex systems was presented, showing the facilities for the creation of FCM, the definition of parameters and options to make the inference process more comprehensible, understanding and used for simulations experiments.

At the end, a real case study was presented, showing a possible Travel Behavior modeling through FCM, and the benefits of the application of a learning method inspired in the PSO metaheuristic, obtaining an improvement on the knowledge structures originally modeled. In this shown example a social and politic repercussion is evident, as we offer to policymakers a framework and real data to play with, in order to study and simulate individual behavior and produce important knowledge to use in the development of city infrastructure and demographic planning.

References

1. Kosko, B.: Neural Networks and Fuzzy systems, a dynamic system approach to machine intelligence, p. 244. Prentice-Hall, Englewood Cliffs (1992)
2. Parpola, P.: Inference in the SOOKAT object-oriented knowledge acquisition tool. Knowledge and Information Systems (2005)
3. Kosko, B.: Fuzzy Cognitive Maps. International Journal of Man-Machine Studies 24, 65–75 (1986)
4. Koulouritios, D.: Efficiently Modeling and Controlling Complex Dynamic Systems using Evolutionary Fuzzy Cognitive Maps. International Journal of Computational Cognition 1, 41–65 (2003)
5. Wei, Z.: Using fuzzy cognitive time maps for modeling and evaluating trust dynamics in the virtual enterprises. Expert Systems with Applications, 1583–1592 (2008)
6. Xirogiannis, G.: Fuzzy Cognitive Maps as a Back End to Knowledge-based Systems in Geographically Dispersed Financial Organizations. Knowledge and Process Management 11, 137–154 (2004)
7. Aguilar, J.: A Dynamic Fuzzy-Cognitive-Map Approach Based on Random Neural Networks. Journal of Computational Cognition 1, 91–107 (2003)
8. Li, X.: Dynamic Knowledge Inference and Learning under Adaptive Fuzzy Petri Net Framework. IEEE Transactions on Systems, Man, and Cybernetics Part C: Applications and reviews (2000)
9. Castillo, E.: Expert Systems and Probabilistic Network Models. Springer, Heidelberg (2003)
10. Intan, R.: Fuzzy conditional probability relations and their applications in fuzzy information systems. Knowledge and Information Systems (2004)

11. Carlsson, C.: Adaptive Fuzzy Cognitive Maps for Hyperknowledge Representation in Strategy Formation Process. In: IAMSR, Abo Akademi University (2005)
12. Stylios, C.: Modeling Complex Systems Using Fuzzy Cognitive Maps. IEEE Transactions on Systems, Man and Cybernetics 34, 155–162 (2004)
13. Mcmichael, J.: Optimizing Fuzzy Cognitive Maps with a Genetic Algorithm AIAA 1st Intelligent Systems Technical Conference. Chicago, Illinois (2004)
14. Mohr, S.: Software Design for a Fuzzy Cognitive Map Modeling Tool. Tensselaer Polytechnic Institute (1997)
15. Contreras, J.: Aplicación de Mapas Cognitivos Difusos Dinámicos a tareas de supervisión y control. Trabajo Final de Grado. Universidad de los Andes. Mérida, Venezuela (2005)
16. Tsadiras, A.: A New Balance Degree for Fuzzy Cognitive Maps. Technical Report. Department of Applied Informatics. University of Macedonia (2007)
17. Gutiérrez, J.: Análisis de los efectos de las infraestructuras de transporte sobre la accesibilidad y la cohesión regional. Estudios de Construcción y Transportes. Ministerio de Fomento. España (2006)
18. Wu, Q.: Multiknowledge for decision making. Knowledge and Information Systems (2005)

Study of Strength Tests with Computer Vision Techniques

Alvaro Rodriguez[1], Juan R. Rabuñal[1,2],
Juan L. Perez[1], and Fernando Martinez-Abella[2]

[1] Dept. of Information and Communications Technologies,
[2] Centre of Technological Innovation in Construction and Civil Engineering
(CITEEC),
University of A Coruña, Campus Elviña s/n 15071, A Coruña
arodriguezta@udc.es, juanra@udc.es, jlperez@udc.es, fmartinez@udc.es

Abstract. Knowing the strain response of materials in strength tests is one of the main issues in construction and engineering fields. In these tests, information about displacements and strains is usually carried out using physical devices attached to the material.

In this paper, the suitability of Computer Vision techniques to analyse strength tests without interfering with the assay is discussed and a new technique is proposed.

This technique measures displacements and deformations from a video sequence of the assay.

With this purpose a Block-Matching Optical Flow algorithm is integrated with a calibration process to extract the vectorfield from the displacement in the material.

To evaluate the proposed technique, a synthetic image set and a real sequence from a strength tests were analysed.

Keywords: Computer Vision, Optical Flow, Block-Matching, Strength Tests.

1 Introduction

Solid materials are composed of molecules separated by empty space and packed in a certain manner with some structure. When forces are applied to the material, this configuration changes, modifying the shape of the material.

In strength tests, monitored loads or strains are applied to a scale model of the meterial. Then, information about the material's behaviour is traditionally obtained using specific devices which are physically linked to the material and which provide information about the length variation of the structure around a given point and in and in a particular direction.

Therefore, these devices can only provide readings in one dimension and they interfere with the experiment generating certain stiffness to displacement.

To overcome this limitations, a new way of analyze strength tests is proposed, using a video camera to record the test and then processing the images to extract the full displacement field of the body.

J.M. Ferrández et al. (Eds.): IWINAC 2011, Part II, LNCS 6687, pp. 257–266, 2011.
© Springer-Verlag Berlin Heidelberg 2011

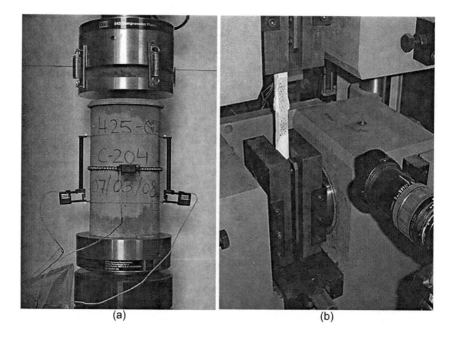

(a)   (b)

Fig. 1. (a) Traditional displacement measurements in strength test, where sensors are attached to the material. (b) Proposed optical system for the analysis strength test.

The general problem of analysing displacements with visual information, is a particular case of the correspondence problem, since vision is not a continuous process, but a discrete one.

Thus, images of the same scene separated by a time interval are analysed, and the correspondence among the objects in the two images must be carried out.

To solve this problem, biological brains analise the visual information in several ways, integrating complex processes of segmentation and labeling of the objects in the scene, analysing similarities, using mental abstractions of the objects and predicting trajectories. The result of this processes is the illusion of the movement.

Using computer processing, the main difficulty for analysing displacement in a non-rigid surface is derived from the fact that the body geometry varies between two moments and thus, it is not possible to find out which part of the information characterises the body and which part may be attributed to deformation.

Optical flow techniques born in the 80s [9]. They provide a flexible approach to extract the motion field of a scene without using any previous knowledge about the displacement of the objects in the image.

Almost at the same time, Computer Vision techniques were applied for the first time in industrial applications and to measure real processes. [4].

The very first approaches to the analysis of non-rigid movements using optical flow methods used gradient-based techniques or those based on the probability distribution of possible displacements [5]. However, the works carried out in the flow analysis field [17] showed that the refinement of the techniques based on region analysis (called Block-Matching techniques) stand for the most robust and flexible approach in this field.

Along the years, several contributions have been made to Block-Matching techniques: For instance, an iterative multi-resolution approach has been widely proposed. [20,2].

Other valuable contributions were the use of Fourier Transforms (FTs) to increase performance and the use one dimensional point estimators to achieve measure- ments below the pixel level [17].

Recently, some new methods have been proposed to include strains in the search process, this is obtained by using some iterative procedure comprising information smoothing and interpolation phases, so that the defined regions are deformable by using the estimate of the displacement in the previous iteration. [18,11,20].

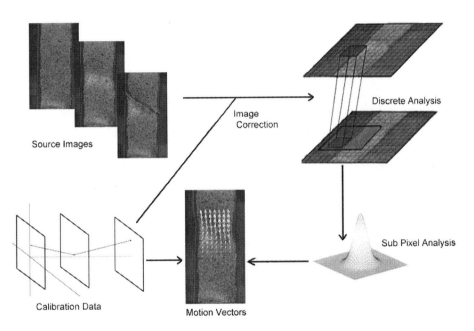

Fig. 2. General scheme of the algorithm purposed, from real images of the stream to obtained results

2   Block-Matching Technique

The proposed algorithm uses the philosophy of the standard Block-Matching procedure. Analysing the statistical similarity of the grey levels in each region (block) of the image. The purpose is to solve the correspondence problem for each region, finding in the next image the region representing the most likely displacement.

In this procedure, a region or block is formally defined as a sub-area of the image of a particular and constant size and shape, which is ideally symmetrical around a central point at which the displacement estimate will be made.

Numerically, the point (i', j') corresponds to (i,j) after applying the displacement $d(i,j)=(dx,dy)$, which may be described as described in (1).

$$(i', j') = (i + x, j + y) \tag{1}$$

This approach is based on the assumption expressed in (2).

$$I(i, j) + N = I'(i + x, j + y) \tag{2}$$

N being a noise factor following a specific statistical distribution, I the original image and I' the deformed one.

The algorithm measures movement using the statistical similarity of the grey levels in each region of the image. In this point, a similarity metric robust to natural variability processes will be used to compute most probable displacement for each point.

With this purpose, the Pearson's correlation quotient defined in (3) is used. Since it is robust in the face of noise and some intensity variations among images.

$$R\left(B_{i,j}, B'_{i',j'}\right) = \frac{\sum\left((I(i,j) - \mu_B) \times I'(i',j') - \mu_{B'}\right)}{N} \tag{3}$$

Where N is the size of the block, μ_B and $\mu_{B'}$ are the average intensity values of the two blocks compared $(B_{i,j}, B'_{i',j'})$ which are centered in the points $I(i,j)$ and $I'(i',j')$ respectively.

This similarity measurement, is limited to the pixel level due to the discrete nature of the images. Therefore the equation 3) provides a statistical measurement of the best discrete displacement.

However, the correlation value itself contains some useful information, given that the correlation values achieved in the pixels surrounding the best value will reflect one part of the displacement located between both pixels.

Thus, in order to achieve these measurements below the pixel level, a numerical adjustment to a continuous function have been made. The two-dimensional Gaussian function defined in (4) is used in this work.

$$f(x, y) = \lambda \times e^{-\left[\frac{(x - \mu_x)^2}{2 \times \sigma_x^2} + \frac{(y - \mu_y)^2}{2 \times \sigma_y^2}\right]} \tag{4}$$

Where $\lambda, \mu_x, \mu_y, \sigma_x$ and σ_y are the parametres of the function to be calculated and μ_x and μ_y, and we have used an iterative adjustment using the Levenberg-Marquardt (LM) method as expressed in (5).

$$\left(J_{5 \times n}^{T(k)} \times J_{n \times 5}^{(k)} + d^{(k)} \times I_5\right) \times Inc_{5 \times 1}^{(k+1)} = J_{n \times 5}^{T(k)} \times E_{n \times 1}^{(k)} \tag{5}$$

With $d \in N^+$ and $d \neq 0$ adjusted for each k iteration of the algorithm. $Inc_{5 \times 1}^{(k+1)}$ is the vector of increments for next iteration, $E^{(k)}$ is the current error matrix and $J^{(k)}$ is the current estimation of the Jacobean matrix of f.

3   Image Correction and Scaling

In order to obtain accurate measurements in a real scale, it must be born in mind the physical aspects which will determine the process how a point in the real space is projected on the image plane in an optical system.

With the assumption that the material can be reduced to a 2 dimensional surface, projection process can be precisely described using *Pin-hole* mathematical model [1,13].

$$\begin{bmatrix} x_{im} \\ y_{im} \\ 1 \end{bmatrix} = \begin{bmatrix} f_x & 0 & c_x \\ 0 & f_y & c_y \\ 0 & 0 & 0 \end{bmatrix}_M * \begin{bmatrix} X_{ref}/Z_{ref} \\ y_{ref}/Z_{ref} \\ 1 \end{bmatrix} \tag{6}$$

Pin-hole model, shown in Fig. 3. is described mathematically in (6), where (x_{im}, y_{im}) are point coordinates in the image and (X_{ref}, Y_{ref}) are the point coordinates in the reference plane, at a distance Z_{ref} from the lens, (f_x, f_y) represent distance from lens to camera sensor (also called focal distance or focal length) and (c_x, c_y) determines the optical centre, establishing the image coordinates where a point is projected through the exact centre of the lens O_c.

Nevertheless, the pin-hole model describes the projection of a point in a perfect optical system. In practice, the real position of the point differs from the ideal one because imperfections in camera lenses introduce some distortions in the image. These distortions follow simple mathematical formulas (7) depending on the distance to optical center and a reduced number of parameters D. Camera matrix M and distortion parameters D determine completely projection equations [12].

$$\begin{aligned} dr_x &= xk_1r^2 + xk_2r^4 \\ dr_y &= yk_1r^2 + yk_2r^4 \\ dt_x &= k_3(r^2 + 2x^2) + 2k_4xy \\ dt_y &= sk_3xy + k_4(r^2 + 2y^2) \\ D &= \begin{bmatrix} k_1 & k_2 & k_3 & k_4 \end{bmatrix} \end{aligned} \tag{7}$$

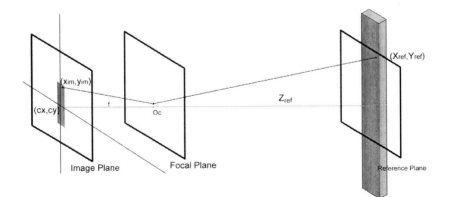

Fig. 3. *Pin-hole* model. The object is projected from the reference plane to the camera lens and then, in a second projection, to the plane were the image will be created.

To calculate the parameters of this model, a standard photometric procedure is used. This consists in analysing a calibration pattern with known geometry in order to solve the equation system.

4   Experimental Results

To evaluate the accuracy and performance of the algorithm, two sets of experiments were carried out.

In each assay the results obtained with the proposed system have been compared with those obtained by two different Block-Matching based techniques:

- The Block-Matching technique provided by the computer vision library OpenCv available at [15].
- The DaVis system, from LA Vision [16]. Currently almost the only commercial application using optical flow techniques for measuring strain in structures. Its algorithm was introduced in [17], and enhanced in [18]. It has been widely used in publications and experiments of various fields [6,10].

The same Block Size was chosen for every technique in each assay. Additionally, the DaVis algorithm used two iterations in each analysis.

4.1   Synthetic Images

The first assay was performed with synthetic images. In this experiment, two test sequences for benchmarking optical flow algorithms were used.

The first used sequence was granted by Otago University, New Zealand. [7]. And it was used in several Computer Vision works [14].

The second one is the classic Yosemite synthetic sequence is a standard test for benchmarking optical flow algorithms. It was created by Lynn Quam [8] and it was widely studied in different works [3].

This experiment was performed analyzing the motion between consecutive frames, calculating the statistics of error according to the true ground data and using to the metrics and methodologies published in [19].

Fig. 4 shows examples of the test sequences. Obtained results are summarized in Table 1

Table 1. Comparative results in the synthetic image sequences

Algorithm	Cube		Yosemite	
	Avg Error	SSD Error	Avg Error	SSD Error
This Work	0.28	0.38	0.14	0.15
DaVis [18]	0.54	0.79	0.16	0.18
Block-Matching [15]	2.14	1.00	0.45	0.45

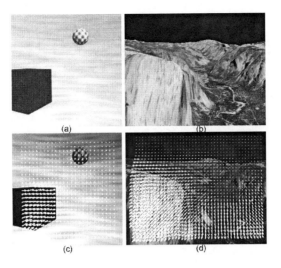

Fig. 4. Synthetic images used in the first assay. (a) Cube and sphere sequence from the sequence granted by the Otago University. (b) Yosemite sequence. (c) and (d) Example of obtained optical flow.

Analysing the results, it may be seen that the proposed technique clearly outperforms regarding the average error and the standard deviation of the error.

In the other hand, the *DaVis* technique had a good performance, and the *Block Matching* one obtained the worst results.

4.2   Strength Test

In the second experiment a real strength test was analysed. In this scenario is not possible to obtain the true ground data. Therefore, the goal will be to determine the validity and potential application of the algorithm in a real scenario and to compare de different techniques in a qualitative way.

With this purpose, the behaviour of a steel bar used in the construction industry as reinforcement of structural concrete was analysed. The material was painted to provide a visual texture and it was subject to traction forces until break-up point. This assay was performed at the Centre of Technological Innovation in Construction and Civil Engineering (CITEEC).

Analysing the obtained results (Fig. 5), it may be seen that only the proposed technique had a good performance during the entire sequence.

Although the three techniques obtain good results in the first half of the sequence, in the second half the *DaVis* technique failed to retrieve the displacement of the scene, and the *Block-Matching* technique obtained several anomalous vectors.

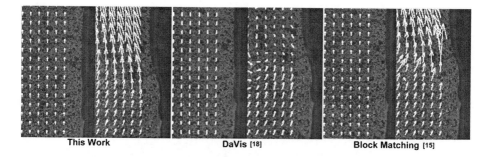

Fig. 5. Results with the different techniques in a real strength test. The same two images are showed for each technique.

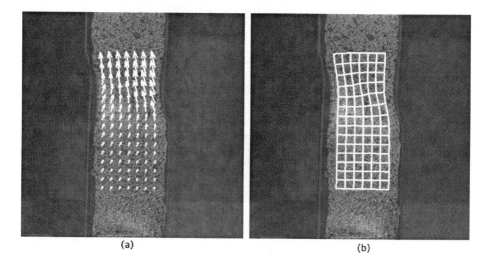

Fig. 6. Results obtained from a real strength test with a steel bar.(a) Displacement Vectors. (b) Virtual deformable grid linked to the surface of the material.

5   Conclussion

Attending to the obtained results this work has probed that strength tests can be analyzed using block matching techniques achieving more flexible measurements than traditional instrumentation.

The present paper introduces as well a new technique to analyze the tests using Block-Matching principles for measuring displacements.

The proposed technique has been compared two different algorithms obtaining the best results in the assays with synthetic images and with the images from a real strength test.

This technique has the advantage of retrieving the complete displacement field of the surface providing information about the global behaviour of the material being tested.

Fig. 6 illustrate examples of output obtained in a real scenario by the algorithm proposed in this work.

Acknowledgment

This work was partially supported by the General Directorate of Research, Development and Innovation (Dirección Xeral de Investigación, Desenvolvemento e Innovación) of the Xunta de Galicia (Ref. 08TMT005CT).

References

1. Abad, F.H., Abad, V.H., Andreu, J.F., Vives, M.O.: Application of Projective Geometry to Synthetic Cameras. In: XIV International Conference of Graphic Engineering (2002)
2. Amiaz, T., Lubetzky, E., Kiryati, N.: Coarse to over-fine optical flow estimation. Pattern Recognition 40(9), 1503–2494 (2007)
3. Austvoll, I.: A Study of the Yosemite Sequence Used as a Test Sequence for Estimation of Optical Flow. In: Kalviainen, H., Parkkinen, J., Kaarna, A. (eds.) SCIA 2005. LNCS, vol. 3540, pp. 659–668. Springer, Heidelberg (2005)
4. Chin, R.T., Harlow, C.A.: Automated visual ispection. IEEE Transactios on Pattern Analysis ad Machine Intelligence 4(6) (1982)
5. Chivers, K., Clocksin, W.: Inspection of Surface Strain in Materials Using Optical Flow. In: British Machine Vision Conference 2000, pp. 392–401 (2000)
6. Deng, Z., Richmond, M.C., Guensch, G.R., Mueller, R.P.: Study of Fish Response Using Particle Image Velocimetry and High-Speed, High-Resolution Imaging. Technical Report. PNNL-14819 (2004)
7. Graphics and Vision Research Laboratory, Department of Computer Science, University of Otago, http://www.cs.otago.ac.nz (accessed November 2010)
8. Heeger, D.: Model for the extraction of image flow. Journal of the Optical Society of America A: Optics, Image Science, and Vision 4, 1455–1471 (1987)
9. Horn, B.K.P., Schunk, B.G.: Determining Optical Flow. Artificial Intelligence 17, 185–203 (1981)
10. Kadem, L.: Particle Image Velocimetry for Fluid Dynamics Measurements. Applied Cardiovascular Fluid Dynamics (Concordia University), Particle Image Velocimetry (2008)
11. Malsch, U., Thieke, C., Huber, P.E., Bendl, R.: An enhanced block matching algorithm for fast elastic registration in adaptive radiotherapy. Phys. Med. Biol. 51, 4789–4806 (2006)
12. Manchado, A.R.: Calibracion de camaras no metricas por el metodo de las lineas rectas. Mapping 51, 74–80 (1999)
13. Martin, N., Perez, B.A., Aguilera, D.G., Lahoz, J.G.: Applied Analysis of Camera Calibration Methods for Photometric Uses. In: VII National Conference of Topography and Cartography (2004)
14. McCane, B., Novins, K., Crannitch, D., Galvin, B.: On Benchmarking Optical Flow. Computer Vision and Image Understanding 84, 126–143 (2001)

15. Open Source Computer Vision, http://opencv.willowgarage.com (accessed November 2010)
16. Particle image Velocimetry, http://www.piv.de (accessed November 2010)
17. Raffel, M., Willert, C., Kompenhans, J.: Particle Image Velocimetry, a Practical Guide. Springer, Berlin (2000)
18. Raffel, M., Willert, C., Kompenhans, J.: Particle Image Velocimetry, a Practical Guide, 2nd edn. Springer, Berlin (2007)
19. Scharstein, D., Baker, S., Lewis, J.P.: A database and evaluation methodology for Optical Flow. In: ICCV (2007)
20. Schwarz, D., Kasparek, T.: Multilevel Block Matching technique with the use of Generalized Partial Interpolation for Nonlinear Intersubject Registration of MRI Brain Images. European Journal for Biomedical Informatics 1, 90–97 (2006)

Scaling Effects in Crossmodal Improvement of Visual Perception

Isabel Gonzalo-Fonrodona[1] and Miguel A. Porras[2]

[1] Departamento de Óptica, Facultad de Ciencias Físicas, Universidad Complutense de Madrid, Ciudad Universitaria s/n. 28040-Madrid, Spain
igonzalo@fis.ucm.es
[2] Departamento de Física Aplicada, ETSIM, Universidad Politécnica de Madrid, Rios Rosas 21, 28003-Madrid, Spain
miguelangel.porras@upm.es

Abstract. Inspired in the work of J. Gonzalo [*Dinámica Cerebral*. Publ. Red Comput. Natural y Artificial, Univ. Santiago de Compostela, Spain 2010] on multisensory effects and crossmodal facilitation of perception in patients with cerebral cortical lesions, we have observed and modelled weaker but similar effects in normal subjects: Moderate and static muscular effort improves visual vernier acuity in ten tested normal subjects, and a scaling power law describes the improvement with the intensity of the effort. This suggests that the mechanism of activation of unspecific (or multispecific) neural mass in the facilitation phenomena in damaged brain is also involved in the normal brain, and that the power law reflects a basic biological scaling with the activated neural mass, inherent to natural networks.

1 Introduction

Summation effects by multisensory interactions, or crossmodal effects, in which the perception of a stimulus is influenced by the presence of another type of stimulus, are briefly referred since very early in the past [1–5]. An extensive study on multisensory interactions, including crossmodal effects improving visual perception, was made by Gonzalo [6–8] in patients with lesions in the parieto-occipital cortex. Nowadays, cross-modal effects and multisensory integration is a highly active research topic. For a review, see for example ref. [9] and references therein, in relation to cross modal influences of sound and touch on visual perception.

There are, however, few studies on the effects of muscular activity on perception. A famous case with cerebral lesion was the *Schn.* case [10], who was able to recognize objects thanks to head movements. The effect of muscular effort on improvement of perception was quantitatively studied by Gonzalo [6–8] in his patients, who presented tilted or almost inverted perception, concentric reduction of the visual field, loss of visual acuity, among many other involvements, which could be corrected by muscular effort or other stimuli. Brief comments by other authors are found on the correction of visual tilted perception by muscular activity, as closing the eyes [11, 12] moving a hand [13], or holding on to a fix

J.M. Ferrández et al. (Eds.): IWINAC 2011, Part II, LNCS 6687, pp. 267–274, 2011.

object [14, 15]. In normal man, a slight improvement of postural equilibrium was found by means of muscular effort by clenching fists [16], and in more recent works, the effect of muscular effort was studied in relation to the perception of distance [17] and near space [18].

Focusing on the work of Gonzalo [6–8], he characterized a bilateral symmetric multi-sensory (SMS) syndrome, associated to a unilateral parieto-occipital cortical lesion equidistant from the visual, tactile and auditory projection areas, causing a general functional depression. A singular feature was the high permeability to crossmodal effects. Muscular effort (as clenching fists, for example) was one of the most efficient ways to improve the perception for any of the sensory systems, the greater the muscular innervation involved the greater the facilitation obtained. It is of fundamental interest the quantitative equivalence found between stimuli as different as, for example, muscular effort and electrical excitation of the retina, to produce minimum phosphene [6]. The improvement by crossmodal effects was greater as the primary stimulus to be perceived was weaker, and as the cerebral lesion was greater, i.e., as the deficit in the cerebral excitation was greater, thus concluding that in normal man, some of the facilitation effects should be very small. The SMS syndrome was interpreted as a scale reduction of the nervous excitability of the cerebral system, which maintains, nevertheless, the same organization as in the normal brain. The permeability to crossmodal effects results from the capability of restoration of the normal scale by recruitment of unspecific (or multispecific) neural mass supplying the neural mass lost. This interpretation arises also from the model of gradients proposed by the same author [8, 19, 20] in which functional specific sensory densities are defined in gradation through the whole cortex, and that could shed some light on multisensory integration also in normal man. This gradients system is related to some works [21–23], has common aspects with the gradients recently found [24, 25] and with findings and proposals of several authors [26–33].

From Gonzalo's data for patients with SMS syndrome, we showed in previous works [20, 34, 35] that perception P improves with the intensity of the stimulus to be perceived, S, following physiological Stevens laws

$$P = pS^m, \tag{1}$$

in the range studied, and that this type of law (with different exponents) also holds for the improvement of perception with the intensity of a facilitating stimulus. Under the assumption [34, 36] that a stimulus S activates a neural mass M_{neur} according to

$$M_{\mathrm{neur}} = \alpha S^\beta, \tag{2}$$

the physiological perception law of equation (1) becomes

$$P = kM_{\mathrm{neur}}^r, \tag{3}$$

with $r = m/\beta$, i.e., is explained as a biological allometric scaling power law with the activated neural mass (unspecific or multispecific neural mass if S is a facilitating stimulus). It is remarkable that many biological observable quantities are statistically seen to follow this type of law with the mass of the organism, with power exponents close to $1/4$ (or multiples), which are supposed to result from geometrical and optimization constrains inherent to the biological networks [37–39]. In cases where the power exponent β in equation (2) is close to unity, then $r \approx m$, and Stevens laws would exhibit quarter powers, as seen in some cases of SMS syndrome [20, 34, 35].

In the present work, we report our observations supporting that the same type of facilitation phenomenon as in the SMS syndrome (but much weaker) takes place in normal man, which entails a substantial generalization. In particular, we detected improvement of visual perception by moderate static muscular effort in ten tested normal subjects, and we observed that the improvement is compatible with a power law with the intensity of the effort (facilitating stimulus). This generalization would imply that, as in the SMS syndrome, the same biological scaling power laws with the recruited mass of the neural network underly general mechanisms of perception in normals.

2   Observation and Modelling of Visual Perception Improvement by Moderate Muscular Effort in Normals

First, we address observations in order to detect variations of visual vernier acuity between two states of the observer: relaxed, and making a moderate static muscular effort, and to characterize the variation as a function of the strength of this effort. Vernier acuity, or hyperacuity, is the capability of discerning misalignment between the two halves of a broken straight line, and is chosen because it is thought to be a process of the cerebral cortex [40] requiring high cerebral excitation and integration, and hence more susceptible to suffer from crossmodal effects [6, 8]. Ten volunteers (age range $21-61$), 8 of them unaware of the purpose of the experiment, gave written informed consent to participate in this study, approved by the ethics committee of the Complutense University of Madrid. All participants had normal cognitive function, no history of brain injury, and normal or corrected-to-normal vision. The Freiburg visual acuity test FrACT [41, 42] version 3.6.3. was presented at the center of a monitor screen and calibrated for an observer at 5 m from the monitor. The vernier target consists of two vertical lines that are slightly offset to the left or to the right. Observers are required to indicate the direction of the offset in a binary decision, using only one eye in central vision, in a maximum time of 6 seconds. Vernier threshold acuity is expressed by means of the minimum angle of resolution (in arcsec), lower values then meaning better acuity. A dark card with a circular aperture covers the screen with the target in the center of the aperture, and a polarizer can be rotated to attenuate the polarized light from the screen as desired to rend the visual stimulus weak enough, as it is required. Each participant made judgement on a total of 72 trials (exceeding the minimum value recommended in the FrACT test) in relaxed state, and 72 trials under moderate muscular effort, performed in series of 24 trials each one with long enough rests between the series in order to avoid fatigue or learning by training, and performed also in random order under the relaxed condition and moderate effort. This effort is determined by holding 14 kg (for males) or 9 kg (for females) in each hand, which are seen to produce significant variations in acuity without diminishing the attention in the visual test or causing fatigue. Other type of stimuli must be avoided.

The results for the ten participants A, B,...J, sorted by increasing age, are shown in Table 1. For each participant, the average obtained for vernier acuity with muscular facilitation, V_{effort}, is lower than the averaged obtained free of muscular effort, V_{free}, i.e., acuities with muscular effort are improved with respect to those in relaxed state. The differences $\Delta V = V_{\text{free}} - V_{\text{effort}}$ are also shown in Table 1 as well as the standard deviations $\sigma_{\Delta V} = \sqrt{\sigma_{\text{free}}^2 + \sigma_{\text{effort}}^2}$, where σ_{free} and σ_{effort} are the standard deviations of V_{free} and V_{effort}. The positive differences ΔV (improvement) for all participants and their ranges of variability $\Delta V \pm \sigma_{\Delta V}$ can

Table 1. For each observer (Obs), mean vernier acuity free of muscular effort, V_{free}, and under static muscular effort, V_{effort}, together with the respective standard deviations σ_{free}, σ_{effort}; and difference $\Delta V = V_{\text{free}} - V_{\text{effort}}$, together with its standard deviation $\sigma_{\Delta V}$. All quantities are expressed in arsec.

Obs	V_{free}	σ_{free}	V_{effort}	σ_{effort}	ΔV	$\sigma_{\Delta V}$
A	10.1	2.8	8.8	6.2	1.3	6.7
B	27.2	1.2	12.9	5.5	14.3	5.6
C	14.8	2.6	10.7	5.7	4.1	6.3
D	28.7	1.1	22.4	5.5	6.2	5.9
E	6.2	2.5	3.0	2.9	3.2	3.8
F	24.6	3.9	21.5	5.0	3.1	6.3
G	23.9	10.4	12.2	0.1	11.7	10.5
H	7.6	4.1	2.5	1.9	5.1	4.6
I	24.8	9.7	18.9	5.5	5.8	11.2
J	35.8	3.4	17.6	8.2	18.2	8.9

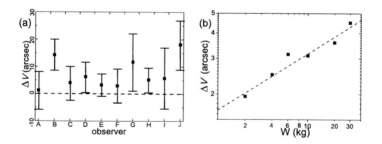

Fig. 1. (a) Squares: Differences ΔV between vernier acuities free of and under muscular effort for all participants. The bars indicate the range of variability given by $\Delta V \pm \sigma_{\Delta V}$. All ΔV are positive, and five of the bars are over zero. (b) Vernier acuity improvement for participant H, as a function of the total weight lifted, in log-log representation. Squares: average experimental values. Dashed line: power law fitting.

be better appreciated in Fig. 1(a). For five of the ten participants, the standard deviation $\sigma_{\Delta V}$ is smaller than the difference ΔV and the range of variability of ΔV does not comprises negative values. These cases can be considered as rather conclusive with regard to the improvement of vernier acuity with the static muscular effort. The oldest participant (J) presents the highest improvement of vernier acuity. This could be due not only to a worse vernier acuity (hence more improvable), but also to an effect of age. Both conditions (here met) seem to favor crossmodal effects, as could be deduced from the studies of Gonzalo [6–8] and Laurienti et al. [43].

The improvement of vernier acuity with the strength of muscular effort can be modelled with a scaling power law. For participant H, Figure 2(b) shows the improvement $\Delta V = V_{\text{free}} - V_{\text{effort}}$ as the participant holds increasing weights from $W = 1$ kg to $W = 15$ kg in each hand. Vernier acuity is seen to present large fluctuations even within the same subject under similar conditions and weight, but the average values show a good correlation of 0.92 to the power law

$$\Delta V \propto W^m \tag{4}$$

[straight line in the log-log plot of Fig. 2(b)] with exponent $m = 0.277 \pm 0.036$.

3   Conclusion

These observations establish a link between the pronounced crossmodal effects observed in patients with symmetric multisensory syndrome [6–8] and the weak crossmodal effects in normal subjects. This link underlines the similarity between the respective neural systems, being related by a scale change in the excitability of the system, as proposed in [6–8].

The improvement by muscular facilitation of visual vernier acuity in normal man can then be interpreted in the same way as was done for the improvement of perception in the symmetric multisensory syndrome, i.e., the mechanism of activation of unspecific (or multispecific) neural mass in multisensory summation effects in deficitary brain, would also apply in normal man. Thus, the found power law (4) perception-facilitating stimulus, of the type of equation (1), could be interpreted, according to expressions (2) and (3), as reflecting the universal allometric scaling power laws of biological properties as functions of the mass. In particular, the fact that the found power exponent $m = 0.277 \pm 0.036$ is compatible with the typical exponent $r = 1/4 = 0.25$ of the biological networks, suggests that the activated neural mass follows an approximately linear relationship [$\beta \simeq 1$ in equation (2)] with the intensity of the facilitating stimulus. Further research to support these ideas is in progress.

Acknowledgements

This work, inspired in the research of Justo Gonzalo [6–8], is dedicated to his memory 25 years after his death.

References

1. Urbantschitsch, V.: Über den Einfluss einer Sinneserregung auf die übrigen Sinnesempfindungen. Pflügers Archiv European J. Physiol. 42, 154–182 (1888)
2. Kravkov, S.V.: Ueber die Abhängigkeit der Sehschärfe vom Schallreiz. Arch. Ophthal. 124, 334–338 (1930)
3. Hartmann, G.W.: Changes in Visual Acuity through Simultaneous Stimulation of Other Sense Organs. J. Exp. Psychol. 16, 393–407 (1933)
4. Hartmann, G.W.: Gestalt Psychology. The Ronald Press, New York (1935)
5. London, I.D.: Research of sensory interaction in the Soviet Union. Psychol. Bull. 51, 531–568 (1954)
6. Gonzalo, J.: Dinámica cerebral. La actividad cerebral en función de las condiciones dinámicas de la excitabilidad nerviosa. Publ. Consejo Superior de Investigaciones Científicas, Inst. S. Ramón y Cajal, vol I (1945) vol II (1950) Madrid. Publ. Red Comput. Natural y Artificial, Univ. Santiago de Compostela, Spain, vols I, II (2010)
7. Gonzalo, J.: La cerebración sensorial y el desarrollo en espiral. Cruzamientos, magnificación, morfogénesis. Trab. Inst. Cajal Invest. Biol. 43, 209–260 (1951)
8. Gonzalo, J.: Las funciones cerebrales humanas según nuevos datos y bases fisiológicas: Una introducción a los estudios de Dinámica Cerebral. Trab. Inst. Cajal Invest. Biol. 44, 95–157 (1952)
9. Shams, L., Kim, R.: Crossmodal influences on visual perception. Phys. Life. Rev. 7, 269–284 (2010)
10. Goldstein, K., Gelb, A.: Psychologische Analysen hirnpathologischer Fälle auf Grund Untersuchungen Hirnverletzer. Z. Gesamte Neurol. Psychiatr. 41, 1–142 (1918)
11. Penta, P.: Due casi di visione capovolta. Il Cervello 25, 377–389 (1949)
12. Klopp, H.: Über Umgekehrt und Verkehrtsehen. Dtsch. Z. Nervenheilkd. 165, 231–260 (1951)
13. River, Y., Ben Hur, T., Steiner, I.: Reversal of vision metamorphopsia. Arch. Neurol. 53, 1362–1368 (1998)
14. Cohen, L., Belee, L., Lacoste, P., Signoret, J.L.: Illusion of visual tilt: a case. Rev. Neurol. 147, 389–391 (1991)
15. Arjona, A., Fernández-Romero, E.: Ilusión de inclinación de la imagen visual. Descripción de dos casos y revisión de la terminología. Neurología 17, 338–341 (2002)
16. Ballus, C.: La maniobra de refuerzo de J. Gonzalo y su objetivación por el test oscilométrico. Anuar. Psicología (Dep. Psicol. Univ. Barcelona) 2, 21–28 (1970)
17. Proffitt, D.R., Stefanucci, J., Banton, T., Epstein, W.: The role of effort in perceiving distance. Psychol. Sci. 14, 106–112 (2003)
18. Lourenco, S.F., Longo, M.R.: The plasticity of near space: Evidence for contraction. Cognition 112, 451–456 (2009)
19. Gonzalo, I., Gonzalo, A.: Functional gradients in cerebral dynamics: The J. Gonzalo theories of the sensorial cortex. In: Moreno-Díaz, R., Mira, J. (eds.) Brain Processes, Theories and Models. An Int. Conf. in honor of W.S. McCulloch 25 years after his death, pp. 78–87. MIT Press, Massachusetts (1996)
20. Gonzalo-Fonrodona, I.: Functional gradients through the cortex, multisensory integration and scaling laws in brain dynamics. Neurocomp. 72, 831–838 (2009)
21. Delgado, A.E.: Modelos Neurocibernéticos de Dinámica Cerebral. Ph.D.Thesis. E.T.S. de Ingenieros de Telecomunicación. Univ. Politécnica, Madrid (1978)

22. Mira, J., Delgado, A.E., Moreno-Díaz, R.: The fuzzy paradigm for knowledge representation in cerebral dynamics. Fuzzy Sets and Systems 23, 315–330 (1987)
23. Mira, J., Manjarrés, A., Ros, S., Delgado, A.E., Álvarez, J.R.: Cooperative Organization of Connectivity Patterns and Receptive Fields in the Visual Pathway: Application to Adaptive Thresholdig. In: Sandoval, F., Mira, J. (eds.) IWANN 1995. LNCS, vol. 930, pp. 15–23. Springer, Heidelberg (1995)
24. Tal, N., Amedi, A.: Multisensory visual-tactile object related network in humans: insights gained using a novel crossmodal adaptation approach. Exp. Brain Res. 198, 165–182 (2009)
25. Hertz, U., Amedi, A.: Disentangling unisensory and multisensory components in audiovisual integration using a novel multifrequency fMRI spectral analysis. NeuroImage 52, 617–632 (2010)
26. Pascual-Leone, A., Hamilton, R.: The metamodal organization of the brain. In: Casanova, C., Ptito, M. (eds.) Progress in Brain Research, vol. 134, pp. 1–19. Elsevier, Amsterdam (2001)
27. Wallace, M.T., Ramachandran, R., Stein, B.E.: A revise view of sensory cortical parcellation. Proc. Natl. Acad. Sci. USA 101, 2167–2172 (2004)
28. Amedi, A., Kriegstein, K., von Atteveldt, N.M., van Beauchamp, M.S., Naumer, M.J.: Functional imaging of human crossmodal identification and object recognition. Exp. Brain. Res. 166, 559–571 (2005)
29. Foxe, J.J., Schroeder, C.E.: The case for feedforward multisensory convergence during early cortical processing. Neuroreport. 16, 419–423 (2005)
30. Pascual-Leone, A., Amedi, A., Fregni, F., Merabet, L.: The plastic human brain cortex. Ann. Rev. Neurosci. 28, 377–401 (2005)
31. Ghazanfar, A.A., Schroeder, C.E.: Is neocortex essentially multisensory? Trends. Cogn. Sci. 10, 278–285 (2006)
32. Martuzzi, R., Murray, M.M., Michel, C.M., Thiran, J.P., Maeder, P.P., Clarke, S., Meuli, R.A.: Multisensory interactions within human primary cortices revealed by BOLD dynamics. Cereb. Cortex 17, 1672–1679 (2007)
33. Stein, B.E., Stanford, T.R., Ramachandran, R., Perrault Jr, T.J., Rowland, B.A.: Challenges in quantifying multisensory integration: alternative criteria, models, and inverse effectiveness. Exp. Brain Res. 198, 113–126 (2009)
34. Gonzalo-Fonrodona, I., Porras, M.A.: Physiological Laws of Sensory Visual System in Relation to Scaling Power Laws in Biological Neural Networks. In: Mira, J., Álvarez, J.R. (eds.) IWINAC 2007. LNCS, vol. 4527, pp. 96–102. Springer, Heidelberg (2007)
35. Gonzalo-Fonrodona, I., Porras, M.A.: Scaling power laws in the restoration of perception with increasing stimulus in deficitary natural neural network. In: Mira, J., Ferrández, J.M., Álvarez, J.R., de la Paz, F., Toledo, F.J. (eds.) IWINAC 2009. LNCS, vol. 5601, pp. 174–183. Springer, Heidelberg (2009)
36. Arthurs, O.J., Stephenson, C.M.E., Rice, K., Lupson, V.C., Spiegelhalter, D.J., Boniface, S.J., Bullmore, E.T.: Dopaminergic effects on electrophysiological and functional MRI measures of human cortical stimulus-response power laws. NeuroImage 21, 540–546 (2004)
37. Anderson, R.B.: The power law as an emergent property. Mem. Cogn. 29, 1061–1068 (2001)
38. West, G.B., Brown, J.H.: A general model for the origin of allometric scalling laws in biology. Science 276, 122–126 (1997)
39. West, G.B., Brown, J.H.: The origin of allometric scaling laws in biology from genomes to ecosystems: towards a quantitative unifying theory of biological structure and organization. J. Exp. Biol. 208, 1575–1592 (2005)

40. Westheimer, G.: Do ocular-dominance columns set spatial limits for hyperacuity processing? Vision Res. 22, 1349–1352 (1982)
41. Bach, M.: The Freiburg Visual Acuity Test – Automatic measurement of visual acuity. Optometry and Vis. Sci. 73, 49–53 (1996)
42. Bach, M.: The Freiburg Visual Acuity Test – Variability unchanged by post-hoc re-analysis. Graefe's Arch. Clin. Exp. Ophthalmol. 245, 965–971 (2007)
43. Laurienti, P.J., Burdette, J.H., Maldjian, J.A., Wallace, M.T.: Enhanced multisensory integration in older adults. Neurobiol. Aging 27, 1155–1163 (2006)

Pattern Recognition Using a Recurrent Neural Network Inspired on the Olfactory Bulb

Lucas Baggio Figueira and Antonio Carlos Roque

Laboratory of Neural Systems, Department of Physics,
FFCLRP, University of São Paulo, Ribeirão Preto, SP, Brazil

Abstract. The olfactory system is a remarkable system capable of discriminating very similar odorant mixtures. This is in part achieved via spatio-temporal activity patterns generated in mitral cells, the principal cells of the olfactory bulb, during odor presentation. In this work, we present a spiking neural network model of the olfactory bulb and evaluate its performance as a pattern recognition system with datasets taken from both artificial and real pattern databases. Our results show that the dynamic activity patterns produced in the mitral cells of the olfactory bulb model by pattern attributes presented to it have a pattern separation capability. This capability can be explored in the construction of high-performance pattern recognition systems.

1 Introduction

The main motivation for the use of artificial neural networks as pattern recognition tools comes from their obvious resemblance with brain circuits. Throughout evolution, biology has found a number of brain organizing principles which are well tuned for interaction with a complex and ever-changing environment. These principles are far from being understood in explicit form but they are gradually being uncovered by the combined efforts of experimental and theoretical neuroscientists [Laurent, 2002], [Abbott, 2008], [Harris et al., 2011], [Jin et al., 2011].

Of the different sensory modalities in the brain, one of the most interesting is the olfactory sense. Through the olfactory sense the brain can perceive complex mixtures of different chemicals as singular entities, the odors, using a relatively small number of processing stages in comparison with other senses [Laurent et al., 2001] [Lledo et al., 2005]. Information from the several odorant molecules present in an odor, after being detected by olfactory receptor neurons (ORNs), is integrated in the first olfactory processing stage – the olfactory bulb (OB) in vertebrates – in the form of a spatio-temporal pattern [Kay and Stopfer, 2006]. Essential odor characteristics like identity and intensity seem to be well represented by this dynamic pattern [Stopfer et al. 2003] [Cleland and Linster 2005].

Considering that patterns are composites of different elements, like odors are constructs of different molecules, one can think of an analogy between pattern recognition and odor identification by the olfactory bulb. We explore this analogy

J.M. Ferrández et al. (Eds.): IWINAC 2011, Part II, LNCS 6687, pp. 275–285, 2011.

in this work and present here a recurrent neural network for pattern recognition inspired on a simplified model of the vertebrate olfactory bulb.

The rest of the paper is organized as follows: Section 2 gives a brief overview of the olfactory bulb. Section 3 presents the model. Section 4 gives results of tests made with our model to recognize patterns taken from public databases. It also contains a discussion of these results. Section 5 has some concluding remarks.

2   Olfactory Bulb

The anatomy and structure of the OB is well documented [Mori et al., 1999], [Shepherd, 2004]. Figure 1 shows a simplified scheme of the OB with its main cell types and their connections. ORNs expressing the same receptor protein converge their axons to the same glomerulus in the OB. Within the glomerulus these axons make synapses with dendrites of mitral cells, which are the output neurons of the OB. There are two inhibitory cell types which make dendrodendritic synapses with mitral cells in the OB, one at the glomerular level and the other at a deeper level, the external plexiform layer (EPL). At the glomerular level, periglomerular (PG) cells whose cell bodies are close to a glomerulus receive inputs from ORNs and mitral cells associated with the glomerulus and in turn inhibit the mitral cells by synapsing onto their apical dendrites inside the glomerulus. These PG cells also send projections to nearby glomeruli, inhibiting the mitral cells associated with them. At the EPL level, mitral cells excite granule cells with their lateral dendrites, and granule cells in turn inhibit mitral cells by synapsing onto their lateral dendrites.

The OB circuitry suggests that it performs some form of contrast enhancement via lateral inhibition, both at the glomerular level mediated by PG cells and at the EPL level mediated by granule cells [Cleland and Linster 2005]. The connectivity pattern in the OB also seems to introduce a temporal component in the odor representation [Laurent et al., 2001]. It is known for decades that the OB exhibits local field potential (LFP) oscillations in response to odor stimulation [Adrian, 1950] [Freeman and Skarda, 1985]. The odor-evoked spiking patterns of mitral cells are correlated with these LFP oscillations, which suggests a functional role for the temporal structure of mitral cells spiking [Kay et al., 2008]. This points to another role for the inhibitory connections that mitral cells receive from PG and granule cells, namely to regulate the spiking times of mitral cells [Davison et al., 2003].

The possible coding mechanisms implemented by the OB neuronal connectivity pattern can only be hypothesized at the present time but evidence strongly point to a spatio-temporal form of odor representation, which involves a significant fraction of all mitral cells [Laurent, 1997], [Laurent et al., 2001], [Cleland and Linster 2005] [Lledo et al., 2005] [Shepherd et al., 2007].

3   The Model

The model consists of a simplified version of the scheme shown in Figure 1. It neglects PG cells and only contains mitral and granule cells with their reciprocal

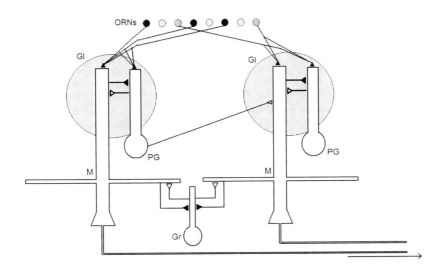

Fig. 1. Simplified scheme of the OB circuitry. Black triangles represent excitatory synapses and white triangles represent inhibitory synapses. ORNs expressing different receptors are indicated by different shades of gray. Cell types are indicated by M (mitral cell), PG (periglomerular cell) and Gr (granule cell). Gl indicates a glomerulus.

synapses. This simplification can be considered minimally sufficient to preserve the two important processing characteristics of the OB circuitry mentioned in the previous section, namely lateral inhibition among mitral cells and temporal regulation of mitral cell spiking via inhibitory inputs [Davison et al., 2003].

The mitral and granule cells were modeled as point neurons according to the Izhikevich's formalism [Izhikevich, 2007] as detailed below.

3.1   Mitral Cell Model

Recordings of mitral cell membrane potentials in slices of rat OB exhibit intrinsic bistability [Heyward et al., 2001]. Mitral cell membrane potential alternates between two resting states: the downstate, which is a hyperpolarized state at $-60mV$, and the upstate, which is a depolarized state at $-48mV$.

We captured this bistability of mitral cell membrane potential by the set of differential equations shown below:

$$40\dot{v} = (v + 55)(v + 50) + 0.5(v_d - v) - u + I$$
$$\dot{v}_d = 0.0125(v - v_d)$$
$$\dot{u} = 0.4(U(v) - u)$$
$$\text{if } v > 35mV \begin{cases} v \leftarrow -60mV \\ u \leftarrow u + 8 \end{cases} \tag{1}$$

In these equations, v is the membrane potential and u represents ionic mechanisms that control membrane potential recovery. The function $U(v) = 0$ when

$v < v_b$ and $U(v) = 20(v - v_b)$ when $v \geq v_b = -48mV$. The variable I represents all external and synaptic input currents.

Figure 2 shows the behavior of the membrane potential implemented by the above equations in response to different input currents. It shows bistability as well as spiking behavior.

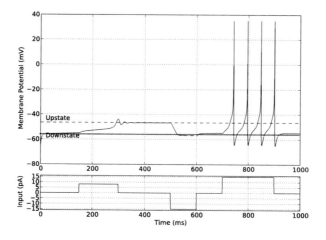

Fig. 2. Mitral cell model membrane potential (upper figure) in response to a time-varying injected current (bottom figure)

3.2   Granule Cell Model

We simplified the conductance-based granule cell model of Davison *et al.* [Davison et al., 2003] and adapted it to the Izhikevich's formalism. The original conductance-based model produces two main behaviors: spike latency and spike adaptation. These behaviors were kept by our model, which is given by the equations below.

$$100\dot{v} = 0.195(v + 65)(v + 50) + 0.5 - u + I$$
$$\dot{u} = A(I)0.7(v - u)$$
$$\text{if } v > 35mV \begin{cases} v \leftarrow -70mV \\ u \leftarrow u + 20 \end{cases} \tag{2}$$

The meanings of v and u in these equations are the same as in the mitral cell model. The current-dependent function $A(I)$ was adjusted so that our model fits well the behavior of the model of Davison *et al.* Its explicit form is $A(I) = 4.722e^{-8}I^3 - 7.883e^{-6}I^2 + 4.552e^{-4}I - 7.633e^{-3}$.

Figure 3 shows the characteristic spike latency and spike adaptation behaviors of the granule cell as captured by our model.

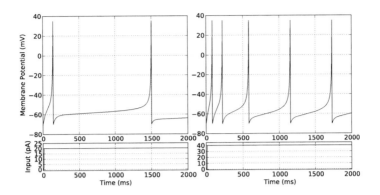

Fig. 3. Spike latency (left) and spike adaptation (right) of our simplified granule cell model in response to step currents of different amplitudes (shown below)

3.3 Connectivity

The connectivity of our model was inspired in previous biophysically detailed OB models [Davison et al., 2003] [Simões-de-Souza and Roque, 2004]. We arranged the mitral and granule cell models into two square arrays (layers) with the same size (Figure 4). The mitral cell layer has 100 cells and the granule cell layer has 400 cells, since the density of granule cells is higher than the density of mitral cells [Shepherd, 2004]. The separation between granule cells is taken as unity. The probability of synaptic connection between a given mitral cell m and a granule cell g was defined as $e^{-(D(m,g)/\lambda)^2}$, where $D(m,g)$ is the Euclidean distance between m and g and λ is a free parameter which controls the average number and range of connections made by the mitral cells. We neglected signal propagation times between mitral and granule cells, regardless of their separation distance.

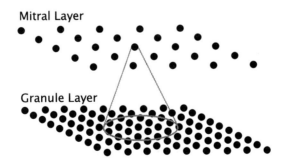

Fig. 4. Scheme of the connectivity between mitral and granule cells. The connection probability between a mitral cell and a granule cell depends solely on the distance between them.

Every time that a connection is established between a mitral and a granule cell two synapses are created. One excitatory from the mitral to the granule cell, and another inhibitory from the granule to the mitral cell. Although our cell models do not have dendrites, these two synapses are used to simulate dendrodendritic synapses.

Each synapse is modeled as an α-function with facilitation and depression modeled by the dynamic synaptic model introduced by Markran et $al.$ [Markram et al., 1998] and Tsodyks et $al.$ [Tsodyks et al., 1998]. Briefly, the magnitude of the synaptic weight μ_k for the kth spike in a train of spike intervals $\Delta_1, \Delta_2, \Delta_3, \cdots, \Delta_{k-1}$ is given by:

$$\begin{aligned}
\mu_k &= AR_k u_k \\
u_k &= U + u_{k-1}(1-U)e^{-\Delta_{k-1}/F} \\
R_k &= 1 + (R_{k-1} - u_{k-1}R_{k-1} - 1)e^{-\Delta_{k-1}/D}
\end{aligned} \tag{3}$$

In these equations, u_k and R_k are variables which measure utilization and availability, respectively, and A is an arbitrary synaptic weight parameter independent of dynamics. The initial value of the utilization parameter is $u_1 = U$ and it varies in time with the facilitation time constant F. The initial value of the availability parameter is $R_k = 1$ and it varies in time with the depression time constant D. The used values of U, F and D are, respectively, 0.5, 125 and 1200.

Each mitral cell in the model is assumed to correspond to a different glomerulus. When a given pattern $\mathbf{x}$ is presented to the system each mitral cell receives a component x_i of $\mathbf{x}$. The input patterns $\mathbf{x}$ have the same size as the mitral layer, i.e. they are 100-component vectors. We assume that each component x_i corresponds to a feature or attribute of real world patterns which the system has to recognize. This means that some sort of feature extraction pre-processing has to be done with patterns to generate vectors $\mathbf{x}$, which will be used as inputs to our model. We will not be concerned here with this pre-processing stage. For this reason we will test our model with patterns from databases which contain only pattern attributes.

The output of the network is a 100-dimensional vector $\mathbf{y(t)}$ whose components give the firing rates of the mitral cells. These were calculated by convolving the spike trains with an exponential kernel with time constant of $30ms$. In all experiments done, patterns $\mathbf{x}$ were presented to the network for $500ms$ and the output vectors $\mathbf{y(t)}$ were measured for the first $475ms$ of this interval. In some cases we used readout modules to transform the dynamic output patterns into stable states [Maass et al., 2002].

4   Results and Discussion

We evaluated how the pattern recognition capacity of our model depends on the combined values of two parameters defined in the previous section: λ and A.

The first is related with network connectivity, so that large values of λ mean large connectivity degrees between mitral and granule cells. The second on is related with the strength of dendrodendritic synapses between mitral and granule cells, so that large values of A mean large synaptic strengths.

Since mitral cells are the input and output cells of our OB model, changes in these parameters reflect the way granule cells interfere in the information flow through OB.

In our first study, for different values of λ and A we submited our model to pairs of fixed 10x10 bidimensional spatial patterns with varying spatial correlation degrees between them. This was done in the following way: we generated two identical copies of our OB model with given values of λ and A. One copy received pattern $\mathbf{u}$ as input while the other received pattern $\mathbf{v}$ as input. The patterns $\mathbf{u}$ and $\mathbf{v}$ do not come from a database. They were artificially generated by us for the first experiment and each one of their components can be considered as a feature of given patterns.

Figure 5 gives the Euclidean distance between $\mathbf{y_u(t)}$ and $\mathbf{y_v(t)}$ for different correlation degrees between $\mathbf{u}$ and $\mathbf{v}$. Each diagram in the figure corresponds to a given combination of parameters λ and A.

One can see that for the case of no coupling between mitral and granule cells (diagram (a) in Figure 5) the distances between patterns remain stable during the entire simulation time. This means that for this case the OB model does not perform any processing in the input patterns. It acts as a relay station that simply propagates what it receives as input.

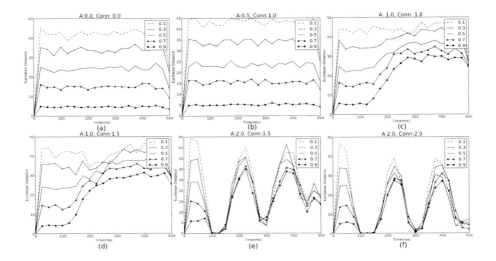

Fig. 5. Euclidean distances between OB output states $\mathbf{y_u(t)}$ and $\mathbf{y_v(t)}$ for input spatial patterns $\mathbf{u}$ and $\mathbf{v}$ with varying correlation degrees between them (given by the insets). Each diagram corresponds to a given pair of parameters λ and A (indicated above the diagram).

For a larger degree of connectivity but small synaptic strength (diagram (b) in Figure 5) the situation is basically the same: the OB model does not seem to perform any significant processing.

When the synaptic strength increases, for the same connectivity degree as in the case of last paragraph (diagram (c) in Figure 5), distances between patterns increase at about $150ms$ and remain high during the rest of simulation. This pattern separation effect is particularly significant for high correlation patterns, meaning that observations of the outputs $\mathbf{y_u(t)}$ and $\mathbf{y_v(t)}$ may be sufficient for identifying patterns $\mathbf{u}$ and $\mathbf{v}$. A similar situation occurs for diagram (d) in Figure 5, which has the same synaptic strength as in diagram (c) but has a higher connectivity degree.

As the values of parameters A and λ increase even more (diagrams (e) and (f) in Figure 5), the distances between patterns oscillate between low and high values. This behavior is a consequence of the strong inhibitory effects from granule cells upon mitral cells. These oscillations may seen undesirable for pattern recognition but a readout module synchronized with the oscillations so that it samples the output signal only at the peaks of the oscillations (like sniffing) would be capable of identifying the patterns.

Next we tested our model with a readout module. The readout module we used is a two-layer perceptron, which receives the outputs of our OB model for given datasets as inputs and offers a classification of patterns in these datasets as output. The outputs of our OB model, used as inputs for the readout, were the values of $\mathbf{y}$ at time $t = 500ms$ after the start of presentation of dataset patterns to the OB model. The same OB model was used in all tests. The two-layer perceptron used as readout module was trained with a simplified version of Parallel Perceptron [Auer et al., 2008], [Auer et al., 2002].

We tested the model with five datasets. Two of them are artificial and were created by us. They will be called A and B. The other three are Iris, BC (breast cancer) and Heart, well known datasets from the UCI database repository [Blake and Merz, 1998]. Datasets A and B have patterns belonging to two classes generated according to a Gaussian distribution. The classes in A have overlap of 20% and the classes in B have overlap of 60%.

Figure 6 shows the performances of the readout module for the five datasets used. The performances were lower for the artificial datasets than for the real ones. For dataset B, which has patterns with overlap of 60%, the readout performance was about 70%. This indicates that our OB model supports high class overlap, i.e. it can sense small differences between instances from different classes.

The readout performances for the real datasets were all above 80%. It is important to mention that the readout parameters were not fine tuned, which suggests that the OB model turns the readout classification task easier. In particular, the readout performance for the Iris dataset was 99%. It is also important to notice that the OB model can handle patterns with missing attributes, since both the BC and the Heart datasets have instances with missed attributes (especially Heart).

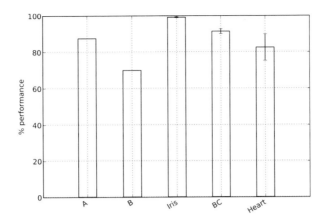

Fig. 6. Readout performances for the five datasets used to test our model (A, B, Iris, BC and Heart). The performance measure is the percent of correct classifications.

5 Conclusion

The results presented here show that a spiking neural network with architecture inspired on the vertebrate OB can be used as a high-performance pattern recognition system, especially if coupled with a readout module.

The main characteristic of the OB is the introduction of the time dimension to code input data, so that input information is represented by a spatio-temporal (dynamic) pattern instead of by just a spatial (static) pattern. This lesson from biology may be the key element for the high-performance of the system described here.

References

[Abbott, 2008] Abbott, L.F.: Theoretical neuroscience rising. Neuron 60, 489–495 (2008)

[Adrian, 1950] Adrian, E.D.: The electrical activity of the olfactory bulb. Electroencephalography and Clinical Neurophysiology 2, 377–388 (1950)

[Auer et al., 2002] Auer, P., Burgsteiner, H., Maass, W.: Reducing communication for distributed learning in neural networks. In: Dorronsoro, J.R. (ed.) ICANN 2002. LNCS, vol. 2415, p. 123. Springer, Heidelberg (2002)

[Auer et al., 2008] Auer, P., Burgsteiner, H., Maass, W.: A learning rule for very simple universal approximators consisting of a single layer of perceptrons. Neural Networks 21(5), 786–795 (2008)

[Blake and Merz, 1998] Blake, C.L., Merz, C.J.: Uci repository of machine learning databases (1998)

[Cleland and Linster 2005] Cleland, T.A., Linster, C.: Computation in the olfactory system. Chemical Senses 30, 801–813 (2005)

[Davison et al., 2003] Davison, A.P., Feng, J., Brown, D.: Dendrodendritic inhibition and simulated odor responses in a detailed olfactory bulb network model. Journal of Neurophysiology 90, 1921–1935 (2003)

[Freeman and Skarda, 1985] Freeman, W.J., Skarda, C.A.: Spatial EEG patterns, non-linear dynamics and perception: the neo-Sherringtonian view. Brain Research 357, 147–175 (1985)

[Heyward et al., 2001] Heyward, P., Ennis, M., Keller, A., Shipley, M.T.: Membrane bistability in olfactory bulb mitral cells. The Journal of Neuroscience 21(14), 5311–5320 (2001)

[Harris et al., 2011] Harris, K.D., Bartho, P., Chadderton, P., Curto, C., de la Rocha, J., Hollender, L., Itskov, V., Luczak, A., Marguet, S.L., Renart, A., Sakata, S.: How do neurons work together? Lessons from auditory cortex. Hearing Research 271, 37–53 (2011)

[Izhikevich, 2007] Izhikevich, E.M.: Dynamical Systems in Neuroscience: Dynamical Systems in Neuroscience: The Geometry of Excitability and Bursting. The MIT Press, Cambridge (2007)

[Jaeger, 2002] Jaeger, H.: Short term memory in echo state networks. GMD-Report 152, GMD - German National Research Institute for Computer Science (2002)

[Jin et al., 2011] Jin, J., Wang, Y., Swadlow, H.A., Alonso, J.M.: Population receptive fields of ON and OFF thalamic inputs to an orientation column in visual cortex. Nature Neuroscience 14, 232–240 (2011)

[Kay and Stopfer, 2006] Kay, L.M., Stopfer, M.: Information processing in the olfactory systems of insects and vertebrates. Seminars in Cell & Developmental Biology 17, 433–442 (2006)

[Kay et al., 2008] Kay, L.M., Beshel, J., Brea, J., Martin, C., Rojas-Líbano, D., Kopell, N.: Olfactory oscillations: the what, how and what for. Trends in Neuroscience 32, 207–214 (2008)

[Laurent, 1997] Laurent, G.: Olfactory processing: maps, time and codes. Current Opinion in Neurobiology 7, 547–553 (1997)

[Laurent, 2002] Laurent, G.: Olfactory network dynamics and the coding of multidimensional signals. Nature Reviews Neuroscience 3, 884–895 (2002)

[Laurent et al., 2001] Laurent, G., Stopfer, M., Friedrich, R.W., Rabinovich, M.I., Volkovskii, A., Abarbanel, H.D.I.: Odor encoding as an active, dynamical process: Experiments, computation, and theory. Annual Review of Neuroscience 24, 263–297 (2001)

[Lledo et al., 2005] Lledo, P.M., Gheusi, G., Vincent, J.D.: Information processing in the mammalian olfactory system. Physiological Reviews 85, 281–317 (2005)

[Markram et al., 1998] Markram, H., Wang, Y., Tsodyks, M.: Differential signaling via the same axon of neocortical pyramidal neurons. Proceedings of the National Academy of Sciences (USA) 95, 5323–5328 (1998)

[Maass et al., 2002] Maass, W., Natschläger, T., Markram, H.: Real-time computing without stable states: A new framework for neural computation based on perturbations. Neural Computation 14(11), 2531–2560 (2002)

[Mori et al., 1999] Mori, K., Nagao, H., Yoshihara, Y.: The olfactory bulb: coding and processing of odor molecule information. Science 286(5440), 711–715 (1999)

[Natschläger et al., 2002] Natschläger, T., Maass, W., Markram, H.: The "liquid computer": A novel strategy for real-time computing on time series. Special Issue on Foundations of Information Processing of TELEMATIK, 8 (2002)

[Shepherd, 2004] Shepherd, G.M.: The synaptic organization of the brain, 5th edn. Oxford University Press, Oxford (2004)

[Shepherd et al., 2007] Shepherd, G.M., Chen, W.R., Willhite, D., Migliore, M., Greer, C.A.: The olfactory granule cell: from classical enigma to central role in olfactory processing. Brain Res. Rev. 55(2), 373–382 (2007)

[Simões-de-Souza and Roque, 2004] Simões-de-Souza, F.M., Roque, A.C.: A biophysical model of vertebrate olfactory epithelium and bulb exhibiting gap junction dependent odor-evoked spatiotemporal patterns of activity. BioSystems 73, 25–43 (2004)

[Stopfer et al. 2003] Stopfer, M., Jayaraman, V., Laurent, G.: Intensity versus identity coding in an olfactory system. Neuron 39, 991–1004 (2003)

[Tsodyks et al., 1998] Tsodyks, M., Pawelzik, K., Markram, H.: Neural networks with dynamic synapses. Neural Computation 10(4), 821–835 (1998)

Experiments on Lattice Independent Component Analysis for Face Recognition

Ion Marqués and Manuel Graña

Computational Intelligence Group, Dept. CCIA, UPV/EHU,
Apdo. 649, 20080 San Sebastian, Spain
www.ehu.es/ccwintco

Abstract. In previous works we have proposed Lattice Independent Component Analysis (LICA) for a variety of image processing tasks. The first step of LICA is to identify strong lattice independent components from the data. The set of strong lattice independent vector are used for linear unmixing of the data, obtaining a vector of abundance coefficients. In this paper we propose to use the resulting abundance values as features for clasification, specifically for face recognition. We report results on two well known benchmark databases.

1 Introduction

Face recognition [3] is one of the most relevant applications of image analysis. It's a true challenge to build an automated system which equals human ability to recognize faces. There are many different industrial applications interested in it, most of them somehow related to security. Face recognition may consist in the authentication of a user, which a binary decision, or in the identification of a user which is a (large) multiclass problem.

Images of faces, represented as high-dimensional pixel arrays, often belong to a manifold of lower dimension. In statistical learning approaches, each image is viewed as a point (vector) in a d-dimensional space. The dimensionality of these data is too high. Therefore, the goal is to choose and apply the right statistical tool for extraction and analysis of the underlying manifold. These tools must define the embedded face space in the image space and extract the basis functions from the face space. This would permit patterns belonging to different classes to occupy disjoint and compacted regions in the feature space. Consequently, we would be able to define a line, curve, plane or hyperplane that separates faces belonging to different classes. The classical approach applied Principal Component Analysis (PCA) for feature extraction [19], other approaches use the variations of the Linear Discriminant Analysis (LDA) [11,22,21,10,13,20,14,2], or the Locality Preserving Projections (LPP) [7]. Other successful statistic tools include Bayesian networks [12], bi-dimensional regression [9], generative models [8], and ensemble-based and other boosting methods [11].

In this paper we report experimental results with a novel feature extraction method based on the notion of lattice independence: Lattice Independent Component Analysis (LICA) [4]. Lattice independent vectors are affine independent

J.M. Ferrández et al. (Eds.): IWINAC 2011, Part II, LNCS 6687, pp. 286–294, 2011.
© Springer-Verlag Berlin Heidelberg 2011

and define a convex polytope. LICA aims to find a set of such lattice independent vectors from the data whose associated convex polytope covers all or most of the data. Feature extraction then consists in the computation of the unmixing process relative to these vectors, which is equivalent to the computation of the convex coordinates relative to them. We explore the performance of this feature extraction process for face recognition over to well known benchmark databases, comparing with Principal Component Analysis (PCA) and Independent Component Analysis (ICA) applied as alternative feature extraction processes.

The paper is organized as follows: Section 2 introduces the LICA approach. Section 3 reports the experimental results. Section 4 gives our conclusions and further work directions.

2   Lattice Independent Component Analysis (LICA)

Lattice Independent Component Analysis is based on the Lattice Independence discovered when dealing with noise robustness in Morphological Associative Memories [16]. Works on finding lattice independent sources (aka endmembers) for linear unmixing started on hyperspectral image processing [6,17]. Since then, it has been also proposed for functional MRI analysis [5] among other.

Under the Linear Mixing Model (LMM) the design matrix is composed of endmembers which define a convex region covering the measured data. The linear coefficients are known as fractional abundance coefficients that give the contribution of each endmember to the observed data:

$$\mathbf{y} = \sum_{i=1}^{M} a_i \mathbf{s}_i + \mathbf{w} = \mathbf{S}\mathbf{a} + \mathbf{w}, \tag{1}$$

where $\mathbf{y}$ is the d-dimension measured vector, $\mathbf{S}$ is the $d \times M$ matrix whose columns are the d-dimension endmembers $\mathbf{s}_i, i = 1, .., M$, $\mathbf{a}$ is the M-dimension abundance vector, and $\mathbf{w}$ is the d-dimension additive observation noise vector. Under this generative model, two constraints on the abundance coefficients hold. First, to be physically meaningful, all abundance coefficients must be non-negative $a_i \geq 0, i = 1, .., M$, because the negative contribution is not possible in the physical sense. Second, to account for the entire composition, they must be fully additive $\sum_{i=1}^{M} a_i = 1$. As a side effect, there is a saturation condition $a_i \leq 1, i = 1, .., M$, because no isolate endmember can account for more than the observed material. From a geometrical point of view, these restrictions mean that we expect the endmembers in $\mathbf{S}$ to be an Affine Independent set of points, and that the convex region defined by them covers *all* the data points.

The *Lattice Independent Component Analysis* (LICA) approach assumes the LMM as expressed in equation 1. Moreover, the equivalence between Affine Independence and Strong Lattice Independence [15] is used to induce from the data the endmembers that compose the matrix $\mathbf{S}$. Briefly, LICA consists of two steps:

1. Use an Endmember Induction Algorithm (EIA) to induce from the data a set of Strongly Lattice Independent vectors. In our works we use the algorithm

---

Algorithm 1. One step of the cross-validation of LICA for face recognition

---

1. Build a training face image matrix $X_{TR} = \{\mathbf{x}_j; j = 1, \ldots, m\} \in \mathbb{R}^{N \times m}$. The testing image matrix is denoted $X_{TE} = \{\mathbf{x}_j; j = 1, \ldots, m/3\} \in \mathbb{R}^{N \times m/3}$.
2. Data preprocessing approaches:

    (a) either perform PCA over X, obtaining $T = \{\mathbf{t}_j; j = 1, \ldots, m\} \in \mathbb{R}^{m \times m}$
    (b) or directly do $T = X_{TR}$.
3. Obtain a set of k endmembers using an EIA over T: $E = \{\mathbf{e}_j; j = 1, \ldots, k\}$ from T. Varying EIA parameters will give different E matrices. The algorithm has been testing with α values ranging from 0 to 10.
4. Unmix train and test data: $Y_{TR} = E^{\#} X_{TR}^T$ and $Y_{TE} = E^{\#} X_{TE}^T$.
5. Nearest Neighbor classification: For each image vector $\mathbf{y}_j \in Y_{TE}$

    (a) calculate the Euclidean distance to each training image $\mathbf{v}_j \in Y_{TR}$.
    (b) assign the class to which $\mathbf{y}_j$ belongs as the class of the nearest $\mathbf{v}_j$.
6. Compute performance statistics: classification accuracy

---

described in [6,5]. These vectors are taken as a set of affine independent vectors that forms the matrix $\mathbf{S}$ of equation 1.
2. Apply the Full Constrained Least Squares estimation to obtain the abundance vector according to the conditions for LMM.

The advantages of this approach are (1) that we are not imposing statistical assumptions to find the sources, (2) that the algorithm is one-pass and very fast because it only uses lattice operators and addition, (3) that it is unsupervised and incremental, and (4) that it can be tuned to detect the number of endmembers by adjusting a noise-filtering related parameter. When $M \ll d$ the computation of the abundance coefficients can be interpreted as a dimension reduction transformation, or a feature extraction process.

2.1   LICA for Face Recognition

Our input is a matrix of face images in the form of column vectors. The induced SLI vectors (endmembers) are selected face images which define the convex polytope covering the data. A face image is defined as a $A_{a \times b}$ matrix composed by $a \cdot b = N$ pixels. Images are stored like row-vectors. Therefore, column-wise the dataset is denoted by $Y = \{\mathbf{y}_j; j = 1, \ldots, N\} \in \mathbb{R}^{n \times N}$, where each $\mathbf{y}_j$ is a pixel vector. Firstly, the set of SLI $X = \{\mathbf{x}_1\} \in \mathbb{R}^{n \times K}$ is initialized with the maximum norm pixel (vector) in the input dataset Y. We chose to use the maximum norm vector as it showed experimentally to be the most successful approach.

We have tested LICA over the original data and over the PCA transformation coefficients. For the PCA we retain all non-null eigenvalue eigenvectors. The maximum number of such eigenvectors is the size of the data sample, because we have much less data samples than the space dimensionality. The classification method performed was a 30 times executed 4-fold cross-validation, randomizing the folds on each iteration; and selecting by euclidean distance the nearest neighbor to decide the class. One step of the cross-validation process is specified in algorithm 1. In this algorithm $E^{\#}$ denotes the pseudo-inverse of the matrix E. Note that we compute the feature extraction process over the training data for each repetition of the data partition into train and test subsamples. When testing PCA as a feature extraction, we retain the eigenvectors with greatest eigenvalues. The algorithm for endmember induction, the EIA, used is the one in [6] which has tolerance parameter α controlling the amount of endmembers detected. In the ensuing experiments we have varied this parameter in order to obtain varying numbers of endmembers on the same data. In other words, in step 3 of algorithm 1 there is implicit an iteration varying the values of α in order to obtain diverse dimensionality reductions.

Fig. 1. An instance of the first 5 eigenfaces (PCA), independent components (ICA) and endmembers (LICA)

3   Experimental Results

The recognition task was performed over the ORL database[18] and the Yalefaces database [1]. We did not perform any image registration or spatial normalization. Neither we did perform any face detection process. Images were taken as given from the databases. On Yalefaces we tested a simple normalization consisting in extracting the mean intensity value of the image to all the pixels (to obtain a zero mean) and adding them the middle value of the gray scale interval. Tests covered dimensionality reduction up to 30 components. For ICA and PCA that was accomplished selecting the desired sources and eigenvectors, respectively. For LICA that implies varying the value of the α parameter and observing the number of endmembers detected. Graphic 5 contains the endmembers obtained depending on the α value. Graphic 4 illustrates the relation between α and hit-rate. Table 1 contains the best cross-validation results obtained for each database and feature extraction process. On the ORL database, LICA obtained better results on the original images than on the result of PCA transformation. LICA improves on ICA, with a greater dimensionality reduction. LICA best result is worse than PCA's on this database. For the Yalefaces, the ICA performs better than the other two and LICA improves over PCA. The normalization of the images introduces some improvement in ICA and LICA based approaches, but not in PCA.

For a better assessment of the algorithm's performance, we show the plots of the recognition accuracy versus the final dimension of the transformed data. These plots represent the average accuracy obtained from the cross-validation repetitions at such dimension reductions. Figure 2 shows the accuracy versus dimension reduction on the ORL database. It can be appreciated that LICA

Table 1. Face recognition results

Method	prep. data	ORL		Yalefaces original		Yalefaces normalized	
		Acc.	Dim.	Acc.	Dim.	Acc.	Dim.
PCA	-	0.94	25	0.70	25	0.70	27
ICA	PCA	0.86	30	0.76	26	0.80	27
LICA	PCA	0.87	24	0.73	10	0.76	30
LICA	-	0.91	15			0.78	30

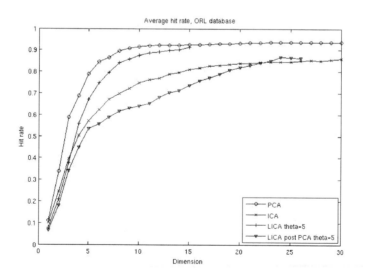

Fig. 2. Plots of accuracy versus dimension on the ORL database

features computed over the original images improve for all dimension over the ICA features and is close to the PCA features. The LICA features computed on the PCA transformed data perform worse than the other approaches for almost all dimensions tested. Figure 3 shows the accuracy versus dimension on the Yalefaces database after the normalization of the images described above. It can be appreciated that PCA performs better for some low dimension but is improved by ICA as the number of dimensions increase. The LICA features on the original images improve steadily with the dimensions approaching the performance of ICA. It's noticeable the good performance obtained over Yalefaces database, taking into account that it includes great illumination variations.

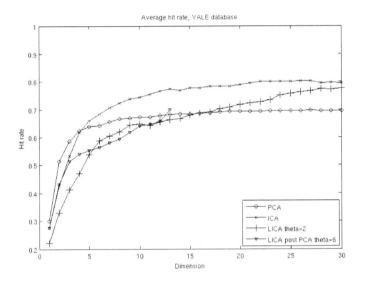

Fig. 3. Plots of accuracy versus dimension on the Yalefaces database

4   Conclusions

We have applied LICA and two well know dimension reduction procedures to feature extraction for face recognition on two well known databases. The results on both databases show that LICA features perform comparable to both linear feature extraction algorithms. This results open a new computational approach to pattern recognition, specially biometric identification problems. However there are some issues on the LICA algorithm: The uncertainty about the amount of endmembers found and therefore the high variance of recognition rates.

Future works will follow these lines:

– Confirm obtained results performing this same experiment over more complex databases like FERET.

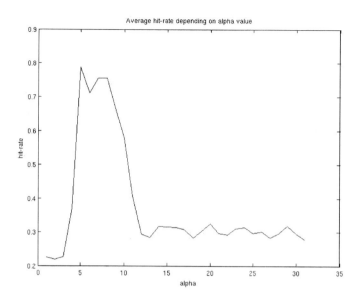

Fig. 4. Accuracy of LICA on the Yalefaces database for different α values

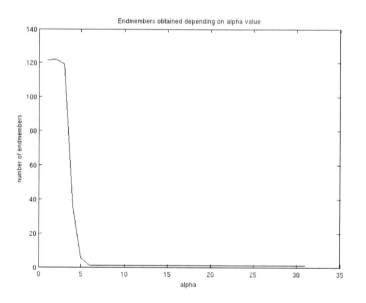

Fig. 5. Number of endmembers retrieved by LICA depending on α

- Combine the non-linear algorithm LICA with other well known statistical tools like PCA, LDA, and other state-of-the art face recognition approaches.
- Work on Lattice Theory mathematical foundations in order to apply energy function-like methods to Lattice Computing implementations that may allow more robust endmember induction.
- Test LICA's capabilities of dealing with face recognition well known problems: Illumination, pose, occlusion, etcetera.

References

1. Bellhumer, P.N., Hespanha, J., Kriegman, D.: Eigenfaces vs. fisherfaces: Recognition using class specific linear projection. IEEE Trans. Pattern Analysis and Machine Intelligence 17(7), 711–720 (1997)
2. Cai, D., He, X., Han, J.: Semi-supervised discriminant analysis. In: IEEE 11th International Conference on Computer Vision, vol. 14, pp. 1–7 (2007)
3. Chellappa, R., Sinha, P., Jonathon Phillips, P.: Face recognition by computers and humans. IEEE Computer 43(2), 46–55 (2010)
4. Graña, M., Chyzhyk, D., García-Sebastián, M., Hernández, C.: Lattice independent component analysis for functional magnetic resonance imaging. Information Sciences (2010) (in Press) Corrected Proof
5. Graña, M., Manhaes-Savio, A., García-Sebastián, M., Fernandez, E.: A lattice computing approach for on-line fmri analysis. Image and Vision Computing 28(7), 1155–1161 (2010)
6. Graña, M., Villaverde, I., Maldonado, J.O., Hernandez, C.: Two lattice computing approaches for the unsupervised segmentation of hyperspectral images. Neurocomputing 72(10-12), 2111–2120 (2009)
7. He, X., Niyogi, P.: Locality preserving projections. In: Proceedings of the Conference on Advances in Nerual Information Processing Systems (2003)
8. Heusch, G.: Sebastien Marcel. A novel statistical generative model dedicated to face recognition. Image and Vision Computing 28(1), 101–110 (2010)
9. Kare, S., Samal, A., Marx, D.: Using bidimensional regression to assess face similarity. Machine Vision and Applications 21(3), 261–274 (2008)
10. Lu, J., Plataniotis, K.N., Venetsanopoulos, A.N.: Face recognition using kernel linear discriminant algorithms. IEEE Trans. on Neural Networks 14(1), 117–126 (2003)
11. Lu, J., Plataniotis, K.N., Venetsanopoulos, A.N., Li, S.Z.: Ensemble-based discriminant learning with boosting for face recognition. IEEE Transactions on Neural Networks 17(1), 166–178 (2006)
12. Nefian, A.V.: Embedded bayesian networks for face recognition. In: Proc. of the IEEE International Conference on Multimedia and Expo., Lusanne, Switzerland, vol. 2, pp. 133–136 (August 2002)
13. Qiao, L., Chen, S., Tan, X.: Sparsity preserving discriminant analysis for single training image face recognition. Pattern Recognition Letters 31(5), 422–429 (2010)
14. Ren, C.-X., Dai, D.-Q.: Incremental learning of bidirectional principal components for face recognition. Pattern Recognition 43(1), 318–330 (2010)
15. Ritter, G.X., Urcid, G.: A lattice matrix method for hyperspectral image unmixing. Information Sciences (2010) (in Press) Corrected Proof
16. Ritter, G.X., Sussner, P., Diaz de Leon, J.L.: Morphological associative memories. IEEE Transactions on Neural Networks 9(2), 281–293 (1998)

17. Ritter, G.X., Urcid, G., Schmalz, M.S.: Autonomous single-pass endmember approximation using lattice auto-associative memories. Neurocomputing 72(10-12), 2101–2110 (2009)
18. Samaria, F.S., Harter, A.C.: Parameterisation of a stochastic model for human face identification. In: Proceedings of the Second IEEE Workshop on Applications of Computer Vision, pp. 138–142 (December 1994)
19. Turk, M., Pentland, A.: Eigenfaces for recognition. Journal of Cognitive Neuroscience 3(1), 71–86 (1991)
20. Yambor, W.S.: Analysis of PCA-based and Fisher Discriminant-Based Image Recognition Algorithms. Technical report cs-00-103, Computer Science Department, Colorado State University (July 2000)
21. Zhou, D., Tang, Z.: Kernel-based improved discriminant analysis and its application to face recognition. Soft Computing - A Fusion of Foundations, Methodologies and Applications 14(2), 103–111 (2009)
22. Zhou, S., Chellappa, R.: Multiple-exemplar discriminant analysis for face recognition. In: Proc. of the 17th International Conference on Pattern Recognition, ICPR 2004, Cambridge, UK, pp. 191–194 (August 2004)

A Hyperheuristic Approach for Dynamic Enumeration Strategy Selection in Constraint Satisfaction

Broderick Crawford[1,2], Ricardo Soto[1], Carlos Castro[2], and Eric Monfroy[2,3]

[1] Pontificia Universidad Católica de Valparaíso, Chile
FirstName.Name@ucv.cl
[2] Universidad Técnica Federico Santa María, Valparaíso, Chile
FirstName.Name@inf.utfsm.cl
[3] LINA, Université de Nantes, France
FirstName.Name@univ-nantes.fr

Abstract. In this work we show a framework for guiding the classical constraint programming resolution process. Such a framework allows one to measure the resolution process state in order to perform an "on the fly" replacement of strategies exhibiting poor performances. The replacement is performed depending on a quality rank, which is computed by means of a choice function. The choice function determines the performance of a given strategy in a given amount of time through a set of indicators and control parameters. The goal is to select promising strategies to achieve efficient resolution processes. The main novelty of our approach is that we reconfigure the search based solely on performance data gathered while solving the current problem. We report encouraging results where our combination of strategies outperforms the use of individual strategies.

Keywords: Constraint Programming, Reactive Search, Heuristic Search.

1 Introduction

Constraint Programming (CP) is a powerful programming paradigm devoted to the efficient resolution of constraint-based problems. It gathers and combines ideas from different domains, among others, from operational research, numerical analysis, artificial intelligence, and programming languages. A main idea under such a paradigm is to capture the variables and properties of the problem in a formal problem representation called Constraint Satisfaction Problem (CSP). This representation mainly consists in a sequence of variables lying in a domain, and a set of relations over such variables, namely constraints restricting the values that variables can adopt from their domains. The goal is to find a variable-value assignment that satisfy the complete set of constraints. The basic CP idea for solving CSPs is to build a tree structure holding the potential solutions by interleaving two main phases: enumeration and propagation. In the enumeration

J.M. Ferrández et al. (Eds.): IWINAC 2011, Part II, LNCS 6687, pp. 295–304, 2011.

phase, a variable and a value from its domain are chosen to create a tree branch. Those decisions are determined by the variable and value ordering heuristics, respectively. In the propagation phase, a consistency property is enforced to prune the tree, i.e., the values that do not lead to any solution are temporarily deleted from domains. In this way, the exploration does not inspect unfeasible instantiations accelerating the whole process. Jointly, the variable and value ordering heuristics constitute what is known as the enumeration strategy. Such a pair of decisions is crucial in the performance of the resolution process, where a correct selection can dramatically reduce the computational cost of finding a solution. However, it is well-known that deciding a priori the correct heuristics is quite hard, as the effects of the strategy can be unpredictable. During the last years, different efforts have been done on determining good strategies based on the information generated through the resolution process. However, deciding what information must be measured and how to redirect the search is an open investigation yet. Following this research direction, we introduce a new framework for the dynamic selection of enumeration strategies. This framework allows one to measure the resolution process state in order to replace strategies exhibiting poor performances. The replacement is carried out "on the fly" by another strategy looking more promising. Promising strategies are selected from a strategy rank which depends on a choice function. The choice function determines the performance of a given strategy in a given amount of time, and it is computed based upon a set of indicators and control parameters. The indicators attempt to reflect the real state of progress in the problem resolution, while the parameters control the relevance of the indicator within the function. Experimental results demonstrate that our framework outperforms in several cases the use of individual strategies. This paper is organized as follows. Section 2 presents the basic notions of CP and CSP solving. Section 3 presents the related work. The framework proposed including the underlying architecture and the choice function are described in Sections 4 and 5. Experimental results obtained are presented in Section 6, followed by conclusions.

2   Preliminaries

In this section, we formally describe the CSPs and we present the basic notions of CSP solving.

2.1   Constraint Satisfaction Problems

Formally, a CSP $\mathcal{P}$ is defined by a triple $\mathcal{P} = \langle \mathcal{X}, \mathcal{D}, \mathcal{C} \rangle$ where:

- $\mathcal{X}$ is a n-tuple of variables $\mathcal{X} = \langle x_1, x_2, ..., x_n \rangle$.
- $\mathcal{D}$ is a corresponding n-tuple of domains $\mathcal{D} = \langle D_1, D_2, ..., D_n \rangle$ such that $x_i \in D_i$, and D_i is a set of values, for $i = 1, ..., n$.
- $\mathcal{C}$ is a m-tuple of constraints $\mathcal{C} = \langle C_1, C_2, ..., C_m \rangle$, and a constraint C_j is defined as a subset of the Cartesian product of domains $D_1 \times ... \times D_n$, for $j = 1, ..., m$.

Then, a constraint C_j is satisfied by a tuple of values $(a_1, ..., a_n)$ if $(a_1, ..., a_n) \in C_j$. The CSP is satisfied when all its constraints are satisfied. If the CSP has a solution, we say that it is consistent; otherwise we say that it is inconsistent.

2.2 CSP Solving

As previously mentioned, the CSP search phase is commonly tackled by building a tree structure by interleaving enumeration and propagation phases. In the enumeration phase, the branches of the tree are created by selecting variables and values from their domains. In the propagation phase, a consistency level is enforced to prune the tree in order to avoid useless tree inspections. Algorithm 1 represents a general procedure for solving CSPs. The goal is to recursively generate partial solutions, backtracking when an inconsistency is detected, until a result is reached. The algorithm uses two data structures: $inst$ and D. The former holds the instantiations while the latter the set of domains. The variable k represents the current level of the tree and $success$ is a boolean variable to be set to true when a solution is found. The $instantiate$ function is responsible for building the partial solutions and assigning them into the $inst$ array. The $consistent$ function decides whether the current instantiation can be extended to a full solution; additionally, it set $success$ to true if the current instantiation is a solution. At the end, $restore$ reinitializes the k variable's domain.

Algorithm 1. A general procedure for solving CSPs

```
 1: procedure solve(k : integer, inst : array) do
 2:     while D[k] ≠ {} and not success do
 3:         a ← choose_value_from(D[k])
 4:         inst ← instantiate(inst, k, a)
 5:         if consistent(inst, k, success) then
 6:             if success then
 7:                 print solution(inst)
 8:             else
 9:                 propagate(k, D, failure)
10:                 if not failure then
11:                     l ← choose_variable()
12:                     solve(l, inst)
13:                 end if
14:             end if
15:         end if
16:     end while
17:     restore(k);
18: end procedure
```

Let us notice that the value and variable selection are performed at line 3 and 11, respectively. The $propagate$ procedure is responsible for pruning the tree by enforcing a consistency property on the constraints of the problem. The most used notion of consistency is the arc-consistency (see Definition 1).

Definition 1 (Arc Consistency). *Let C be a constraint over a set of variables $\{x_1, \ldots, x_n\}$ and let k be an integer, $k \in \{1, \ldots, n\}$. We say that C is arc consistent iff:*

$$\forall a_k \in D_k : \exists a_1 \in D_1, \ldots, \exists a_{k-1} \in D_{k-1}, \exists a_{k+1} \in D_{k+1}, \ldots, \exists a_n \in D_n \text{ such that}$$
$$(a_1, \ldots, a_n) \in C$$

A constraint is said to be arc consistent if it is arc consistent wrt. to all its variables. A CSP is said to be arc consistent if all its constraints are arc consistent.

3   Related Work

The study of enumeration strategies has been focus of research during many years mainly since their well-known impact on the solving process. From the 70's, there exist different studies concerning strategies. For instance, preliminary studies were focused on defining general criteria, e.g. the smaller domain for variable selection, and its minimum, maximum, or a random value. There is also work focused on defining strategies for specific class of problems, e.g. for job shop scheduling [7] and for configuration design [4]. We can also find research focused on determining the best strategy based in some static criterion [2,8]. However, it turns out that taking an a priori decision is quite difficult as their effect are hard to predict. During the last years there is a trend to analyze the state of progress of the solving process in order to automatically identify good-performing strategies (or a combination of them). For instance, the Adaptive Constraint Engine (ACE) [5] is a framework that learns ordering heuristics by gathering the experience from problem solving processes. The main idea is to manage a set of advisors that recommend in the form of comment a given action to perform e.g. "choose the variable with maximum domain size". Another interesting approach following a similar goal is the weighted degree heuristic [3]. The idea is to associate weights to constraints, which are incremented during arc-consistency propagation whenever this causes a domain wipeout. The sum of weights is computed for each variable involved in constraints and the variable with the largest sum is selected. The random probing method [9] adress two drawbacks of the weighted degree heuristic. On one hand, the initial choices are made without information on edge weights, and on the other, the weighted degree is biased by the path of the search. This makes the approach too sensitive to local instead of to global conditions of failure. The random probing method proposes to perform sampling during an initial gathering phase arguing that initial choices are often the most important. Preliminary results demonstrates that random probing performs better that weighted degree heuristic. In our framework, we also collect information during the search to make correct decisions. However, the aforementioned approaches are mainly focused on sampling and learning good strategies after solving a problem or a set of problems. Our main goal here is to react "on the fly" to allow an early replacement of bad-performance strategies without waiting the entire resolution process or an exhaustive analysis of a given class of problems.

4   Architecture

Our framework is supported by the architecture proposed in [6]. This architecture consists in 4 components: SOLVE, OBSERVATION, ANALYSIS and UPDATE. We reuse the three first components, while the later is completely new (see Sect. 5). The SOLVE component runs a generic CSP solving algorithm performing a depth-first search by alternating constraint propagation with enumeration phases. SOLVE has a set of basic enumeration strategies each one characterized by a priority that evolves during computation: the UPDATE component evaluate strategies and update their priorities. For each enumeration, the dynamic enumeration strategy selects the basic strategy to be used based on the attached priorities. The OBSERVATION component aims at regarding and recording some information about the current search tree, i.e., it spies the resolution process in the SOLVE component. The ANALYSIS component studies the snapshots taken by the OBSERVATION. It evaluates the different strategies, and provides indicators to the UPDATE component. The UPDATE component makes decisions using the indicators computed by ANALYSIS, it interprets the indicators, and then updates the enumeration strategies priorities.

The indicators as well variable and value ordering heuristics used in this implementation are depicted in Table 1, 2, and 3, respectively.

Table 1. Search process indicators

Name	Description
$T_n(S_j)$	Number of steps since the last time that an enumeration strategy S_j was used until step n^{th}
SB	Number of Shallow Backtracks [1]
B	Number of Backtracks
N	Number of Visited Nodes by a procedure
In1	Represents a Variation of the Maximum Depth calculated as Current Maximum Depth - Previous Maximum Depth
In2	Current Depth - Previous Depth. A positive value means that the current node is deeper than the one explored at the previous step

5   A Hyperheuristic Approach for the UPDATE Component

A hyperheuristic approach is a heuristic that operates at a higher level of abstraction than the CSP solver. The hyperheuristic is responsible for deciding which enumeration strategy to apply at each decision step during the search. It manages a portfolio of enumeration strategies and has no prior problem specific knowledge. The hyperheuristic is the core of the new UPDATE component

Table 2. Variable Ordering Heuristics

Name	Description
First (F)	The first variable of the list is selected
Minimum Remaining Values (MRV)	At each step, the variable with the smallest domain size is selected
Anti Minimum Remaining Values (AMRV)	At each step, the variable with the largest domain size is selected
Occurrence (O)	The variable with the largest number of attached constraints is selected

Table 3. Value Ordering Heuristics

Name	Description
In Domain (ID)	It starts with the smallest element and upon backtracking tries successive elements until the entire domain has been explored.
In Domain Max (IDM)	It starts the enumeration from the largest value downwards.

of the architecture. To allow the hyperheuristic operating, we define a choice function which adaptively ranks the enumeration strategies. The choice function provides guidance to the hyperheuristic by indicating which enumeration strategy should be applied next based upon the information of the search process. The choice function is defined as a weighted sum of indicators expressing the recent improvement produced by the enumeration strategy had been called.

5.1   Choice Function

The choice function attempts to capture the correspondence between the historical performance of each enumeration strategy and the decision point currently being investigated. Here, a decision point or step is every time the solver is invoked to fix a variable by enumeration. The choice function is used to rank and choose between different enumeration strategies at each step. For any enumeration strategy S_j, the choice function f in step n for S_j is defined by equation 1, where l is the number of indicators considered and α is a parameter to control the relevance of the indicator within the choice function.

$$f_n(S_j) = \sum_{i=1}^{l} \alpha_i f_{in}(S_j) \tag{1}$$

Additionaly, to control the relevance of an indicator i for an strategy S_j in a period of time, we use a popular statistical technique for producing smoothed time series called exponential smoothing. The idea is to associate, for some indicators, greater importance to recent performance by exponentially decreasing weights to older observations. In this way, recent observations give relatively more weight that older ones. The exponential smoothing is applied to the computation of $f_{in}(S_j)$, which is defined by equations 2 and 3, where x_0 is the value of the indicator i for the strategy S_j in time 1, n is a given step of the process, β is the smoothing factor, and $0 < \beta < 1$

$$f_{i1}(S_j) = x_0 \qquad (2)$$

$$f_{in}(S_j) = \beta_i x_{n-1} + (1 - \beta_i) f_{in-1}(S_j) \qquad (3)$$

Let us note that the speed at which the older observations are smoothed (dampened) depends on β. When β is close to 0, dampening is quick and when it is close to 1, dampening is slow. The general solving procedure including the choice function can be seen in Algorithm 2. Three new function calls have been included: for calculating the indicators (line 10), the choice functions (line 11), and for choosing promising strategies (line 12), that is, the ones with highest choice function[1]. They are called after constraint propagation to compute the real effects of the strategy (some indicators may be impacted by the propagation). At lines 3 and 14, the procedures for selecting variables and values have been modified to respond to the dynamic replacement of strategies. Let us notice, that the initial selection of strategy enumeration is performed randomically outside the procedure.

Algorithm 2. A procedure for solving CSPs including the new approach

---

```
 1: procedure solve(k : integer, inst : array) do
 2:     while D[k] ≠ {} and not success do
 3:         a ← choose_value_from2(D[k])
 4:         inst ← instantiate(inst, k, a)
 5:         if consistent(inst, k, success) then
 6:             if success then
 7:                 print solution(inst)
 8:             else
 9:                 propagate(k, D, failure)
10:                 calculate_indicators()
11:                 calculate_choice_functions()
12:                 choose_promising_enum_strategy()
13:                 if not failure then
14:                     l ← choose_variable2()
15:                     solve(l, inst)
16:                 end if
17:             end if
18:         end if
19:     end while
20:     restore(k);
21: end procedure
```

---

[1] When strategies have the same score, one is selected randomly.

5.2   Choice Function Tuning with a Multilevel Structure

In order to determine the most appropriate set of parameters α_i for the choice function a multilevel approach is used. The parameters are fine-tuned by a Genetic Algorithm (GA) which trains the choice function carrying out a sampling phase. Sampling occurs during an initial information gathering phase where the search is run repeatedly to a fix cutoff (i.e. until a fixed number of variables instantiated, visited nodes or backtracks). After sampling, the problem is solved with the most promising set of parameter values for the choice function. The GA evaluates and evolves different combinations of parameters, relieving the task of manual parameterization. Each member of the population encodes the parameters of a choice function. Then, these individuals are used in order to create a choice function instance. Each choice function instantiated (each chromosome) is evaluated in a sampling phase trying to solve partially the problem at hand to a fixed cutoff. As an evaluation value for the chromosome, an indicator of performance process is used (number of backtracks). After each chromosome of the population is evaluated, selection, crossover and mutation are used to breed a new population of choice functions. As noted above, the multilevel approach is used to tuning the choice function, the resulting choice function is applied to solve the CSP problem. A population size of 10 is used. The domain of parameters α_i is [-100, 100]. The crossover operator randomly selects two chromosomes from the population and mates them by randomly picking a gene and then swapping that gene and all subsequent genes between the two chromosomes. The two modified chromosomes are then added to the list of candidate chromosomes. The crossover operator uses a fixed crossover rates, this operation is performed 0.5 as many times as there are chromosomes in the population. The mutation operator runs through the genes in each of the chromosomes in the population and mutates them in statistical accordance to the given mutation rate (0.1). Mutated chromosomes are then added to the list of candidate chromosomes destined for the natural selection process.

6   Experimental Results

Our implementation has been written in the ECL^{i}PS^e Constraint Programming System version 5.10. Tests have been performed on a 2.33GHZ Intel Core2 Duo with 2GB RAM running Windows XP. The cut-off is 65535 steps for each experiment and the problem used was N-queens (NQ). Table 4 presents the results measured in terms of number of backtracks and table 5 presents the results in terms of number of visited nodes. For both evaluations, we consider 8 enumeration strategies (F+ID, AMRV + ID, MRV + ID, O + ID, F + IDM, AMRV + IDM, MRV + IDM, and O + IDM) and the dynamic approach with $\alpha = 1.0$ and $\beta = 0.6$. Results show that the dynamic approach gains very good position in the global ranking.

Table 4. Number of Backtracks solving different instances of the N-Queens problem with different strategies

Strategy	NQ n=8	NQ n=10	NQ n=12	NQ n=15	NQ n=20	NQ n=50	NQ n=75
F + ID	10	6	15	73	10026	>27406	>26979
AMRV + ID	11	12	11	808	2539	>39232	>36672
MRV + ID	10	4	16	1	11	177	818
O + ID	10	6	15	73	10026	>26405	>26323
F + IDM	10	6	15	73	10026	>27406	>26979
AMRV + IDM	11	12	11	808	2539	>39232	>36672
MRV + IDM	10	4	16	1	11	177	818
O + IDM	10	6	15	73	10026	>26405	>26323
Dynamic	4	8	4	1	4	19	593

Table 5. Number of Visited Nodes solving different instances of the N-Queens problem with different strategies

Strategy	NQ n=8	NQ n=10	NQ n=12	NQ n=15	NQ n=20	NQ n=50	NQ n=75
F + ID	24	19	43	166	23893	>65535	>65535
AMRV + ID	21	25	30	1395	4331	>65535	>65535
MRV + ID	25	16	45	17	51	591	2345
O + ID	25	19	46	169	24308	>65535	>65535
F + IDM	24	19	43	166	23893	>65535	>65535
AMRV + IDM	21	25	30	1395	4331	>65535	>65535
MRV + IDM	25	16	45	17	51	591	2345
O + IDM	25	19	46	169	24308	>65535	>65535
Dynamic	14	24	18	16	27	95	1857

7   Conclusion

In this work, we have presented a framework for the dynamic selection of enumeration strategies. Based on a set of indicators, the framework measures the resolution process state to allow the replacement of strategies exhibiting poor performances. A main element of the framework is the choice function, which is responsible for determining the quality of strategies. The choice function is calculated based upon a set of indicators and control parameters, while the adjustment of parameters is handled by a genetic algorithm. We have applied our approach to solve a benchmark of CSP, the results demonstrate that in several cases the dynamic selection outperforms the use of classic enumeration strategies. The framework introduced here is ongoing work, and we believe there is a considerable scope for future work, for instance, the addition of new combinations of enumeration strategies, analysis of the control parameters, as well as the study of new statistical methods for improving the choice function.

References

1. Barták, R., Rudová, H.: Limited assignments: A new cutoff strategy for incomplete depth-first search. In: Proceedings of the 20th ACM Symposium on Applied Computing (SAC), pp. 388–392 (2005)
2. Beck, J.C., Prosser, P., Wallace, R.J.: Trying again to fail-first. In: Faltings, B.V., Petcu, A., Fages, F., Rossi, F. (eds.) CSCLP 2004. LNCS (LNAI), vol. 3419, pp. 41–55. Springer, Heidelberg (2005)
3. Boussemart, F., Hemery, F., Lecoutre, C., Sais, L.: Boosting systematic search by weighting constraints. In: Proceedings of the 16th Eureopean Conference on Artificial Intelligence (ECAI), pp. 146–150. IOS Press, Amsterdam (2004)
4. Chenouard, R., Granvilliers, L., Sebastian, P.: Search heuristics for constraint-aided embodiment design. AI EDAM 23(2), 175–195 (2009)
5. Epstein, S.L., Freuder, E.C., Wallace, R.J., Morozov, A., Samuels, B.: The adaptive constraint engine. In: Van Hentenryck, P. (ed.) CP 2002. LNCS, vol. 2470, pp. 525–542. Springer, Heidelberg (2002)
6. Monfroy, E., Castro, C., Crawford, B.: Adaptive enumeration strategies and metabacktracks for constraint solving. In: Yakhno, T., Neuhold, E.J. (eds.) ADVIS 2006. LNCS, vol. 4243, pp. 354–363. Springer, Heidelberg (2006)
7. Sadeh, N.M., Fox, M.S.: Variable and value ordering heuristics for the job shop scheduling constraint satisfaction problem. Artif. Intell. 86(1), 1–41 (1996)
8. Sturdy, P.: Learning good variable orderings. In: Rossi, F. (ed.) CP 2003. LNCS, vol. 2833, p. 997. Springer, Heidelberg (2003)
9. Wallace, R.J., Grimes, D.: Experimental studies of variable selection strategies based on constraint weights. J. Algorithms 63(1-3), 114–129 (2008)

Genetic Algorithm for Job-Shop Scheduling with Operators

Raúl Mencía, María R. Sierra, Carlos Mencía, and Ramiro Varela

Department of Computer Science,
University of Oviedo, Campus de Viesques s/n, Gijón, 33271, Spain
http://www.di.uniovi.es/tc

Abstract. We face the job-shop scheduling problem with operators. To solve this problem we propose a new approach that combines a genetic algorithm with a new schedule generation scheme. We report results from an experimental study across conventional benchmark instances showing that our approach outperforms some current state-of-the-art methods.

1 Introduction

In this paper we propose a Genetic Algorithm (GA) to solve the job-shop scheduling problem with operators. This problem has been recently proposed in [1] and it is motivated by manufacturing processes in which part of the work is done by human operators sharing the same set of tools. The problem is formalized as a classical job-shop problem in which the processing of an operation on a given machine requires the assistance of one of the p available operators. In [1] the problem is studied and the minimal NP-hard cases are established. Also, a number of exact and approximate algorithms to cope with this problem are proposed and evaluated on a set of instances generated from that minimal relevant cases. The result of their experimental study makes it clear that instances with 3 jobs, 3 machines, 2 operators and a number of 30 operations per job are hard to solve to optimality.

We combine a genetic algorithm with a new schedule generation scheme which is inspired in the $G\&T$ algorithm proposed in [5] for the classical job-shop scheduling problem and conducted an experimental study across instances of different sizes and characteristics. The results of this study show that our approach outperforms the approximate state-of-the-art algorithms, that as far as we know are those proposed in [1].

The remaining of the paper is organized as follows. In Section 2 we define the problem and propose a disjunctive model for it. In Section 3 we describe and study the schedule generation scheme termed $OG\&T$. Section 4 presents the main components of the GA used. Section 5 is devoted to the experimental study. Finally, we summarize the main conclusions of the paper and propose some ideas for future research in Section 6.

J.M. Ferrández et al. (Eds.): IWINAC 2011, Part II, LNCS 6687, pp. 305–314, 2011.

2   Description of the Problem

Formally the job-shop scheduling problem with operators can be defined as follows. We are given a set of n jobs $\{J_1, \ldots, J_n\}$, a set of m resources or machines $\{R_1, \ldots, R_m\}$ and a set of p operators $\{O_1, \ldots, O_p\}$. Each job J_i consists of a sequence of v_i operations or tasks $(\theta_{i1}, \ldots, \theta_{iv_i})$. Each task θ_{il} has a single resource requirement $R_{\theta_{il}}$, an integer duration $p_{\theta_{il}}$ and a start time $st_{\theta_{il}}$ to be determined. A feasible schedule is a complete assignment of starting times and operators to operations that satisfies the following constraints: (i) the operations of each job are sequentially scheduled, (ii) each machine can process at most one operation at any time, (iii) no preemption is allowed and (iv) each operation is assisted by one operator and one operator cannot assist more than one operation at a time. The objective is finding a feasible schedule that minimizes the completion time of all the operations, i.e. the makespan. This problem was first defined in [1] and is denoted as $JSO(n, p)$. The significant cases of this problem are those with $p < min(n, m)$, otherwise the problem is a standard job-shop problem denoted as $J||C_{max}$.

We use the following disjunctive model for the $JSO(n, p)$. A problem instance is represented by a directed graph $G = (V, A \cup E \cup I \cup O)$. Each node in the set V represents either an actual operation, or any of the fictitious operations introduced with the purpose of giving the graph a particular structure: starting and finishing operations for each operator i, denoted O_i^{start} and O_i^{end} respectively, and the the dummy operations $start$ and end.

The arcs in A are called $conjunctive$ $arcs$ and represent precedence constraints among operations of the same job. The arcs in E are called $disjunctive$ $arcs$ and represent capacity constraints. E is partitioned into subsets E_i with $E = \cup_{\{i=1,\ldots,M\}}E_i$. E_i includes an arc (v, w) for each pair of operations requiring the resource R_i. The set O of $operator$ $arcs$ includes three types of arcs: one arc (u, v) for each pair of operations of the problem, and arcs (O_i^{start}, u) and (u, O_i^{end}) for each operator node and operation. The set I includes arcs connecting node $start$ to each node O_i^{start} and arcs connecting each node O_i^{end} to node end. The arcs are weighted with the processing time of the operation at the source node.

From this representation, building a solution can be viewed as a process of fixing disjunctive and operator arcs. A disjunctive arc between operations u and v gets fixed when one of (u, v) or (v, u) is selected and consequently the other one is discarded. An operator arc between u and v is fixed when (u, v), (v, u) or none of them is selected, and fixing the arc (O_i^{start}, u) means discarding (O_i^{start}, v) for any operation v other than u. Analogously for (u, O_i^{end}).

Therefore, a feasible schedule S is represented by an acyclic subgraph of G, of the form $G_S = (V, A \cup H \cup I \cup Q)$, where H expresses the processing order of operations on the machines and Q expresses the sequences of operations that are assisted by each operator. The makespan is the cost of a $critical$ $path$ in G_S. A critical path is a longest cost path from node $start$ to node end.

Figure 1 shows a solution graph for an instance with 3 jobs, 3 machines and 2 operators. Discontinuous arrows represent operator arcs. So, the sequences of operations assisted by operators O_1 and O_2 are $(\theta_{21}, \theta_{11}, \theta_{32}, \theta_{12}, \theta_{13})$ and

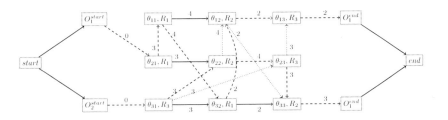

Fig. 1. A feasible schedule to a problem with 3 jobs, 3 machines and 2 operators

$(\theta_{31}, \theta_{22}, \theta_{23}, \theta_{33})$ respectively. In order to simplify the picture, only the operator arc is drawn when there are two arcs between the same pair of nodes. Continuous arrows represent conjunctive arcs and doted arrows represent disjunctive arcs; in these cases only arcs not overlapping with operator arcs are drawn. In this example, the critical path is given by the sequence $(\theta_{21}, \theta_{11}, \theta_{32}, \theta_{12}, \theta_{33})$, so the makespan is 14.

In order to simplify expressions, we define the following notation for a feasible schedule. The *head* r_v of an operation v is the cost of the longest path from node *start* to node v, i.e. it is the value of st_v. The *tail* q_v is defined so as the value $q_v + p_v$ is the cost of the longest path from v to *end*. Hence, $r_v + p_v + q_v$ is the makespan if v is in a critical path, otherwise, it is a lower bound. PM_v and SM_v denote the predecessor and successor of v respectively on the machine sequence, PJ_v and SJ_v denote the predecessor and successor operations of v respectively on the job sequence and PO_v and SO_v denote the predecessor and successor operations of v respectively on the operator of v.

A partial schedule is given by a subgraph of G where some of the disjunctive and operator arcs are not fixed yet. In such a schedule, heads and tails can be estimated as

$$r_v = \max\{ \max_{J \subseteq P(v)} \{\min_{j \in J} r_j + \sum_{j \in J} p_j\}, \tag{1}$$

$$\max_{J \subseteq PO(v)} \{\min_{j \in J} r_j + \sum_{j \in J} p_j\}, r_{PJ_v} + p_{PJ_v}\}$$

$$q_v = \max\{ \max_{J \subseteq S(v)} \{\sum_{j \in J} p_j + \min_{j \in J} q_j\}, \tag{2}$$

$$\max_{J \subseteq SO(v)} \{\sum_{j \in J} p_j + \min_{j \in J} q_j\}, p_{SJ_v} + q_{SJ_v}\}$$

with $r_{start} = q_{end} = r_{O_i^{start}} = q_{O_i^{end}} = 0$ and where $P(v)$ denotes the disjunctive predecessors of v, so as for all $w \in P(v)$, $R_w = R_v$ and the disjunctive arc (w, v) is already fixed (analogously, $S(v)$ denotes the disjunctive successors of v). $PO(v)$ denotes the operator predecessors of v, i.e $w \in PO(v)$ if it is already established that $O_w = O_v$ and w is processed before v, so as the operator arc (w, v) is fixed (analogously, $SO(v)$ are the operator successors of v).

3   Schedule Generation Schemes

In [5], the authors proposed a schedule generation scheme for the $J||C_{max}$ problem termed $G\&T$ algorithm. This algorithm has been used in a variety of settings for job-shop scheduling problems. For example, in [4] it was used to devise a greedy algorithm, in [3] and [8] it was exploited as a decoder for genetic algorithms, and in [9] it was used to define the state space for a best-first search algorithm. Unfortunately, for other scheduling problems the original $G\&T$ algorithm is not longer suitable, but it has inspired the design of similar schedule builders such as the $EG\&T$ algorithm proposed in [2] for the job-shop scheduling problems with sequence dependent setup times. In this section we propose to use a schedule generation scheme for the $JSO(n,p)$ which is also inspired in the $G\&T$ algorithm.

As it is done by the $G\&T$ algorithm, the operations are scheduled one at a time in sequential order within each job. When an operation u is scheduled, it is assigned a starting time st_u and an operator O_i, $1 \leq i \leq p$. Let SC be the set of scheduled operations at a given time and G_{SC} the partial solution graph built so far. For each operation u in SC, all operations in $P(u)$ or $PO(u)$ are scheduled as well. So, there is a path in G_{SC} from the node O_i^{start} to u through all operations in $PO(u)$ and a path from $start$ to u through all operations in $P(u)$. Let A be the set that includes the first unscheduled operation of each job that has at least one unscheduled operation, i.e.

$$A = \{u \notin SC; \nexists PJ_u \vee PJ_u \in SC\} \tag{3}$$

For each operation u in A, r_u is the starting time of u if u is selected to be scheduled next. In accordance with expression (1), r_u is greater or at least equal than both the completion time of PJ_u and the completion time of the last operation scheduled on the machine R_u. Moreover, the value of r_u also depends on the availability of operators at this time, hence r_v is not lower than the earliest time an operator is available. Let t_i, $1 \leq i \leq p$, be the time at which the operator i is available, then

$$r_u = \max\{r_{PJ_u} + p_{PJ_u}, r_v + p_v, \min_{1 \leq i \leq p} t_i\} \tag{4}$$

where v denotes the last operation scheduled having $R_v = R_u$. In general, a number of operations in A could be scheduled simultaneously at their current heads, however it is clear that not all of them could start processing at these times due to both capacity constraints and operators availability. So, a straightforward schedule generation scheme is obtained if each one of the operations in A is considered as candidate to be scheduled next.

If the selected operation is u, it is scheduled at the time $st_u = r_u$ and all the disjunctive arcs of the form (u, v), for all $v \notin P(u)$, get fixed. Operator arcs should also be fixed from the set of scheduled operations assisted by any of the operators. In principle, any operator i with $t_i \leq r_u$ can be selected for the operation u. So, if we start from the set A containing the first operation of each job and iterate until all the operations get scheduled, no matter what operation is selected in each step, we will finally have a feasible schedule, and there is a sequence of choices eventually leading to an optimal schedule.

Let us now consider the operation v^* with the earliest completion time if it is selected from the set A, i.e.

$$v^* = \arg\min\{r_u + p_u; u \in A\} \tag{5}$$

and the set A' given by the operations in A that can start before the earliest completion of v^*, i.e.

$$A' = \{u \in A; r_u < r_{v^*} + p_{v^*}\} \tag{6}$$

If we restrict the selection, in each step of the previous scheduling algorithm, to an operation in A', at least one of the schedules that may be eventually reached is optimal. The reason for this is that for every operation u in A not considered in A', the possibility of being scheduled at a time $st_u = r_u^A$ remains in the next step.

Moreover, if the number of operators available is large enough, it is not necessary to take all the operations in the set A' as candidate selections. Let $[T, C)$ be a time interval, where $C = r_{v^*} + p_{v^*}$ and $T = min\{r_u; u \in A'\}$, and the set of machines $R_{A'} = \{R_u; u \in A'\}$. If we consider the simplified situation where $r_u = T$, for all $u \in A'$ we can do the following reasoning: if, for instance, the number of machines in $R_{A'}$ is two and there are two or more operators available along $[T, C)$, then the set A' can be reduced to the operations requiring the machine R_{v^*}. In other words, we can do the same as it is done in the $G\&T$ algorithm for the classical job-shop problem. The reason for this is that after selecting an operation v requiring R_{v^*} to be scheduled, every operation $u \in A'$ requiring the other machine can still be scheduled at the same starting time as if it were scheduled before v, so as this machine may not be considered in the current step. However, if there is only one operator available along $[T, C)$ then A' may not be reduced, otherwise the operations removed from A' will no longer have the possibility of being processed at their current heads.

The reasoning above can be extended to the case where p' operators are available along $[T, C)$ and the number of machines in $R_{A'}$ is $m' \geq p'$. In this case A' can be reduced to maintain the operations of only $m' - p' + 1$ machines in order to guarantee that all the operations in A' have the opportunity to get scheduled at their heads in G_{SC}.

In general, the situation is more complex as the heads of operations in A' are distributed along the interval $[T, C)$ as well as the times at which the operators get available. Let $\tau_0 < \cdots < \tau_{p'}$ be the times given by the head of some operation in A' or the time at which some operator becomes available along the interval $[T, C)$. It is clear that $\tau_0 = T$ and $\tau_{p'} < C$. Let NO_{τ_i}, $i \leq p'$, be the number of operators available in the interval $[\tau_i, \tau_{i+1})$, with $\tau_{p'+1} = C$, R_{τ_i} the set of machines that are required before τ_{i+1}, i.e $R_{\tau_i} = \{R_u; r_u \leq \tau_i\}$ and NR_{τ_i} be the number of machines in R_{τ_i}.

We now consider the time intervals from $[\tau_{p'}, C)$ backwards to $[T, \tau_1)$. If $NO_{\tau_i} \geq NR_{\tau_i}$ then only the operations of just one of the machines in R_{τ_i} should be maintained in A' due to the interval $[\tau_i, \tau_{i+1})$. Otherwise, i.e. if $NO_{\tau_i} < NR_{\tau_i}$, then the operations from at least $NR_{\tau_i} - NO_{\tau_i} + 1$ machines must be maintained from A' in order to guarantee that any operation requiring a machine in R_{τ_i} could be eventually processed along the interval $[\tau_i, \tau_{i+1})$. In other words, in this way we guarantee that any combination of NO_{τ_i} operations requiring different machines can be processed along the interval $[\tau_i, \tau_{i+1})$, as at least one operation requiring each of the $NR_{\tau_i} - NO_{\tau_i} + 1$ machines will appear in such combination of operations.

Here, it is important the following remark. As it was pointed, in principle any operator available at time r_u is suitable for u. However, we now have to be aware of leaving available operators for the operations that are removed from A'. In order to do that we select an operator i available at the latest time t_i such that $t_i \leq r_u$. In this way, we maximize the availability of operators for the operations to be scheduled in the next steps. To be more precise, we guarantee that any operation removed from A' can be scheduled at its current head in a subsequent step.

From the interval $[\tau_{p'}, C)$ at least the operation v^* is maintained and then some new operations are added from the remaining intervals. The set of operations obtained in this way is termed B and it is clear that $|B| \leq |A'| \leq |A|$. An important property of this schedule generation scheme is that if the number of operators is large enough, in particular if $p \geq \min(n, m)$ so as $JSO(n, p)$ becomes $J||C_{max}$, it is equivalent to the G&T algorithm. So, we call this new algorithm OG&T (Operators G&T). From the reasoning above the following result can be established.

Theorem 1. *Let $\mathcal{P}$ be a $JSO(n, p)$ instance. The set $\mathcal{S}$ of schedules that can be obtained by the OG&T algorithm to $\mathcal{P}$ is dominant, i.e. $\mathcal{S}$ contains at least one optimal schedule.*

4   Genetic Algorithm for the JSO(n,p)

The GA used here is inspired in that proposed in [6]. The coding schema is based on permutations with repetition as it was proposed in [3]. A chromosome is a permutation of the set of operations that represents a tentative ordering to schedule them, each one being represented by its job number. For example, the sequence (2 1 1 3 2 3 1 2 3) is a valid chromosome for a problem with 3 jobs and 3 machines. As it was shown in [10], this encoding has a number of interesting properties for the classic job-shop scheduling problem; for example, it tends to represent orders of operations as they appear in good solutions. So, it is expected that these characteristics are to be good for the $JSO(n, p)$ as well.

For chromosome mating, the GA uses the Job-based Order Crossover (JOX) described in [3]. Given two parents, JOX selects a random subset of jobs and copies their genes to the offspring in the same positions as they are in the first parent, then the remaining genes are taken from the second parent so as they maintain their relative ordering. We clarify how JOX works in the next example. Let us consider the following two parents

Parent1 (**2** 1 1 3 **2** 3 1 **2** 3)   Parent2 (3 3 1 2 1 3 2 2 1)

If the selected subset of jobs from the first parent just includes the job 2, the generated offspring is

Offspring (**2** 3 3 1 **2** 1 3 **2** 1).

Hence, operator JOX maintains for each machine a subsequence of operations in the same order as they are in Parent1 and the remaining in the same order as they are in Parent2.

To evaluate chromosomes, the GA uses the algorithm $OG\&T$ described in the previous section, so that the non-deterministic choice is done by looking at the chromosome: the operation in B which is in the leftmost position in the chromosome sequence is selected to be scheduled next.

The remaining elements of GA are rather conventional. To create a new generation, all chromosomes from the current one are organized into couples which are mated and then mutated to produce two offsprings in accordance with the crossover and mutation probabilities respectively. Finally, tournament replacement among every couple of parents and their offsprings is done to obtain the next generation.

5   Experimental Study

The purpose of the experimental study is to assess our approach and to compare it with the state-of-the-art methods, namely the dynamic programming and the heuristic algorithms given in [1]. We have experimented across two sets of instances. The first one is that proposed in [1], all these instances have $n = 3$ and $p = 2$ and are characterized by the number of machines (m), the number of operations per job (v_{max}) and the range of processing times (p_i). A set of small instances was generated combining the values $m = 3, 5, 7$; $v_{max} = 5, 7, 10$ and $p_i = [1, 10], [1, 50], [1, 100]$ and a set of larger instances was generated with $m = 3$, combining $v_{max} = 20, 25, 30$ and $p_i = [1, 50], [1, 100], [1, 200]$; 10 instances were considered from each combination. The sets of small instances are identified by numbers from 1 to 27: set 1 corresponds to triplet $3 - 5 - 10$, the second is $3 - 5 - 50$ and so on. The large instances are identified analogously by labels from $L1$ to $L9$. There are 360 instances in all. The second set includes a number of instances from the OR-library. Firstly, small instances with $m = 5$ and different number of jobs: $LA01 - 05$ with $n = 10$, $LA06 - 10$ with $n = 15$ and $LA11 - 15$ and $FT20$ with $n = 20$. Then larger instances with 10 jobs and 10 machines: $FT10$, $LA16 - 20$, $ABZ5, 6$ and $ORB01 - 10$. For each instance, all values of p in $[1, \min(n, m)]$ are considered.

In all the experiments GA was parameterized with a population of 100 chromosomes, a number of 140 generations, crossover probability of 0,7 and mutation probability of 0,2; so as about 10000 chromosomes are evaluated in each run. The GA was run 30 times for each instance. The algorithm was coded in C++ and the target machine was Intel Core2 Duo E7500 at 2,93GHz. 4 GB RAM.

For all experiments, we report the mean relative error in percent of the best and average solutions (*Ebest* and *Eavg* respectively) reached by GA, w.r.t. the best solutions obtained with a prototype implementation of an exact method based in the A* algorithm. This algorithm was able to solve optimally all the instances taken from [1] and most of the instances $LA01 - 15$ and $FT20$. For the remaining ones it has given a suboptimal solution in a time limit of 3600 s, or after the memory of the target machine (Intel Xeon 2.26 GHz. 24 GB RAM) got exhausted. We also report the time taken in average and the Pearson coefficient of variation in percent of the solutions, denoted CV and calculated as the ratio of standard deviation and the average of the solutions obtained.

Table 1 reports the results across the first set, together with the results from the best exact and approximate algorithms given in [1], namely the dynamic programming algorithm (DP) and the heuristic algorithms (*Heur* and *Heur+*). DP was able to solve optimally all sets of instances but a number of instances from sets L4-9. In this cases, DP was stopped after 3600 s and the incumbent solution is taken as reference for calculating the gap of the heuristic algorithms. For these algorithms the gap is defined as the error in percent w.r.t. the optimal solution, averaged for all instances. For the sets L4-9, the instances that were not solved by DP are not considered. For the GA, we report *Ebest*, *Eavg*, *CV* and $T(s)$. In all cases the results are averaged for subsets of instances with the same number of operations per job v_{max}. The first conclusion we can draw from these experiments is that GA clearly outperforms the heuristic methods *Heur* and *Heur+*; not only the error is much lower for GA than is is for both heuristic methods, but also the time taken is much lower for GA, in particular for the largest instances. It is also remarkable that GA is not able to solve optimally all these instances in the time given and that the errors are quite similar for all the instances disregarding their size. Even the largest errors are obtained for instances S1-9. As the time given is really small, new experiments should be conducted with larger run times in order to establish a clear comparison with DP.

Table 2 reports the results across instances with 5 machines and 10, 15 and 20 jobs. In this case, GA is able to reach optimal solutions for many instances. Even for some instances that A* was not solve optimally, GA reached better solutions taking less time.

Table 1. Summary of results from instances with 3 jobs and 2 operators

| | | GA | | | | T.(s) | | | Gap | |
Instances	*Ebest*	*Eavg*	*CV*	T.(s)	*DP*	*Heur*	*Heur+*	*Heur*	*Heur+*
1–9	4,50	4,51	0,03	0,18	0,03	0,01	0,04	15,33	11,22
SMALL 10–18	1,55	1,88	0,12	0,18	1,29	0,04	0,47	14,81	10,98
19–27	1,07	1,28	0,23	0,26	14,93	0,47	4,86	15,31	11,12
L1-L3	1,43	2,37	0,50	0,50	654,10	9,40	91,47	17,80	12,90
LARGE L4-L6	1,82	2,97	0,58	0,64	1683,60	35,77	366,73	15,40	11,97
L7-L9	2,53	3,96	0,68	0,73	4351,00	97,00	823,97	16,50	11,57

Table 2. Summary of results from instances with 5 machines and 10, 15 and 20 jobs

#Op.	10*jobs* (T=0,5 s)			15*jobs* (T=1,5 s)			20*jobs* (T=2,3 s)		
	Ebest	*Eavg*	*CV*	*Ebest*	*Eavg*	*CV*	*Ebest*	*Eavg*	*CV*
1	0,00	0,00	0,00	0,00	0,00	0,00	0,00	0,00	0,00
2	0,00	0,00	0,00	0,00	0,00	0,00	0,00	0,00	0,00
3	0,00	0,00	0,00	0,00	0,07	0,04	-0,07	0,04	0,07
4	0,83	2,04	0,55	0,27	0,66	0,23	-0,41	0,35	0,30
5	0,24	1,19	0,61	0,00	0,00	0,00	0,52	1,12	0,24

Table 3. Summary of results from instances with 10 jobs and 10 machines

#Op.	*Ebest*	*Eavg*	*CV*	T.(s)	#Op.	*Ebest*	*Eavg*	*CV*	T.(s)
1	0,00	0,00	0,00	0,86	6	-0,25	1,83	0,94	2,27
2	0,00	0,00	0,00	1,41	7	0,81	3,13	1,10	2,23
3	-0,18	-0,05	0,07	1,70	8	1,37	3,86	1,18	2,21
4	-0,02	0,56	0,27	2,06	9	1,60	4,01	1,13	2,01
5	-0,25	0,90	0,54	2,24	10	1,22	3,76	1,19	1,80

Finally, Table 3 shows results for instances with 10 jobs and 10 machines which are hard to solve. In these experiments, GA has reached the optimal solutions for all trivial instances with 1 and 2 operators. It has improved the solutions for many instances with 3-6 operators which A^* was not able to solve to optimality. And for the hardest instances, it has not improved the solutions reached by A^*. Nevertheless, we have to be aware of the small amount of time given to GA w.r.t. that given to A^*.

From this experimental study we can conclude that the proposed GA outperforms the approximate algorithms proposed in [1] and that at difference of all methods proposed in that paper, it can be efficiently scaled to solve instances with more than 3 jobs and more than 2 operators. From our experimental study we have observed that the $JSO(n, p)$ is harder to solve than the classic job-shop scheduling problem when the number of operators available takes an intermediate value in between $n/2$ and n.

6   Conclusions

The proposed GA outperforms or at least is quite competitive with the state-of-the-art methods for the job-shop scheduling problem with operators proposed in [1]. We have used here a powerful schedule generation scheme and a simple GA implementation. Also, in the experiments we have given GA a small amount of time (no more than 3s. per run), so there is even much room to improving. As future work we plan to do an exhaustive experimental study across larger instances giving GA more time and also to improve the GA. To do that we will try to devise local search methods and combine them with GA. As it is usual, we will intend to adapt methods from the classic job-shop problem, as it was done for example in [6] for the job-shop scheduling problem with sequence-dependent setup times and in [7] for job-shop scheduling with uncertain durations.

Acknowledgements

This work has been supported by the Spanish Ministry of Science and Innovation under research project MICINN-FEDER TIN2010-20976-C02-02 and by FICYT under grant BP09105.

References

1. Agnetis, A., Flamini, M., Nicosia, G., Pacifici, A.: A job-shop problem with one additional resource type. Journal of Scheduling (2010), doi:10.1007/s10951-010-0162-4
2. Artigues, C., Lopez, P., Ayache, P.: Schedule generation schemes for the job shop problem with sequence-dependent setup times: Dominance properties and computational analysis. Annals of Operations Research 138, 21–52 (2005)
3. Bierwirth, C.: A generalized permutation approach to jobshop scheduling with genetic algorithms. OR Spectrum 17, 87–92 (1995)
4. Brucker, P., Jurisch, B., Sievers, B.: A branch and bound algorithm for the job-shop scheduling problem. Discrete Applied Mathematics 49, 107–127 (1994)
5. Giffler, B., Thompson, G.L.: Algorithms for solving production scheduling problems. Operations Research 8, 487–503 (1960)
6. González, M.A., Vela, C.R., Varela, R.: A new hybrid genetic algorithm for the job shop scheduling problem with setup times. In: Proceedings of the Eighteenth International Conference on Automated Planning and Scheduling (ICAPS 2008). AAAI Press, Sidney (2008)
7. González Rodríguez, I., Vela, C.R., Puente, J., Varela, R.: A new local search for the job shop problem with uncertain durations. In: Proceedings of the Eighteenth International Conference on Automated Planning and Scheduling (ICAPS 2008). AAAI Press, Sidney (2008)
8. Mattfeld, D.C.: Evolutionary Search and the Job Shop Investigations on Genetic Algorithms for Production Scheduling. Springer, Heidelberg (1995)
9. Sierra, M.R., Varela, R.: Pruning by dominance in best-first search for the job shop scheduling problem with total flow time. Journal of Intelligent Manufacturing 21(1), 111–119 (2010)
10. Varela, R., Serrano, D., Sierra, M.: New codification schemas for scheduling with genetic algorithms. In: Mira, J., Álvarez, J.R. (eds.) IWINAC 2005. LNCS, vol. 3562, pp. 11–20. Springer, Heidelberg (2005)

Bio-inspired System in Automatic Speaker Recognition

Lina Rosique–López[1] and Vicente Garcerán–Hernández[2]

[1] Hospital Rafael Mendez, Lorca, Murcia, España
lrosiquel@hotmail.com
[2] Universidad Politécnica de Cartagena,
Antiguo Cuartel de Antiguones (Campus de la Muralla),
Cartagena 30202, Murcia, España
vicente.garceran@upct.es

Abstract. Automatic speaker recognition determines whether a person is who he/she claims to be without the intervention of another human being, from their voiceprint. In recent years different methods of voice analysis to fulfill this objective have been developed. At the moment, the development of computational auditory models that emulate the physiology and function of the inner ear have added a new tool in the field of speaker recognition, including the Triple Resonance Nonlinear filter. This paper studies the behavior of a bio-inspired model of inner ear applied to the speakers's voice analysis. Their ability on speaker recognition tasks was statistically analyzed. We conclude, given the results obtained, that this system is an excellent tool, which shows a high rate of sensitivity and specificity in speaker recognition.

1 Introduction

Hearing is the ability to capture and interpret the sound, defined as periodic waves that cause vibrations of the molecules in the environment that is external to the individual. These vibrations are collected by the pinna, and conveyed by the ear's canal to the tympanic membrane. This, in turn, vibrates and transmits sound to the tiny bones (malleus, incus and stapes), located in the middle ear. The stapes will be responsible for transmitting sound and amplifying the vibrations to adapt the air environment to the inner ear liquid, where contain the sensory apparatus is contained. The inner ear consists of the organ of Corti and basilar membrane; this is where it transforms the sound wave to electrophysiological signals that are sent to the central nervous system, which will finally be understood. This complex system is what allows us to recognize sounds, such as the voice of an individual [1].

Automatic speaker recognition is defined as the process of recognizing, without the intervention of any human figure, who is speaking based on personal characteristics included in the voice signals. Most of the investigations carried out have the voice signal as their starting point. It is analyzed by sophisticated

J.M. Ferrández et al. (Eds.): IWINAC 2011, Part II, LNCS 6687, pp. 315–323, 2011.

algorithms that try to extract the characteristics that are assumed as individual to each voice, then calculating the degree of similarity between the different voices to decide whether this is same speaker or a different one [2].

Similarly, in recent years a series of software or computer models have been developed, whose role has been to mimic the physiology of the auditory system, by analyzing the sound waves just like human ear would. The emergence of computer systems is driving the developing of new theories that explain the operating principle of the cochlea [3]. From these theories a series of models are developed, among which are the mechanical and phenomenological, which try to mimic as much as possible the functions of the human cochlea, taking into account an element of great importance such as the functioning of basilar membrane (BM),[4].

Ludwig Hermann von Helmholtz (1821-1894) provided the first mechanical model of the cochlea based on physical and physiological variables. That is represented by a bank of harmonic oscillators tuned to different frequencies. Thus, he was the first to imagine the cochlea as a Fourier analyzer, providing a map that spatially locates each fundamental characteristic frequency. The widening of the Helmholtz model took account of the middle ear and the effect of oscillators forced by a pressure wave in the presence of a fluid [5].

Phenomenological models, in turn, seek to reproduce the response of the BM without the need to develop equations based on properties and physical and physiological variables. They are based on analog signal processing theory and/or digital. The first works that inspired the development of existing phenomenological models [6,7,8]using analog band pass filters were designed in order to reproduce the discharge of the nerve fiber auditory pattern and reflect the suppressive effects of a second tone, giving origin to BPNL models (Band Pass Nonlinear) [6,8] included Lyon, Seneff and Kates [9]. The implementation of two branches, one with Multiple Band Pass Nonlinear (MBPNL) and one with a band pass filter, improved the previous results [7,10].

Finally the current computational auditory models come, among which we cite those models that use a Dual Resonance Nonlinear (DRNL) filter of [11], development of Pfeiffer's architecture [8] and Goldstein[7,10]. The computational model of Meddis [11] was published in 1986. It describes how the beginning of sound transduction begins at the hair cells and at the auditory nerve, keeping a close resemblance to the physiology of hearing. The model consists of four stages [12]:

- Permeability of the inner hair cells, which depends on range of motion of the cilia of the inner hair cells, and ease of passage of neurotransmitter from the cell into the synaptic gap.
- Number of neurotransmitter in the inner hair cells.
- Number of neurotransmitter in the synaptic cleft.
- Probability of producing an action potential (spike).

Like previous models, DRNL filter has been used to provide the input to models of higher stages of the auditory pathway, for example, to simulate the response of the auditory nerve of the guinea pig or to reproduce psychoacoustic logs to create an equivalent bank filters to those of the human cochlea [11]. This model had some shortcomings that were solved with Triple Resonance Nonlinear (TRNL filter), which will be applied in our investigation. These filters have been investigated by Dr. López Poveda at the Department of Hearing and Psychoacoustics Computing Laboratory at the University of Salamanca, and they have been applied to complex computer programs for use; the biologically-inspired inner ear model is the result.

TRNL filter bank can reproduce the basilar membrane response of regions for which we have no experimental data, and simulate the excitement of the basilar membrane activated with complex stimuli, largely mimicking the human inner ear. The model structure consists of two stages in cascaded. The first of these is a filter that simulates the function of the middle ear and the second is a filter that simulates the response of a region on the BM. The model input signal varies over time and is assumed in pressure units (Pa). This signal is filtered in the first stage filter and its answer is expressed in units of oscillating stapes speed (m/s). In turn, this signal is used as input to the TRNL filter, whose response is interpreted in units of oscillation BM speed (m/s). TRNL filter has three parallel branches: Two of them (one linear and the other nonlinear) are the filter DRNL per se [11]. The third one is formed for a linear gain and an all-pass filter. The output signal is the sum of the responses of the three branches. The input signal is expressed as velocity of the stapes (m/s) and the output is interpreted as the oscillating speed of a specific region of the BM (m/s). The middle ear is an essential stage in the model that determines the characteristics of the TRNL filter response, especially in front of clicks.

It has been shown that, therefore, it is key to successfully simulate the temporal variation of the instantaneous frequency of the impulse response. TRNL filter reasonably reproduces the experimental response of the basilar membrane with pure tones, both in amplitude and phase for a wide range of frequencies and sound levels of stimulation, and reproduces the region of the plateau observed in the experimental response. In our work we will perform the analysis of the voice signal through Triple Resonance Nonlinear filter, which form what we called *"biologically-inspired model of inner ear"* which simulates the human inner ear function. This form of voice analysis is originally in the field of speaker recognition.

2   Methods and Materials

2.1   Speaker Database and Locutions

The aim of this work is to validate the bio-inspired model of inner ear through statistical analysis as a useful method in the automatic speaker recognition. In the preparation of programs for the speaker identification, we used the database 'Ahumada' as input, consisting of recordings of 100 male speakers [13]. For the work recordings in normal duration of about 5 ms will be used, in which the speaker says the digit "uno" (which means "one" in Spanish). The collection of fragments of a definite duration is defined by the program's ability to process the signal; this procedure is carried out by the PRAAT program (a free scientific

software program for the analysis of speech in phonetics), [14]. For each speaker
we will get two fragment of different vocal recording sessions, both saying the
digit "uno". With the help of the program PRAAT, we select voice fragments
suitable for further analysis. Has been taken into account that both, frequency
and intensity of those fragments do not have much deviation. When we made
comparisons of the two vocal fragments from the same speaker, we obtained
the so-called intra-speaker (within-speaker) comparisons, and we got a number
of 100 comparisons. When the comparison was made from a voice fragment of
a speaker with the vocal portion of the rest of the speakers of the database,
we obtained the so-called inter-speaker comparisons, yielding a total of 5,250.
Voice fragments previously obtained were processed with bio-inspired model and
by MatlLab "Duque.m" program, [15]. We obtained a graphical representation:
The cochleogram diagram. In this cochleogram the movement or stimulation that
occurs at 30 specific areas of the basilar membrane are represented. Channel V1
(100 Hz) is located at the apex or end of the basilar membrane, and it represents
low frequencies. Channel V30 (7,000 Hz) is located at the beginning or base of
the basilar membrane and it represents high frequencies, Figure 1.

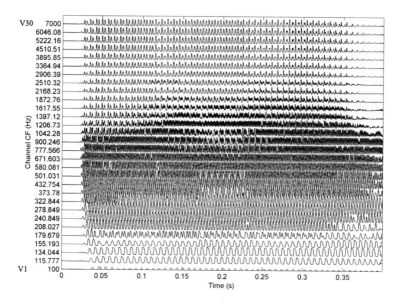

Fig. 1. Representing 30 channels of bio-inspired model, using the Spanish word "uno"
pronounced by the speaker 00

Those thirty channels we have obtained will be what we will call "variables"
throughout the study, so we get V1, V2, V3 ... V30, and a total of thirty variables.
Each of these 30 variables can be computed and therefore represented by a
numeric expression, at each instant throughout the time. The value expressed,
as indicated above, refers to movement or stimulation of the basilar membrane
in different parts of it, when stimulated with a sound wave as phonation "uno".

The number of points recorded for each variable, varies slightly, depending on the duration of phonation to be analyzed, with an average of 4,500 points per 5ms of phonation duration. The cochleogram can be represented in a table. The rows represent the thirty variables and the columns are the measure of time in a number of points for each variable. Thus we see that at each point in time the variable takes on a different value, depending on the motion to the basilar membrane at the right time, Table 1.

Table 1. Representation of ten variables in rows, and the time values in columns.(10^{-9})

	V1	V2	V3	V4	V5	V6	V7	V8	V9	V10	...	V30
1	0.296	0.296	0.296	0.296	0.296	0.296	0.296	0.296	0.296	0.296	...	
2	0.555	0.555	0.555	0.555	0.555	0.555	0.555	0.555	0.555	0.555	...	
3	0.781	0.781	0.781	0.781	0.781	0.781	0.781	0.781	0.781	0.781	...	
4	0.98	0.98	0.98	0.98	0.98	0.98	0.98	0.98	0.98	0.98	...	
5	1.15	1.15	1.15	1.15	1.15	1.15	1.15	1.15	1.15	1.15	...	
6	1.31	1.31	1.31	1.31	1.31	1.31	1.31	1.31	1.31	1.31	...	
7	1.44	1.44	1.44	1.44	1.44	1.44	1.44	1.44	1.44	1.43	...	
8	1.55	1.55	1.55	1.55	1.55	1.55	1.55	1.55	1.55	1.54	...	
9	1.64	1.64	1.64	1.64	1.64	1.64	1.64	1.64	1.63	1.62	...	
10	1.72	1.72	1.72	1.72	1.72	1.72	1.71	1.71	1.70	1.65	...	

Subsequently we carried out a data reduction method, such as factor analysis, allowing us to observe the different grouping of variables, focusing on those of greater weight or importance in speaker recognition. This process showed that the most important variables are between V20 to V30. For this reason, we summarize our results in the variables V20, V25 and V30.

3   Results

The method of Pearson distances is the statistical method used for comparisons between variables. Previously, we obtain the periodograms from each of the variables to compare. The Pearson distance expresses the relationship between pairs of variables to compare. If the result of this is close or equal to unity, we conclude that we are comparing two variables belonging to the same speaker. However, if the result of this distance is close or equal to zero, we compared variables of different speakers. Figures 2 and 3 show the outline for intra-speaker and iter-speaker comparisons.

A summary of the results obtained, following the Pearson distances, is shown in Tables 2 and 3. Table 2 shows the results obtained for intra-speaker comparisons for eleven principal variables. As we can observe, we obtain values close to unity, as was expected when the speaker compare himself. Table 3 shows some of the results of inter-speaker comparison between speaker number 001 with the rest of the database speakers. We observe how the results are of less value than the previous ones, some close to zero.

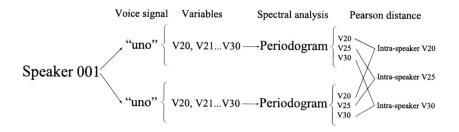

Fig. 2. Method of calculating the Pearson distance between two periodograms to obtain intra-speaker values between variables, V20, V25, V30

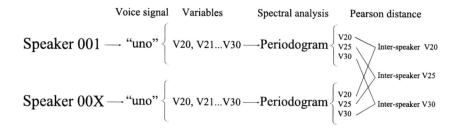

Fig. 3. Method of calculating the Pearson distance between two periodograms to obtain inter-speaker values between variables, V20, V25 and V30

Table 2. Intra-speaker distance matrix between the variables V20-V30, by a hundred speakers pronouncing the digit "uno"

	V20	V21	V22	V23	V24	V25	V26	V27	V28	V29	V30
001-001	0.973	0.965	0.976	0.978	0.983	0.980	0.979	0.971	0.972	0.973	0.971
002-002	0.821	0.832	0.836	0.853	0.784	0.816	0.825	0.824	0.830	0.833	0.834
003-003	0.915	0.894	0.891	0.861	0.881	0.861	.845	0.844	0.844	0.844	0.842
004-004	0.803	0.793	0.776	0.823	0.837	0.841	0.857	0.859	0.862	0.865	0.864
005-005	0.892	0.889	0.886	0.879	0.898	0.914	0.913	0.916	0.922	0.925	0.919
006-006	0.865	0.830	0.911	0.871	0.842	0.819	0.813	0.808	0.805	0.803	0.805
007-007	0.884	0.800	0.834	0.875	0.873	0.845	0.815	0.819	0.823	0.825	0.825
008-008	0.876	0.875	0.889	0.903	0.882	0.877	0.869	0.864	0.860	0.858	0.864
009-009	0.771	0.799	0.809	0.778	0.823	0.846	0.854	0.864	0.872	0.876	0.868
010-010	0.951	0.946	0.942	0.923	0.928	0.927	0.917	0.917	0.914	0.913	0.908
011-011	0.893	0.842	0.909	0.802	0.817	0.846	0.848	0.844	0.847	0.850	0.853
...	...	...	...	...	...	...	...	...	...	...	...
100-100	...	...	...	...	...	...	...	...	...	...	...

Table 3. Inter-speaker distance matrix, for the comparison between speaker number 001 with the rest, pronouncing the digit "uno". The variables V20-V30 are represented.

	V20	V21	V22	V23	V24	V25	V26	V27	V28	V29	V30
001-002	0.521	0.251	0.058	0.095	0.008	0.004	0.003	0.003	0.003	0.003	0.003
001-003	0.240	0.085	0.027	0.037	0.022	0.012	0.011	0.011	0.012	0.012	0.011
001-004	0.364	0.095	0.066	0.085	0.061	0.045	0.047	0.045	0.050	0.052	0.049
001-005	0.204	0.084	0.038	0.026	0.036	0.036	0.026	0.025	0.024	0.024	0.024
001-006	0.253	0.128	0.091	0.083	0.043	0.040	0.033	0.033	0.034	0.035	0.034
001-007	0.470	0.196	0.066	0.099	0.024	0.014	0.008	0.008	0.008	0.008	0.008
001-008	0.477	0.266	0.152	0.187	0.139	0.128	0.129	0.130	0.133	0.133	0.132
001-009	0.311	0.222	0.124	0.141	0.101	0.099	0.094	0.093	0.093	0.093	0.094
001-010	0.344	0.147	0.117	0.168	0.102	0.076	0.071	0.068	0.068	0.068	0.068
001-011	0.260	0.173	0.095	0.083	0.062	0.047	0.048	0.049	0.050	0.052	0.050
001-012	0.228	0.237	0.120	0.197	0.102	0.076	0.078	0.084	0.093	0.099	0.088
...	...	...	...	...	...	...	...	...	...	...	...
001-100	...	...	...	...	...	...	...	...	...	...	...

For the assessment of the recognition rate, that was obtained with our analysis of voice using the *"bio-inspired model of inner ear"*. We used the ROC curve (Receiver Operating Characteristic) applied to the results obtained from the Pearson distance. From that, we obtain the parameters of specificity and sensitivity of the new method of speaker recognition. The ROC curves provide different cutoff points, each with a different sensitivity and specificity, allowing a dichotomous classification of test values, higher or lower depending on the value chosen as the cutoff. Let us recall that the sensitivity in our model to assess is the ability of the test to classify as positive (compared intra-speaker) among all the existing intra-speaker comparisons (true positive rate, TPR = TP / (TP + FN)), being TP = True Positive and FN = False Negative. In contrast to our study, the specificity is the ability to classify as negative (inter-speaker comparison) among all the values inter-speaker, (true negative rate, TNR = TN / (TN + FP)), being TN = True Negative and FP = False Positive. Once the most appropriate cutoff point has been chosen (in our study we choose the highest sensitivity possible), we will take as values intra-speaker those that are above, and as inter-speaker values those below. Some values have been wrongly classified (FP and FN). We summarize the results obtained using the method of the ROC curves into three variables: V20, V25 and V30. Thus, for the cutoff value of variable V20 we choose 0.6575. Those values that are above this value will be taken as intra-speaker and that those that remain below will be considered as inter-speaker comparisons. With this cutoff point for classification of variable V20, we get a 98% of sensitivity and 98% of specificity. The cutoff chosen for V25 was 0.6455, which showed 100% of sensitivity and 99.9% of specificity. The cutoff chosen for V30 was 0.6710, which showed 100% of sensitivity and 99.9% of specificity.

4   Conclusions and Future Work

The results obtained are close to 100% of sensitivity and 99.9% of specificity for the variables studied, making the bio-inspired model of inner ear in an excellent way to test for speaker identification by their voices. This work provides a new tool in forensic acoustics applications, where the environment is noisy, and in pathological voices applications, proving its effectiveness in extreme conditions, as opposed to the problems that previous recognition systems have had.

Acknowledgements

Acknowledgement: This project is funded by the Spanish Ministry of Education and Science (Profit ref. CIT-3900-2005-4), Spain.

References

1. Gil Loyzaga, P.: Estructura y función de la corteza auditiva. Bases de la vía ascendente auditiva. In: Perello, E., Salesa, E. (eds.) Tratado de audiología. Barcelona: Masson: Capítulo 2 (2005)
2. Esteve, C.: Reconocimiento de locutor dependiente de texto mediante adaptación de modelos ocultos de Markov fonéticos (Final work). Universidad autónoma de Madrid (2007)
3. Martínez Rams, E.A., Cano Ortiz, S.D., Garcerán-Hernández, V.: Implantes cocleares: Desarrollo y perspectivas. Revista mexicana de ingeniería biomédica 27(1), 45–54 (2006)
4. López Poveda, E.A., Barrios, L.F., Alves Pinto, A.: Psychophisical stimates of level- dependent best- frequency shifts in the apical region of the human basilar membrane. Acoustic. Soc. Am. J. 121(6), 2654–3646 (2007)
5. Fletcher, H.: On the dinamics of the cochlea. Acoustic Soc. Am. J. 23(6), 637–646 (1951)
6. Goblick Jr., T.J., Pfeiffer, Y.: Time-domain measurements of cochlear nonlinearities using combination click stimuli. J. Acoust. Soc. Am. 46, 924–938 (1969)
7. Goldstein, J.L.: Modeling rapid waveform compression on the basilar membrane as multiple- bandpass-nonlinearity filtering. Hear. Res. 49, 39–60 (1990)
8. Pfeiffer, R.R.: A model for two-tone inhibition of single cochlearnerve Fibers. Acoust. Soc. Am. J. 48(suppl.) (1970)
9. Dos Santos Perdigao, F.M.: Modelos do sistema auditivo periférico no reconocimiento automático de fala (Thesisl). University of Coimbra (1997)
10. Goldstein, J.L.: Relations among compression, suppression, and combination tones in mechanical responses of the basilar membrane: data and MBPNL mode. Hear. Res. 89, 52–68 (1995)
11. Meddis, R., O'Mard, L.P., Lopez-Poveda, E.A.: A computational algorithm for computing nonlinear auditory frequency selectivity. J. Acoust. Soc. Am. 109, 2852–2861 (2001)
12. Christiansen, T.U.: The Meddis Inner Hair-Cell Model. In: Vestergaard, M.D. (ed.), Collection Volume, Papers from the first seminar on Auditory Models, DTU, Acoustic Technology. Technical University of Denmark, pp. 113–128 (2001); ISSN-1395-5985

13. Ortega-Garcia, J., González-Rodriguez, J., Marrero-Aguiar, V., et al.: Ahumada: A large speech corpus in Spanish for speaker identification and verification. Speech Communication 31(2-3), 255–264 (2004)
14. Boersma, P., Weenink, D.: Praat. Program for speech analysis and synthesis. The University of Amsterdam, Department of Phonetics
15. Martínez-Rams, E., Garcerán-Hernández, V.: A Speaker Recognition System based on an Auditory Model and Neural Nets: performance at different levels of Sound Pressure and of Gaussian White Noise. In: Ferrández, J.M., et al. (eds.) IWINAC 2011. LNCS, vol. 6687, pp. 157–166. Springer, Heidelberg (2011)

Independent Component Analysis:
A Low-Complexity Technique

Rubén Martín-Clemente*, Susana Hornillo-Mellado,
and José Luis Camargo-Olivares

Departamento de Teoría de la Señal y Comunicaciones,
Escuela S. de Ingenieros, Avda. de los Descubrimientos, s/n.,
41092 Seville, Spain
University of Seville, Spain
ruben@us.es, susanah@us.es, jlcamargo@yahoo.es

Abstract. This paper presents a new algorithm to solve the Independent Component Analysis (ICA) problem that has a very low computational complexity. The most remarkable feature of the proposed algorithm is that it does not need to compute higher-order statistics (HOS). In fact, the algorithm is based on trying to guess the sign of the independent components, after which it approximates the rest of the values.

Keywords: Independent Component Analysis, Blind Signal Separation.

1 Introduction

Independent Component Analysis (ICA) is a method of finding latent structure in high dimensional data. In ICA, we express a set of multidimensional observations as a linear combination of unknown latent variables. These underlying latent variables are called sources or independent components, and they are assumed to be statistically independent of each other. The mathematical ICA model, which is assumed to be consistent with experimental data, in matrix form is written as:

$$\mathbf{x}(n) = \mathbf{A}\,\mathbf{s}(n) \tag{1}$$

where $n \in \mathbb{Z}$ is the discrete time, $\mathbf{x}(n)$ is the $N \times 1$ vector of observed signals, $\mathbf{A} \in \mathbb{R}^{N \times N}$ is an invertible unknown matrix, and $\mathbf{s}(n)$ is an $N \times 1$ vector of the sources, where this source vector $\mathbf{s}(n) = \langle s_1(n), \ldots, s_N(n) \rangle'$ is made of unknown mutually independent elements. The goal of ICA techniques is to estimate the components $s_i(n)$ of $\mathbf{s}(n)$ from the sole observation of $\mathbf{x}(n)$.

ICA has been applied to problems in fields as diverse as speech processing, brain imaging, electrical brain signals (e.g., EEG signals), telecommunications, and stock market prediction [1,2,3,4]. Specially, it has been suggested that ICA mimics the biological processing in the simple cells of the mammalian primary

---

* Corresponding author.

J.M. Ferrández et al. (Eds.): IWINAC 2011, Part II, LNCS 6687, pp. 324–332, 2011.
© Springer-Verlag Berlin Heidelberg 2011

visual cortex [1,2]. The similarity of the ICA latent variables to optimal responses measured for neurons in the primary visual cortex using single-cell recordings suggest that these neurons perform some form of redundancy reduction, i.e., a representation with independent components, as proposed by Barlow (1989). It seems likely that information processing strategies successful in the primary visual cortex would also be useful in higher visual processing, and indeed in processing of other sensory signals; thus it seems probable that ICA or related methods could be applied in modeling these functions as well.

Most algorithms solve the ICA problem by evaluating higher-order statistics (HOS), e.g. by estimating fourth-order cumulants. However, in general, very large data records are required to obtain accurate estimates of HOS and the computational burden of HOS-based algorithms is often severe. These difficulties severely limit the practical application of ICA to realistic problems.

In this paper we propose a low-cost algorithm to find the solutions to ICA. The most remarkable feature is that it does not require to compute HOS. The paper is organized as follows: in the following Subsection, a pre-processing known as 'whitening' is reviewed. In Section II we present our main results. Section III illustrates the performance of the method via experiments. Finally, Section IV is devoted to the conclusions.

1.1 Preprocessing

Given an ensemble of T data points $\mathbf{x}(1), \ldots, \mathbf{x}(T)$ of zero sample mean, i.e.,

$$\frac{1}{T} \sum_{n=1}^{T} \mathbf{x}(n) = \mathbf{0},$$

ICA is typically carried in two-step. First, whitening yields a set of signals

$$\mathbf{z}(n) = \mathbf{C}_x^{-1/2} \mathbf{x}(n), \quad n = 1, \ldots, T \tag{2}$$

where $\mathbf{C}_x^{1/2}$ is any square root of the sample covariance matrix

$$\mathbf{C}_x = \frac{1}{T} \sum_{n=1}^{T} \mathbf{x}(n) \mathbf{x}(n)'$$

It holds that the components of $\mathbf{z}(n)$ are of *zero mean, unit variance,* and are also *uncorrelated* among themselves, i.e.

$$\frac{1}{T} \sum_{n=1}^{T} \mathbf{z}(n) = \mathbf{0} \tag{3}$$

$$\mathbf{C}_z = \frac{1}{T} \sum_{n=1}^{T} \mathbf{z}(n) \mathbf{z}(n)' = \mathbf{I} \tag{4}$$

As a consequence, the source and whitened vectors are related through an orthogonal transformation:

$$\mathbf{z}(n) = \mathbf{Q}\,\mathbf{s}(n) \tag{5}$$

The ICA problem thus reduces to the computation of the orthogonal matrix $\mathbf{Q}$, which is accomplished in the second step.

2   Proposed Approach

The proposed algorithm is as follows:

Algorithm 1. Estimation of a single independent component

Data: $\mathbf{z}$(n)
Result: $\hat{s}(n)$
    // $\hat{s}(n)$ is the estimate of one independent component
1   initialize $\hat{s}(n)$ at random;
2   **repeat**
3      $\mathbf{w} \leftarrow \mathbf{0}$;   // $\mathbf{w} \in \mathbb{R}^N$ is the extracting vector
       // Main loop
4   **for** $n \leftarrow 1$ **to** T **do**
5   **if** $\hat{s}(n) \geq 0$ **then**
6            $\mathbf{w} \leftarrow \mathbf{w} + \mathbf{z}(n)$;
7   **end**
8   **end**
       // End of main loop
9      $\mathbf{w} \leftarrow \dfrac{\mathbf{w}}{\|\mathbf{w}\|}$;   // normalize $\mathbf{w}$ to have unit-norm
10     $\hat{s}(n) \leftarrow \mathbf{w}'\,\mathbf{z}(n)$;   // calculate $\hat{s}(n)$ for all n
11 **until** *until convergence*;

The algorithm has a very low computational complexity since, unlike the usual ICA techniques, it does not need to compute higher-order statistics. *Note that this is the most interesting feature of the proposed method.* To estimate more than one independent component, we can use the procedure described in Chapter 4 of book [2]. Basically, we remove $\hat{s}(n)$ from $\mathbf{z}(n)$ by $\hat{\mathbf{z}}(n) = \mathbf{z}(n) - \mathbf{w}\,\hat{s}(n)$. Then we reduce the dimensionality of the data in one unit and apply whitening again. The algorithm is repeated until all the independent components are recovered.

2.1   Justification of the Proposed Method

To fix ideas, and without any loss of generality, consider for the moment the problem of estimating the first independent component $s_1(n)$. Let

$$\mathbb{S} = \{n \mid s_1(n) \geq 0\}$$

be the set of indexes corresponding to $s_1(n)$ greater than zero. Now, define:

$$\hat{s}(n) = \mathbf{w}' \, \mathbf{z}(n) \tag{6}$$

where

$$\mathbf{w} = \frac{1}{Z} \sum_{i \in \mathbb{S}} \mathbf{z}(i)' \tag{7}$$

$$Z = \left\| \sum_{i \in \mathbb{S}} \mathbf{z}(i) \right\| \tag{8}$$

Next consider the following property, whose proof will be published in a forthcoming paper:

Lemma 1. *Under mild conditions, (6) is an estimate of the first independent component (up to sign).*

Equivalently, we may also say that vector $\mathbf{w}$ is the first row of the matrix $\mathbf{Q}$ which was defined in (5).

The exact elements of the set $\mathbb{S}$ are of course unknown in practice. In the above algorithm, we have actually substituted $\mathbb{S}$ with

$$\mathbb{S}^* = \{n \mid \hat{s}(n) \geq 0\}$$

(this is the purpose of lines 4 – 9). Then, we iterate the procedure to produce successive approximations to the independent component (lines 2 – 11). The algorithm is consistent in the sense that the solution $\hat{s}(n) = s_1(n)$ satisfies it. To see it, note that, using (5) and the fact that $\mathbf{Q}$ is an orthogonal matrix, (6) can be rewritten as

$$\hat{s}(n) = \left[\frac{1}{Z} \sum_{i \in \mathbb{S}^*} \mathbf{s}(i) \right]' \mathbf{s}(n) \tag{9}$$

The independent components can be assumed to be zero-mean without loss of generality, i.e.,

$$\frac{1}{T} \sum_{i=1}^{T} \mathbf{s}(n) = \mathbf{0} \Rightarrow \sum_{i \in \mathbb{S}^*} \mathbf{s}(i)' = - \sum_{i \notin \mathbb{S}^*} \mathbf{s}(i)'$$

Substituting in (9) we get

$$\hat{s}(n) = \mathbf{g}' \, \mathbf{s}(n) \tag{10}$$

where

$$\mathbf{g} = (2Z)^{-1} \sum_{i=1}^{T} \text{sign}\{\hat{s}(i)\} \mathbf{s}(i) \tag{11}$$

The nth component of vector $\mathbf{g}$ equals

$$g_n = (2Z)^{-1} \sum_{i=1}^{T} \text{sign}\{\hat{s}(i)\} s_n(i) \tag{12}$$

Substituting $\hat{s}(n) = s_1(n)$, we get

$$g_n = (2Z)^{-1} \sum_{i=1}^{T} \text{sign}\{s_1(i)\} s_n(i) \tag{13}$$

Note that the sum looks like a mathematical expectation (up to scale factors). Since the sources are statistically independent among themselves, for $n \neq 1$ we get

$$\sum_{i=1}^{T} \text{sign}\{s_1(i)\} s_n(i) \propto \left[\sum_{i=1}^{T} \text{sign}\{s_1(i)\}\right] \left[\sum_{i=1}^{T} s_n(i)\right] = 0 \tag{14}$$

In conclusion, only the first element of $\mathbf{g}$ is nonzero (i.e. $\hat{s}(n) = s_1(n)$).   □

The stability of the algorithm (i.e., the conditions under which the algorithm converges) will be also studied in detail in a forthcoming paper. Having said that, observe that (13) can be rewritten in the framework of the algorithm as follows:

$$g_n^{(k+1)} \propto \sum_{i=1}^{T} \text{sign}\left\{\hat{s}(i)^{(k)}\right\} s_n(i) \tag{15}$$

where $g_n^{(k)}$ and $\hat{s}(i)^{(k)}$ are, respectively, the k-th coefficient of vector $\mathbf{g}$ and the estimated independent component after the k-th iteration. Furthermore,

$$\hat{s}(i)^{(k)} = \sum_{p=1}^{N} g_p^{(k)} s_p(i) \tag{16}$$

from which it follows that

$$g_n^{(k+1)} \propto \sum_{i=1}^{T} \text{sign}\left\{\sum_{p=1}^{N} g_p^{(k)} s_p(i)\right\} s_n(i) \tag{17}$$

The key for convergence is that $\text{sign}\left\{\sum_{p=1}^{N} g_p^{(k)} s_p(i)\right\} = \text{sign}\{s_1(i)\}$. If we assume that the algorithm is near to convergence (i.e., $g_p^{(k)} \approx 0$ for $p \neq 1$), this condition is intuitively satisfied when the independent components $s_p(i)$ do not take large values as compared to $s_1(i)$ (i.e., they should not have large spikes).

3   Computer Experiment

In this experiment $s_1(n)$ was the discrete-time sequence obtained by sampling the Spanish phrase "ahora sí", spoken by a male speaker. The sampling frequency was 44100 Hz. The samples of the other independent components, $s_2(n), \ldots, s_5(n)$ were drawn from an uniform distribution ($T = 4000$ samples each), and matrix $\mathbf{A}$ was generated at random. Specifically, we got:

$$\mathbf{A} = \begin{bmatrix} -0.23 & 0.62 & 0.24 & 0.39 & -0.95 \\ 0.12 & 0.8 & -1 & 0.088 & 0.78 \\ 0.31 & 0.94 & -0.74 & -0.64 & 0.57 \\ 1.4 & -0.99 & 1.1 & -0.56 & -0.82 \\ -0.35 & 0.21 & -0.13 & 0.44 & -0.27 \end{bmatrix}$$

Let $\hat{\mathbf{s}}(n) = \langle \hat{s}_1(n), \ldots, \hat{s}_5(n) \rangle'$ be the vector that contains the estimated independent components. Both $\mathbf{s}(n)$ and $\hat{\mathbf{s}}(n)$ are related by a simple linear transformation, viz.

$$\hat{\mathbf{s}}(n) = \mathbf{G}\,\mathbf{s}(n)$$

In this experiment, we got using the proposed algorithm:

$$\mathbf{G} = \begin{bmatrix} 0.029 & -0.035 & -1 & -0.04 & -0.00044 \\ -0.024 & -0.013 & 0.024 & -1 & 0.014 \\ -0.039 & 0.0015 & -0.0023 & 0.012 & -1 \\ -0.07 & -1 & 0.03 & 0.02 & 0.029 \\ 1 & -0.044 & 0.0026 & -0.028 & -0.065 \end{bmatrix}$$

e.g., $\hat{s}_1(n) \approx -s_3(n)$ or $\hat{s}_5(n) \approx s_1(n)$. Observe that the approximation is very accurate. To have a basis for comparison, FastICA [5], which is a celebrated HOS-based algorithm for ICA, produced the following matrix:

$$\mathbf{G}_{\text{FastICA}} = \begin{bmatrix} -0.0072 & 0.0047 & 1 & 0.0042 & -0.00079 \\ -0.025 & -0.0077 & 0.013 & 1 & -6.6e-05 \\ -0.0015 & 1 & 0.001 & 0.0021 & -0.0024 \\ -0.99 & -0.029 & 0.021 & -0.021 & 0.15 \\ -0.12 & 0.021 & -0.0014 & 0.02 & -0.99 \end{bmatrix}$$

which is similar in terms of accuracy. However, FastICA took about two times more CPU time than the proposed algorithm. Figures 1–3 show the waveforms of the different signals that have appeared in the experiment.

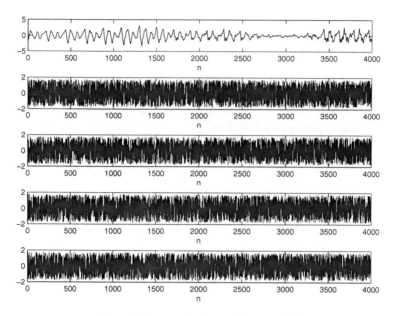

Fig. 1. Source signals $s_1(n), \ldots, s_5(n)$

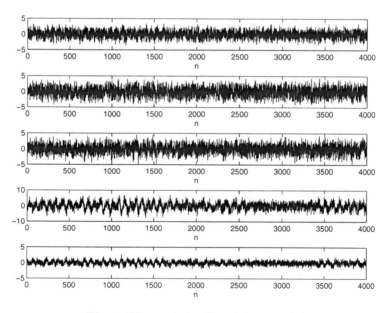

Fig. 2. Observed signals $x_1(n), \ldots, x_5(n)$

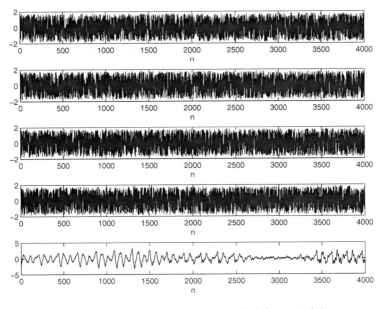

Fig. 3. Estimated source signals $\hat{s}_1(n), \ldots, \hat{s}_5(n)$

4 Conclusions

We have presented a new iterative algorithm to solve the ICA problem. Even though it is true that there already exist a plethora of solutions for ICA, the proposed method has a remarkable very low complexity since, unlike what it is usual in ICA, it does not to compute higher-order statistics. Furthermore, the algorithm exploits a new property (i.e. see Lemma 1) that, as far as we know, has not been described in the literature before. Even though the research is still in its early stages, experiments show that the proposed approach is promising.

Acknowledgements

This work was supported by a grant from the 'Junta de Andalucía' (Spain) with reference P07-TIC-02865.

References

1. Comon, P., Jutten, C. (eds.): Handbook of blind source separation, independent componet analysis and applications. Elsevier, Amsterdam (2010)
2. Cichocki, A., Amari, S.I.: Adaptive blind signal and image processing: learning algorithms and applications. Wiley, Chichester (2002)
3. Zarzoso, V., Nandi, A.K.: Noninvasive fetal electrocardiogram extraction: blind separation versus adaptive noise cancelation. IEEE Tr. on Biomedical Engineering 48 (1), 12–18 (2001)

4. Camargo-Olivares, J.L., Martín-Clemente, R., Hornillo, S., Román, I.: Improving independent component analysis for fetal ecg estimation using a reference signal. IEEE Signal Processing Letters (2011) (accepted for publication)
5. Hyvärinen, A.: Fast and robust fixed-point algorithms for independent component analysis. IEEE Transactions on Neural Networks 10 (3), 626–634 (1999)

AdaBoost Face Detection on the GPU Using Haar-Like Features

M. Martínez-Zarzuela, F.J. Díaz-Pernas, M. Antón-Rodríguez,
F. Perozo-Rondón, and D. González-Ortega

Higher School of Telecommunications Engineering, Paseo de Belén, 15 47007
Valladolid, Spain
marmar@tel.uva.es,
http://www.gti.tel.uva.es/moodle

Abstract. Face detection is a time consuming task in computer vision
applications. In this article, an approach for AdaBoost face detection
using Haar-like features on the GPU is proposed. The GPU adapted
version of the algorithm manages to speed-up the detection process when
compared with the detection performance of the CPU using a well-known
computer vision library. An overall speed-up of $\times$ 3.3 is obtained on the
GPU for video resolutions of 640x480 px when compared with the CPU
implementation. Moreover, since the CPU is idle during face detection,
it can be used simultaneously for other computer vision tasks.

Keywords: Face Detection, Adaboost, Haar-like features, GPU, CUDA,
OpenGL.

1 Introduction

In the field of Computer Vision (CV), detecting a specific object in an image is a computationally expensive task. Face detection can be addressed using
feature-based approaches, without machine learning "a priori", or appearance-
based approaches, with machine learning inside. Feature-based approaches advantage is they make an explicit use of face knowledge: like local features of the
face (nose, mouth, eyes,...) and the structural relationship between them. They
are generally used for one single face localization and are very robust to illumination conditions, occlusions or viewpoints. However, good quality pictures
are required and algorithms are computationally expensive. On the other hand,
the appearance-based approaches consider face detection as a two-class pattern
recognition problem that rely on statistical learning methods to build a face/non
face classifier from training samples; they can be used for multiple face detection in even low resolution images. In practice, appearance-based approaches
have proven to be more successful and robust than featured-based approaches.

Different appearance-based methods mainly differ in the choice of the classifier: support vector machines, neural networks [6], Bayesian classifiers [3] or Hidden Markov Models (HMMs) [12]. A well-known method developed for frontal

J.M. Ferrández et al. (Eds.): IWINAC 2011, Part II, LNCS 6687, pp. 333–342, 2011.
© Springer-Verlag Berlin Heidelberg 2011

face detection was independently introduced by Viola and Jones [16], by Romd-hani [13], and by Keren [10]. All of these algorithms use a 20x20 pixel patch (searching window) around the pixel to be classified. The main difference be-tween these approaches lies in the manner in which a cascade of hierarchical filters devised to classify the window as a face or a non-face are obtained, and more specifically, the criterion used for performance optimization during train-ing. Facing this problem as seen in [15], computation time is an important factor that will set the need of a implementation method that reduce this time to a minimum.

In this paper we identify different parallelizable steps of the AdaBoost algo-rithm for face detection introduced in [16] and propose how to translate it for its execution under the GPU. This way, not only the face detection is faster than when it is done on the CPU, but also the CPU remains idle and can be used for other computations simultaneously.

1.1   Motivation

Using boosting algorithms reduces the number of computations needed to classify a window as face/non-face. The introduction of these algorithms made possible to detect faces in real time on CPUs (Central Processing Units). However, the speed of the detection is still not very fast and performance falls down when the resolution of the images increases. Also, some ad-hoc hardware developments and software implementations for specific architectures different from CPU have been proposed. Masayuki Hirimoto et al. studied the requirements of a special-ized processor suitable for AdaBoost detection using Haar-like features [9] and Yuehua Shi et al. developed a cell array architecture using a multipipeline for speeding up its computation [14]. Some researchers work optimizing the well-known open source computer vision library OpenCV to run not only under Intel platforms, but also on the Cell BE processor. For face detection using Haar-like features and AdaBoost algorithm, their implementation speeds-up computation between ×3 and ×11 times for 640x480 px video resolutions [1]. Ghorayeb et al. proposed an hybrid CPU and GPU implementation of AdaBoost for face detection, not using Haar-like features but Control Points features, achieving a classification speed of 15 fps for video resolutions 415x255 px on an Athlon 64 3500+ at 2.21GHz and a 6600GT GPU [5].

For many years, CPU software immediately run faster on new generations of microprocessors due to a significant increases in clock frequencies. However, the new trend in microprocessors design is not increasing the clock frequency, but the number of cores inside a die. Sequential implementations of algorithms have to be redesigned so that the workload is efficiently delivered to CPUs equipped with 2, 4 or 8 cores. On the other hand, GPU computing has proved to be a nice technique for speeding-up the execution of algorithms using not CPUs, but GPUs (Graphics Processing Units), massively parallel processors with hundreds of cores hidden on commodity graphics cards. GPU computing deals with the translation of algorithms into data parallel operations and their implementation on GPUs.

In this article, some approaches for AdaBoost face detection using Haar-like features on the GPU are proposed. Section 2 gives an overview of the algorithm. Section 3 describes algorithm implementation on the GPU, using CUDA and a combination of CUDA and OpenGL/Cg. In section 4 experimental performance tests, comparing CPU and GPU performance are detailed. Finally, section 5 draws the main conclusions obtained from this research.

2   AdaBoost Face Detection with Haar-Like Features

AdaBoost learning [4], combined with Haar-like features [7] computation is one of the most employed algorithms for face detection. The AdaBoost learning process is able to significantly reduce the space of Haar features needed to classify a window on an image as containing or not containing a face. Selected Haar-like features are disposed on a classifier made up of a cascade of hierarchical stages. A scanning window is used over the input image at different locations and scales, and then the contents of the window are labeled along the cascade as a potential face or discarded as a non-face. Haar-like features are widely used in face searching and numerous prototypes have been trained to accurately represent human faces through the AdaBoost learning algorithm. Results of some of these trainings are available through the open source OpenCV library [11], in which is possible to find XML descriptions of cascades of classifiers for frontal or partially rotated faces. Viola and Jones proposed four basic types of scalar features for face detection [16]. In this paper we use five different types of Haar-like features, that are shown in figure 1. Every feature can be located on any subregion inside the searching window and vary in shape (aspect ratio) and size. Therefore, for a window of size 20x20 pixels, there can be tens of thousands of different features.

Haar-like features in figure 1 are computed using equation (1), where $h_j(x,y)$ is the Haar-feature j computed over coordinates (x,y) on the window, relative to the position of the window on the original image i in grey scale. The sum of the pixels over positions (m,n) inside every rectangle r conforming the Haar-feature, is weighted according to a w_{jr} factor. Selection of h_j and associated weights are decided by the boosting procedure during training. Stages in the AdaBoost cascade are comprised of a growing number of J Haar-features, as stated in equation (2) . Probability of face or non-face depends on a threshold θ, computed during the training process. Different Haar features contribute with a different weight to the final decision of the strong classifier. Thresholds have to be chosen to maximize the amount of correct detections while maintaining a low number of false positives.

$$h_j(x,y) = \sum_{r=1}^{R} \left[w_{jr} \cdot \sum_{(m,n)\in r} i(m,n) \right] \tag{1}$$

$$H(x,y) = \begin{cases} 1 & \sum_{j=1}^{J} h_j(x,y) < \theta \\ 0 & \text{otherwise} \end{cases} \qquad (2)$$

3   AdaBoost Haar-Like Face Detection on the GPU

3.1   Parallel Computation of the Integral Image

Within a detection window, computing a stage in the cascade of classifiers implies scanning several Haar features at different scales and positions. Each Haar feature is a weak classifier and its evaluation is based on a sum of intensities, thus requiring fetching every pixel under the feature area. This involves lots of lookups, which is undesirable. A preprocessing stage can be added to speed-up weak classifiers computation, by generating an *Integral Image* (II) [2]. In the integral image, the value stored on a pixel is the sum of those pixel intesities above and to the left in the original input image, as described in (3), being $i(x,y)$ the value on the input image and $ii(x,y)$ the value on the integral image.

$$ii(x,y) = \sum_{x' \leq x, y' \leq y} i(x',y') \qquad (3)$$

Once the II image is calculated, computing a Haar-like feature needs just four memory fetches per rectangle. The sum-up of the intensities under a rectangle at any location and scale can be computed in constant time using equation (4), as described in figure Features comprised of three different rectangles in figure 1 can be expressed using only two opposite weighted rectangles.

$$S = ii(x_C, y_c) + ii(x_A, y_A) - ii(x_B, y_B) - ii(x_D, y_D) \qquad (4)$$

Fig. 1. Five types of rectangular Haar wavelet-like features. A feature is a scalar calculated by summing up the pixels in the white region and subtracting those in the dark region.

3.2   Parallel Computation of the Cascade of Classifiers: CUDA Naïve Implementation

Detection is divided into several CUDA kernels, pieces of code that are executed in parallel by many GPU threads. Blocks of threads and grid data organization are mapped around a mask vector of integers, which contains information about the window being processed or 0 if the window has been discarded on a previous stage. Initial CUDA grid is defined unidimensional, and has the length of the mask. Each thread classifies one single window at a given stage and scale, sequentially computing all necessary weak classifiers. In a block of threads, the feature description is uploaded to shared memory before it is computed. Figure 2 shows mask vector and how processing is delivered to blocks of threads and a grid of blocks for an image of size MxN pixels. Each integer in the mask identifies a single sliding window over the original image, placed at coordinates $(x, y) = (v\%M, v/M)$, where v is the value read from the mask vector and $\%$ is the modulo operation.

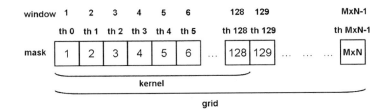

Fig. 2. Data organization for parallel computation (phase 1)

Once the first stage in the cascade of classifiers has been calculated, several values in the mask will be 0. Corresponding windows will not have to be processed in subsequent stages. However, when a kernel is mapped to a grid and block for computation, hardware executes a minimum group of threads in parallel *(warp of threads)*. A thread reading a 0 value from the mask will not have any operation to do, so it will be idle waiting for the rest of the threads in the warp. A way to avoid idle threads is to compact the mask, putting the values different from zero all together at the beginning of the mask, as it is shown in figure 3. The algorithm to do this in parallel is described in [8]. The size of the mask, therefore the size of the grid, after compactation will be $M'xN'$. It is possible to dynamically adjust the size of the kernel on the fly for maximum performance. Although performance of compactations on the GPU is larger than on the CPU only for large masks and input images, doing the compactation on the CPU is not practical, as it would involve two extra memory copies. One of the advantages of compactation process is that it returns the number of valid points in the mask.

Knowing the number of surviving windows for the next stage allows for another optimization. On the first stages of the cascade, there is a small number of features to compute and many windows have to be computed. Along the cascade, the number of candidate windows is reduced exponentially and the number of features grows linearly.Thus, it is possible to divide computation of the cascade into two different phases: parallelizing the computation of windows (Kernel 1) and parallelizing the computation of features inside a window (Kernel 2).

Kernel 1 is used along the cascade first stages. The number of windows is big and the number of features to compute inside a window is smaller than the warp size. Every thread sequentially computes the features inside the window. The size of the block is dynamically adjusted depending on the number of windows that survive the strong classifier.

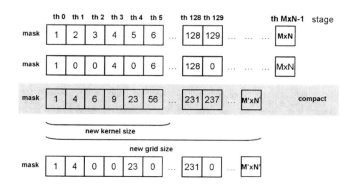

Fig. 3. Modification of mask along the cascade of classifiers

Kernel 2 can be used instead of Kernel 1 when the number of windows becomes small enough and the number of features inside a window is large, thus is worthwhile to use a block of threads for parallel computing the features inside a window. Figure 4 shows how computation of a window is mapped to a block of threads. Size of the block depends on the number of features that have to be computed in a the given stage and is chosen to be a multiple of the warp size. Each thread stores in shared memory the result of the computation of a single feature. Then this values are modified depending on each feature threshold and summed up using a parallel reduction. Different blocks are executed on different multiprocessors, so that surviving windows are still computed in parallel. The final result is compared to the strong classifier and the window is discarded as necessary.

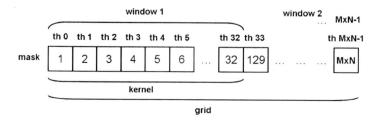

Fig. 4. Data organization for parallel computation (phase 2)

3.3   Further Optimizations

Processing all the possible windows over an input image is a really expensive computational task. On the implementation considered in section 3.2, all the pixels of the image are considered as potential window origins. Hence, for robustness, all of them are classified as face/non-face regions, using a brute force approach. However, it is possible to increase detection speed by increasing the distance between windows. For a better performance, CPU implementations use an offset of $y = 2$ px between windows, so that only odd rows of the input are processed. An offset of $x = 1$ px is used when the previous window has been classified as a potential face and $x = 2$ px in other case. For the CUDA implementation, it is possible to define the mask to be of size $(M/2)x(N/2)$, so that it will contain one fourth of the possible windows. On odd and even frames of a video sequence it is possible to use an overlapping mask of the same size, but displaced an offset of $x = 1$ px. For the given masks, the coordinates of a window can be calculated as $(x, y) = (v\%M, v/M) * (offset_x, offset_y)$.

Moreover, re-scaling the sliding windows and associated Haar-like features is needed when looking for an object in different sizes. The same integral image can be used, but lookup accesses to textures become more dispersed, what provokes cache fails and slows down the process. Another improvement consists on resizing the input image and keeping the sliding window size, since resizing the input image on the GPU can be done almost for free.

4 Performance Tests

The GPU-based face detector developments described in section 3 have two interfaces able to process static images or video sequences captured from a video webcam. Several tests were done in order to test the performance of the GPU implementations, using different sized images and video resolutions on a 3.2 GHz Pentium 4 with 4GB RAM, and using a NVIDIA GeForce GTX285 1GB GPU. The GPU performance was tested against the CPU implementation of the algorithm delivered with the OpenCV library [11]. Both CPU and GPU implementations use the same XML cascade of classifiers for detecting frontal faces and detection rate was exactly the same on the tests that are presented here.

Figure 5 shows a graph comparing the time needed to detect only *one face* on the CPU and on the GPU. Figure 5(a), shows times for detection on static images depending on their resolution. Times given are average times after 10 executions over different images for two different GPU implementations: the naïve version of the algorithm and the algorithm optimized as detailed in section 3.3. The naïve implementation on the GPU runs more or less at the same rate than the CPU OpenCV version. The GPU optimized code gives a speed-up of ×1.6 for images of size 800x800 px.

Figure 5(b) shows a comparison of times for *one face* detection on video sequences. Again, the GPU optimized implementation is the fastest. The speed-up obtained with respect to the OpenCV implementation is up to ×3.3 times for video resolutions of 640x480. The speed-up obtained is greater for video than for static images due to the initialization processes that take place on the GPU when a kernel or a shader is launched for the first time. This initialization is only necessary for the first frame when the face detector is launched with the video interface. Memory and algorithm initialization makes also face detection on a single image slower than on a single video frame when using OpenCV.

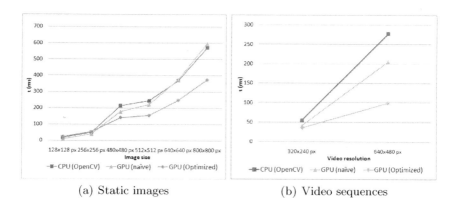

(a) Static images   (b) Video sequences

Fig. 5. Performance test for only one face detection on static images and video sequences at different resolutions

Finally, we analyzed the performance of the different implementations depending on the *number of faces* that have to be detected on a single image. Figure 6(a) shows an image from the CMU/VASC database of size 512x512 px. Figures 6(b) to 6(d) have the same size, but incorporate an increasing number of faces to be detected. The number of windows that survive along the cascade of classifiers varies depending on the number of faces on the image. For the image containing 9 faces, the number of windows that have to be evaluated after the first stage is ×1.5 times greater than on an image containing only one face. In the last stage, the number of windows grows up by a factor of ×9. On a sequential implementation on the CPU, the performance of the algorithm can be significantly hurt with an increasing number of faces on the image. However on a parallel GPU-based implementation, the performance its only reduced by a factor of ×1.3 on a GPU GTX285, containing 240 stream processors.

(a) Original   (b) Two faces   (c) Four faces   (d) Nine faces

Fig. 6. Original image from CMU/VASC database with one face and modifications including two, four and nine faces

5   Conclusions

In this paper, different GPU-based approaches for implementing AdaBoost face detection using Haar-like features were described and tested. In the AdaBoost algorithm, detection speed is increased through the evaluation of a cascade of hierarchical filters: windows easy to discriminate as not containing the target object are classified by simple and fast filters and pixels that resemble the object of interest are classified by more involved and slower filters. The main limitation we found when translating the algorithm was the high level of branching imposed, what makes difficult data parallelization. For GPU performance, a cascade of classifiers comprised of a smaller number of stages or even just one computationally expensive stage would perform rather fast. Mapping the algorithm to the GPU to evaluate simultaneously as many windows as possible in the first stages (large number of windows including a small number of features) and parallelize the computation of features inside a window on the next stages (small number of windows including a large number of features) demonstrated to be more convenient to take advantage of the underlying parallel hardware.

For static images, a maximum speed-up of ×1.6 was obtained with the fastest GPU implementation at resolutions of 800x800 px. For video resolutions of 640x480 px, face detection was ×3.3 times faster than the OpenCV library version for CPU. A cascade of well-known classifiers was employed. Although this cascade was designed for its execution on sequential hardware, nice figures of speedup are obtained on the GPU.

Acknowledgements

This work has been partially supported by the *Spanish Ministry of Science and Innovation* under project TIN2010-20529.

References

1. CellCV: Opencv on the cell. adaboost face detection using haar-like features optimization for the cell. code downloading and performance comparisons (2009), http://cell.fixstars.com/opencv/index.php/Facedetect (last Visit February 2009)
2. Crow, F.C.: Summed-area tables for texture mapping. In: SIGGRAPH 1984: Proceedings of the 11th Annual Conference on Computer Graphics and Interactive Techniques, pp. 207–212. ACM Press, New York (1984)
3. Elkan, C.: Boosting and naive bayesian learning. Tech. rep. (1997)
4. Freund, Y., Schapire, R.E.: A decision-theoretic generalization of on-line learning and an application to boosting. Journal of Computer and System Sciences 55(1), 119–139 (1997)
5. Ghorayeb, H., Steux, B., Laurgeau, C.: Boosted algorithms for visual object detection on graphics processing units. In: Narayanan, P.J., Nayar, S.K., Shum, H.-Y. (eds.) ACCV 2006. LNCS, vol. 3852, pp. 254–263. Springer, Heidelberg (2006)
6. Rowley, H.A., Baluja, S., Kanade, T.: Neural network-based face detection. IEEE Transactions on PAMI (1998)
7. Haar, A.: Zur theorie der orthogonalen funktionensysteme. Math. Annalen. 69, 331–371 (1910)
8. Harris, M.: Parallel prefix sum (scan) with cuda. In: Nguyen, H. (ed.) GPU Gems 3, ch. 39, pp. 851–876. Addison Wesley Professional, Reading (2007)

9. Hiromoto, M., Nakahara, K., Sugano, H., Nakamura, Y., Miyamoto, R.: A specialized processor suitable for adaboost-based detection with haar-like features. In: IEEE Conference on Computer Vision and Pattern Recognition, CVPR 2007, pp. 1–8 (June 2007)

10. Keren, D., Osadchy, M., Gotsman, C.: Antifaces: A novel, fast method for image detection. IEEE Transactions on Pattern Analysis and Machine Intelligence 23(7), 747–761 (2001)

11. OpenCV: Open source computer vision library (2009), http://sourceforge.net/projects/opencvlibrary (last visit February 2009)

12. Rabiner, L.R.: A tutorial on hidden markov models and selected applications in speech recognition, pp. 267–296 (1990)

13. Romdhani, S., Torr, P., Scholkopf, B., Blake, A.: Computationally efficient face detection. In: IEEE International Conference on Computer Vision, vol. 2, p. 695 (2001)

14. Shi, Y., Zhao, F., Zhang, Z.: Hardware implementation of adaboost algorithm and verification. In: 22nd International Conference on Advanced Information Networking and Applications - Workshops, AINAW 2008, pp. 343–346 (March 2008)

15. Vaillant, R., Monrocq, C., Le Cun, Y.: Original approach for the localization of objects in images. In: IEEE Proceedings of Vision, Image and Signal Processing, vol. 141(4), pp. 245–250 (August 1994)

16. Viola, P., Jones, M.: Robust real-time object detection. International Journal of Computer Vision 57(2), 137–154 (2002)

Fuzzy ARTMAP Based Neural Networks on the GPU for High-Performance Pattern Recognition

M. Martínez-Zarzuela, F.J. Díaz-Pernas, A. Tejero de Pablos,
F. Perozo-Rondón, M. Antón-Rodríguez, and D. González-Ortega

Higher School of Telecommunications Engineering,
Paseo de Belén, 15 47007
Valladolid, Spain
marmar@tel.uva.es
http://www.gti.tel.uva.es/moodle

Abstract. In this paper we introduce, to the best of our knowledge, the first adaptation of the Fuzzy ARTMAP neural network for its execution on a GPU, together with a self-designed neural network based on ART models called SOON. The full VisTex database, containing 167 texture images, is proved to be classified in a very short time using these GPU-based neural networks. The Fuzzy ARTMAP neural network implemented on the GPU performs up to ×100 times faster than the equivalent CPU version, while the SOON neural network is speeded-up by ×70 times. Also, using the same texture patterns the Fuzzy ARTMAP neural network obtains a success rate of 48% and SOON of 82% for texture classification.

Keywords: Fuzzy ARTMAP, SOON, GPU Computing, CUDA, Neural Networks, Texture recognition.

1 Introduction

Among the various methods for pattern recognition, statistical and Neural Network (NN) approaches have received an important attention. One of the main disadvantages of the neural network techniques is the large time required for pattern classification. Even when neural network models have an inherent data-parallel structure, most implementations have been developed to be executed on single-core CPUs. A new trend in microprocessor manufacturing is including many cores under a single die of silicon, and this has led to an increased interest in translating existing algorithms and developing new ones to exploit new parallel processing opportunities. Parallel processing can be done using task-parallel, data-parallel or a mixture of both approaches. Multi-core CPUs are well suited for the former, but not for data-parallel execution. On the other hand, commodity GPUs *(Graphics Processing Units)*, comprised of a high number of symmetric multi core processors, have most of their transistors oriented to computations, thus are well suited for SIMD *(Single Instruction Multiple Data)* processing.

J.M. Ferrández et al. (Eds.): IWINAC 2011, Part II, LNCS 6687, pp. 343–352, 2011.

Several researchers have obtained impressive performance results with their GPU-based artificial neural network proposals, speeding up the execution dozens of times with respect to their CPU versions [15,12,4,3,11]. Moreover, a personal computer can be used as a heterogeneous computing system, being possible to combine the power of the CPU and the GPU for speeding up neural networks execution [10].

In this paper, we propose to implement on GPU, using CUDA technology, SOON [2] and Fuzzy ARTMAP neural networks [5]. These implementations are also employed to compare color texture recognition success.

1.1 Motivation

In the neural networks research field, an important contribution can be found in the Adaptive Resonance Theory (ART). This theory on brain behavior has favored the development of an important number of neural models capable of unsupervised [6], supervised [5] and semi-supervised learning [1], using incremental learning, in contrast to other neural networks such as MLP, based on prediction error measurement for training. The authors of this paper introduced the first GPU-based implementations of Fuzzy ART, an unsupervised NN, using OpenGL [13] and CUDA [14]. In [2] we proposed a texture recognition neural system comprised of a module for color segmentation and a module for categorization which employed a network called SOON (Supervised OrientatiOnal invariant Neural network) inspired on ARTMAP theory.

The proposed architecture, CPU implemented, was trained and employed to recognize up to 30 image textures from the VisTex database, improving recognition rates obtained using two other architectures: the HSG-MLP system [7], based on log-Gabor filtering and Multilayer Perceptron classification; and the ARTEX system [8], based on oriented filtering and ARTMAP theory. However, one of the limitations of the proposed neural architecture running on the CPU is that it requires a lot of time for pattern classification. Requirements of time are heavily increased with the number of textures to recognize. In order to classify a larger number of texture patterns in a reasonable quantity of time we decided to explore the implementation of Fuzzy ARTMAP and SOON on the GPU using CUDA technology. This paper describes the results of that study. Measurements of the speed-up obtained with respect to the original CPU versions of the algorithms are given. Moreover, experimental results include a performance comparison of Fuzzy ARTMAP and SOON classifiers for texture recognition of the whole VisTex database, containing 167 image textures, using pattern extraction described in [2].

Section 2 focuses on the parallelization of the algorithms and their implementation on the GPU. In section 3 exhaustive tests are done in order to measure the performance of both approaches running on the GPU and in terms of recognition rate using the same input features. Finally, section 4 draws the main conclusions and puts forward future research tasks.

2   Fuzzy ARTMAP and SOON Implementations Using CUDA

In this section, we describe how to adapt the Fuzzy ARTMAP and SOON algorithms for data-parallel computation and main issues to bear in mind when implementing the code in CUDA.

2.1   Fuzzy ARTMAP and SOON Principles

Fuzzy ART-based supervised neural networks are self-organizing algorithms capable of incremental learning [6]. These networks are comprised of an ART_A module which analyzes input patterns and generates long-term memory (LTM) traces, another ART_B module which controls supervision input and a MapField (MF) which communicates the two aforementioned modules and maps LTM traces to the class selected by an expert.

This kind of NNs can be used to clusterize a stream of P input patterns, belonging to C different classes, into N different categories, generated during a supervised training phase. Each category is associated to a given class via the MF. In Fuzzy ARTMAP and SOON, the first layer of neurons F_1 receives the input pattern and neurons in the upper layer F_2 represent a specific category from those emerged during the self-organizing training phase. F_1 activity vector is denoted by $\boldsymbol{I_p} = (I_1, \cdots, I_M)$ where each component I_i is within the $[0, 1]$ interval. F_2 neurons synapses are weighted by the LTM traces denoted by $\boldsymbol{w_j} = (w_{j1}, \cdots, w_{jM})$. Activity of these neurons is computed as $T_j(\boldsymbol{I}) = |\boldsymbol{I} \wedge \boldsymbol{w_j}|/(\alpha + |\boldsymbol{w_j}|)$. SOON NN only generates activities of neurons whose class matches the current class of the input vector (supervision input) and introduces an additional step consisting on computing $T_j(I)$ for O different possible orientations of the input patterns, being $T_j(\boldsymbol{I}) = \max(T_j(\boldsymbol{I_r}) : r = 1 \cdots O)$.

In Fuzzy ARTMAP, activities vector $\boldsymbol{T} = (T_1, \cdots, T_N)$ is organized decreasingly, by activation value and each component is verified to meet two conditions in this order: the associated *match function* meets the *vigilance criterion* $(|\boldsymbol{I} \wedge \boldsymbol{w_J}|)/|\boldsymbol{I}| \geq \rho$ and the MF value associated to the weight activation has to be equal to the current supervision input. If this last condition is not met, ρ is increased $\rho^{new} = $ *match function* $+ \epsilon$, being ϵ a very low value, enough to prevent the commitment of a similar category. The system enters in *resonance* only if an indexed category J meets these terms.

In SOON, only those activities elicited through LTM traces whose class matches the supervision input are generated. The category choice is indexed by J, where $T_J = \max(T_j : j = 1 \cdots N)$. The system enters in *resonance* only if the *match function* meets the *vigilance criterion*. When this occurs and learning is enabled $(\beta \neq 0)$, associated $\boldsymbol{w_J}$ is updated $\boldsymbol{w_J^{new}} = \beta(\boldsymbol{I} \wedge \boldsymbol{w_J^{old}}) + (1 - \beta)\boldsymbol{w_J^{old}}$, being $\boldsymbol{I}$ in SOON the rotation of current input pattern which generates the chosen activity $(\boldsymbol{I_r})$. If no neuron j is found to meet these requirements, a new neuron is committed in F_2 and and its associated MF value is stored.

2.2   Parallelization of Fuzzy ARTMAP

The pseudocode in Figure 1(a) details the different steps of the algorithm. Parallel operations are indexed by a κ that indicates a *kernel* in CUDA, while labels *Non-Parallel (NP)* and *Checking (Ch)* indicate some must-be sequential data operations.

In Fuzzy ARTMAP training phase, every input vector I_p has to be sequentially analyzed, since learning process can modify the weights associated to previous committed categories. First of all, supervision input is searched in the MF. This process can be easily parallelized in a kernel where each thread checks a MF element (kernel κ_1). In case the supervision input is not found, a new neuron is generated in F_2 (kernel κ_2). Although input patterns are presented serially, it is possible to simultaneously compute the match criterion and the activity of every output neuron (kernel κ_3).

Figure 1(b) describes parallelization of the testing phase (kernel κ_5). In this case every input vector I_p can be processed in parallel and the activity of nodes in F_2 are sequentially computed using a single kernel call. This is possible now because the network is not altered during the process and, since for many applications the number of input vectors is much larger than the number of categories. Besides, only activities which met vigilance are calculated and it is not necessary to analyze them after, but just choose the higher. Then, its respective MF value will determine in which class are categorized input patterns.

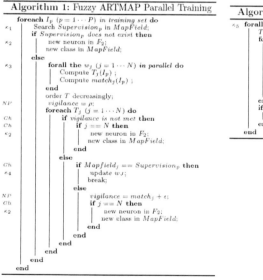

(a) Training   (b) Classification

Fig. 1. Parallel training and classification in ARTMAP

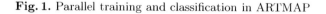

2.3   Parallelization of SOON

Figure 2(a) details parallelization of SOON algorithm for parallel training. The main feature of SOON neural network is the inclusion of pattern rotations. This implies calculating O match values for each input pattern, being O the number of pattern orientations encoded inside the input patterns. But whereas Fuzzy ARTMAP algorithm needs to compute all activities and match values in order to perform a sequential analysis of them, in SOON it is only necessary to calculate those activities when vigilance criterion is met (kernel κ_1). Moreover, each vector is only confronted with the LTM traces of the same class, which means avoiding to check if the class associated to each activated neuron (MF) corresponds with the current class of the input pattern.

Instead of a sorting function, a reduction algorithm [9][14] is executed (kernel κ_2) to find the neuron in F_2 with the highest activity value (T_J) and, if it exists, update the corresponding LTM trace using the rotation of I_p which elicited the activity T_J (kernel κ_3). If $T_J == 0$, then another neuron in F_2 is generated (kernel κ_3).

Figure 2(b) shows the parallelization of the algorithm during classification. Like in Fuzzy ARTMAP version, testing phase is implemented on a single kernel (kernel κ_4), which works with all the input patterns at the same time. This phase also operates with their different orientations. Then, every match criterion and activities are calculated using the rotation of I_p that offers the greatest value. This difference with respect to Fuzzy ARTMAP algorithm makes SOON classification testing phase more complex.

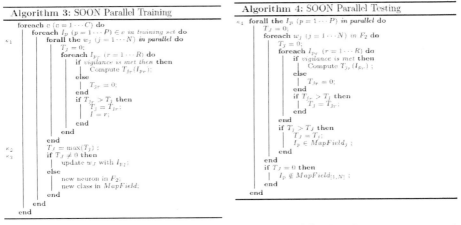

(a) Training   (b) Classification

Fig. 2. Parallel training and classification in SOON

2.4   Common Implementation Aspects

The main advantage of using a graphics card is the possibility to execute the same operation on various vectors simultaneously. In figure 3 data organization

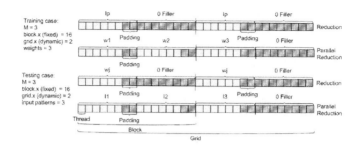

Fig. 3. Grid and kernel organization for training and classification processes

in CUDA is depicted. Single-dimension blocks and grid are used in order to simplify vector accesses. A grid of blocks is generated to simultaneously deal with neural network weights on training and with the input patterns during classification, as described in pseudocodes of Figures 1(a), 1(b), 2(a) and 2(b) including various vectors on each block. Dimensions of the grid are dynamically computed during the training, so that depending on the number of committed categories by the neural network, more blocks of threads are generated.

As a single block can operate on various vectors, specialized parallel operations have to be programmed. In SOON testing phase, various input vectors have to be rotated in the same block at the same time, so a parallel rotation algorithm has to be implemented. In order to compute simultaneously the norm of every vector indepently, a kind of parallel reduction is utilized [14]. One remarkable technique which helps speeding-up reductions is *unrolling kernel loops* [9]. When applied for reduction operations, the number of threads used in each iteration of the algorithm can be adjusted to use only those needed. There are some restrictions when using this technique though [9][14].

GPU kernels are designed to associate each thread with a vector component and, due to the kind of operations performed in these kernels (as reductions programmed in order to compute the norm of a vector) the number of threads needs to be a power of two. But since the number of components of both input patterns and categories utilized in Fuzzy ARTMAP does not, it is necessary to work with zero-padded vectors.

3   Experimental Results

This section includes performance comparisons among the Fuzzy ARTMAP and SOON implementations using CUDA and their respective CPU C++ implementations over a Windows 7 operating system. Timings were taken on a dual-core 3.2 GHz Pentium 4 with 4 GB RAM, equipped with a GPU NVIDIA GeForce GTX285 and include times for copying GPU results to the CPU. In the tests, input patterns of 26 components within sets of 16, 64 and 167 different classes were used. Patterns from the same class are extracted from the image of a texture in the VisTex database [16]. The training set is formed by 30 random patterns of each class and neural networks were trained once. Classification is performed on 8836 patterns from every class. The block size has been chosen to get the higher occupancy on the GPU. More occupancy means less idle threads while executing the program. Given the number of registers and shared memory needed for the kernels in the program, a block size of 128 is the optimal in most of the cases, and has been chosen for all the tests presented in this section.

Figure 4 compares the time needed for training the network on the CPU and on the GPU for 16, 64 and 167 textures. Training SOON is faster than Fuzzy ARTMAP for the same amount of input patterns. Training on the GPU is useful when the number of committed neurons is very large, and so is the number of LTM traces; otherwise, most of the time there will be a large number of idle threads. In the Fuzzy ARTMAP GPU implementation is ×2 times slower than CPU for 16 textures, but ×2 times faster for 167 textures.

Classification, on the other hand, is always much faster on the GPU than on the CPU in both neural networks, because it is always possible to process in parallel a huge amount of input patterns. Figure 5 shows testing performance under GPU and CPU platforms. The largest speed-up achieved for the SOON GPU implementation with respect to CPU is ×70 and ×100 for Fuzzy ARTMAP.

Fuzzy ARTMAP presents a faster overall execution time. However, being Fuzzy ARTMAP implementation executed faster than SOON, working with the same original patterns, error classification rates are greater, as it is showed on Figure 6. SOON generates more accurate categories due to the inclusion of input pattern orientation analysis in both classification and training phases. Learning a single random input, the NN collects information of similar points on the image that could appear rotated. Fuzzy ARTMAP higher error rates imply that this algorithm needs more than one epoch (training iteration) and more patterns to generate better quality LTM traces.

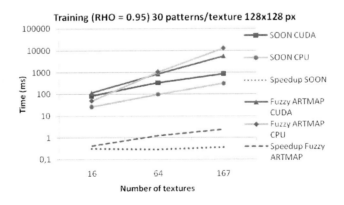

Fig. 4. Fuzzy ARTMAP and SOON neural networks execution time for training. Only 30 patterns per texture in the dataset are learned.

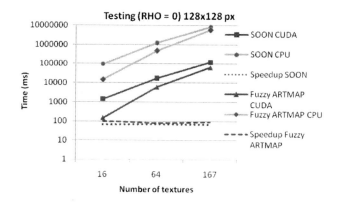

Fig. 5. Fuzzy ARTMAP and SOON testing execution times

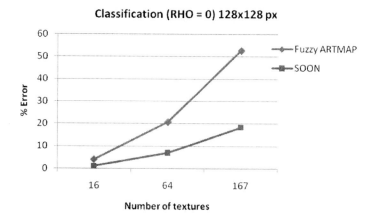

Fig. 6. Fuzzy ARTMAP and SOON error rates

4   Conclusions and Future Work

In this paper, parallelization of Fuzzy ARTMAP and SOON neural networks for training and classification was presented. Implementations of the algorithms on CUDA were briefly detailed and tested against equivalent CPU implementations. Unfortunately, the lack of space avoided the authors to give more details on the techniques used for GPU implementation and on the experiments performed. A speed-up of ×70 for SOON and ×100 for Fuzzy ARTMAP is achieved for pattern classification against the sequential version of the algorithm running on the CPU. In training, when the number of neurons committed in the Fuzzy ARTMAP NN is large enough to take advantage of all the GPU hardware resources, the relative speed-up between GPU and CPU grows exponentially whereas the GPU SOON implementation remains slower.

When comparing both neural networks, it can be observed that SOON rotation operations in training and testing phases involve the generation of more accurate LTM traces and more precision during input patterns classification. Also, having Fuzzy ARTMAP a less parallel training method, SOON exhibits better time results. However, classification phase on Fuzzy ARTMAP only requires the computation of activities when vigilance criterion is met and because of the absence of rotations, it is the fastest.

Although Fuzzy ARTMAP presented clear sequential-based operations, a hybrid implementation was approached bearing in mind the forthcoming heterogeneous microprocessors. Possible memory bottlenecks between CPU and GPU will not be a problem anymore and further research can be done using these heterogeneous processors to combine the strength of data-parallel and task-parallel processing. Using the large parallelization capability of GPUs, neural networks with complex algorithms like SOON can be programmed in order to obtain more accurate results, without sacrificing acceptable execution times.

Acknowledgements

This work has been partially supported by the *Spanish Ministry of Science and Innovation* under project TIN2010-20529.

References

1. Amis, G.P., Carpenter, G.A.: Self-supervised artmap. Neural Networks 23(2), 265–282 (2010)
2. Antón-Rodríguez, M., Pernas, F.J.D., Higuera, J.F.D., Martínez-Zarzuela, M., Ortega, D.G., Boto-Giralda, D.: Recognition of coloured and textured images through a multi-scale neural architecture with orientational filtering and chromatic diffusion. Neurocomputing 72(16-18), 3713–3725 (2009)
3. Bernhard, F., Keriven, R.: Spiking neurons on gPUs. In: Alexandrov, V.N., van Albada, G.D., Sloot, P.M.A., Dongarra, J. (eds.) ICCS 2006. LNCS, vol. 3994, pp. 236–243. Springer, Heidelberg (2006)
4. Campbell, A., Berglund, E., Streit, A.: Graphics hardware implementation of the parameter-less self-organising map. In: Gallagher, M., Hogan, J.P., Maire, F. (eds.) IDEAL 2005. LNCS, vol. 3578, pp. 343–350. Springer, Heidelberg (2005)
5. Carpenter, G.A., Grossberg, S., Markuzon, N., Reynolds, J.H., Rosen, D.B.: Fuzzy ARTMAP: A neural network architecture for incremental supervised learning of analog multidimensional maps. IEEE Trans. Neural Networks 3(5), 698–712 (1992)
6. Carpenter, G.A., Grossberg, S., Rosen, D.B.: Fuzzy ART: Fast stable learning and categorization of analog patterns by an adaptive resonance system. Neural Networks 4(6), 759–771 (1991)
7. Greenspan, H.: Non-parametric texture learning (1996)
8. Grossberg, S., Williamson, J.: A self-organizing neural system for leaning to recognize textured scenes. Vision Research (39), 1385–1406 (1999)
9. Harris, M.: Parallel prefix sum (scan) with cuda. In: Nguyen, H. (ed.) GPU Gems 3, ch. 39, pp. 851–876. Addison Wesley Professional, Reading (2007)

10. Ho, T.Y., Park, A., Jung, K.: Parallelization of cellular neural networks on gpu. Pattern Recogn. 41(8), 2684–2692 (2008)
11. Jang, H., Park, A., Jung, K.: Neural network implementation using cuda and openmp. In: DICTA 2008: Proceedings of the 2008 Digital Image Computing: Techniques and Applications, pp. 155–161. IEEE Computer Society, Washington, DC, USA (2008)
12. Luo, Z., Liu, H., Wu, X.: Artificial neural network computation on graphic process unit. In: IJCNN 2005: Proceedings of the 2005 IEEE International Joint Conference on Neural Networks, Montreal, Canada, pp. 622–626 (August 2005)
13. Martínez-Zarzuela, M., Díaz Pernas, F.J., Díez Higuera, J.F., Rodríguez, M.A.: Fuzzy ART neural network parallel computing on the GPU. In: Sandoval, F., Prieto, A.G., Cabestany, J., Graña, M. (eds.) IWANN 2007. LNCS, vol. 4507, pp. 463–470. Springer, Heidelberg (2007)
14. Martínez-Zarzuela, M., Pernas, F.J.D., de Pablos, A.T., Rodríguez, M.A., Higuera, J.F.D., Giralda, D.B., Ortega, D.G.: Adaptive resonance theory fuzzy networks parallel computation using CUDA. In: Cabestany, J., Sandoval, F., Prieto, A., Corchado, J.M. (eds.) IWANN 2009. LNCS, vol. 5517, pp. 149–156. Springer, Heidelberg (2009)
15. Oh, K., Jung, K.: Gpu implementation of neural networks. Pattern Recognition 37(6), 1311–1314 (2004)
16. VisTex: Vision texture database massachusetts institute of technology (1995), http://vismod.media.mit.edu/vismod/imagery/VisionTexture/vistex.html (last visit June 2010)

Bio-inspired Color Image Segmentation on the GPU (BioSPCIS)

M. Martínez-Zarzuela, F.J. Díaz-Pernas, M. Antón-Rodríguez,
F. Perozo-Rondón, and D. González-Ortega

Higher School of Telecommunications Engineering, Paseo de Belén,
15 47007 Valladolid, Spain
marmar@tel.uva.es
http://www.gti.tel.uva.es/moodle

Abstract. In this paper we introduce a neural architecture for multiple scale color image segmentation on a Graphics Processing Unit (GPU): the BioSPCIS *(Bio-Inspired Stream Processing Color Image Segmentation)* architecture. BioSPCIS has been designed according to the physiological organization of the cells on the mammalian visual system and psychophysical studies about the interaction of these cells for image segmentation. Quality of the segmentation was measured against hand-labelled segmentations from the Berkeley Segmentation Dataset. Using a stream processing model and hardware suitable for its execution, we are able to compute the activity of several neurons in the visual path system simultaneously. All the 100 test images in the Berkeley database can be processed in 5 minutes using this architecture.

1 Introduction

About one-fourth of the brain is involved in vertebrates sensory visual processing system, more than for all other senses [3]. Optical imaging techniques have made it possible to visualize how response properties, such as the selectivity for edge orientation, are mapped across the cortical surface [15]. These physiological studies proved that data processing along the visual system is a parallel process comprised of several stages and the knowledge acquired through these studies can be used for the development of bio-inspired image processing models, that try to imitate the behavior of the human visual system when performing image processing operations [7] [14]. However, while image segmentation and pre-attentive recognition are really fast processes when performed by the brain, most of the equivalent software algorithms are very long-time consuming, as they have been traditionally executed on a Central Processing Unit (CPU), based on a SISD (Single Instruction Single Data) processing model.

Vision and other neural processes are achieved through a massively interconnected neural parallel system, thus sequential hardware should not be selected for simulating neural interactions within the visual pathway. Today, almost every personal computer is equipped with an external graphics card hiding an incredible computational power under a many-core processor called Graphics

J.M. Ferrández et al. (Eds.): IWINAC 2011, Part II, LNCS 6687, pp. 353–362, 2011.

Processing Unit (GPU), designed according to SIMD (Single Instruction Multiple Data) and MIMD (Multiple Instruction Multiple Data) processing models. During last years *GPU Computing* or *General Purpose Computing on the GPU* (GPGPU) has been established as a well-accepted application acceleration technique. GPUs are being considered in many fields of computations, including numerical simulation, signal and image processing or computer vision [16]. In the field of image segmentation some developments have been done to specifically run on the GPU [19] [17] and even basic retina simulations have been performed on a GPU, achieving computation speed-ups 20 times greater with respect to CPU conventional programming [6]. In previous works, we demonstrated how GPUs, originally devoted to computer graphics, were more adequate than CPUs to solve certain pattern recognition neural models [13] [12].

1.1   Motivation

In the field of color image segmentation, we proposed a new architecture inspired in the behavior of the mammalian visual path system [5]. In the field of image recognition, we proposed bio-inspired architectures for multiple scale feature extraction and recognition in [1] [4]. These architectures demonstrated great performance for texture recognition, using image properties such as texture patterns and texture color information using retina and LGN cells opponent color codification, to feed-up complex neural networks invariant to texture patterns orientation. Although these architectures demonstrated to be very adequate for texture recognition and provided very high texture classification rates, their CPU coded implementations were very slow and only a small amount of texture recognition tests were possible. In a further work, we decided to modify the original architecture so that a full GPU implementation, for parallel texture recognition using a stream processing model, was possible [11]. This new implementations run up to ×25 times faster than its CPU-based counterpart, what allowed us to better adjust the neural network parameters under a large number of tests and successfully classify the whole VisTex texture database.

In the present paper, a GPU-based evolution of the architecture presented in [5] is described. The new design improves the architecture performance, as more processing layers of neurons can be incorporated to the model for a better response, and more tests can be done to tune more than one hundred parameters that control the behavior of the segmentation process. This GPU-based architecture is called BioSPCIS *(Bio-Inspired Stream Processing Color Image Segmentation)* and provides image segmentation robust to lighting variations in the scene, as it is inspired in both the physiological and psycophysical principles of the mammalian visual system. This new development will be incorporated to all our neural models, so that also image recognition processes will be accelerated.

The rest of the paper is organized as follows. Section 2 briefly describes the proposed architecture and includes a description on how to map the neural architecture to be executed on a GPU following a stream processing model. Section 3 exposes different image processing tests performed with the BioSPCIS architecture, giving quantitative measurements of performance. Finally, section 4 concludes this paper and puts forward some of the related future work.

2   Description of the BioSPCIS Architecture

2.1   Overview and Relation to the Primary Visual Path System

Figure 1 shows the different processing stages of the BioSPCIS architecture and the visual path system areas in which these computations are inspired. Perception of a visual scene starts when rods and cones, placed in the innermost

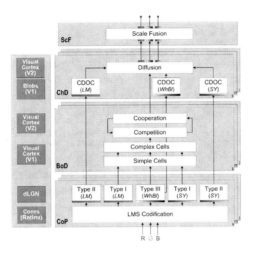

Fig. 1. BioSPCIS architecture

layer of the eye, turn light into electrical signals that modulate the response of retinal cells. There are three types of cones, that react to long(L), medium(M) or short(S) wavelengths. The axons of most retinal ganglion cells synapse on the neurons that form the dorsal Lateral Geniculate Nucleus (dLGN) through the *optic nerve*. In the BioSPCIS architecture (Figure 1) the *color Opponency* (CoP) stage simulates color-opponent interactions that take place on the retina and the dLGN receptive fields. At this stage, the input image is parallel processed and decomposed into five different channels of information. Four of these channels contain chromatic information (LM and SY cell types I and II), while the fifth one contains luminance information (WhBl type III cells). Primary target site for projections from the dLGN is the striate cortex, where fundamental aspects of visual analysis occur [18]. The V1 zone is densely packed with cells in many layers. Some neurons within these layers present patterns of responsiveness similar to those observed in the retina and the dLGN, while most of them present elongated receptive fields that are most fired when information coming from the dLGN contains bars or edges [9]. The first kind of cells are named CDOC *Chromatic Double Opponent Cells*. They are located in the *blobs* and are involved in double opponency processes that are fundamental for visual color processing. In the BioSPCIS architecture, CDOC cells located in the *Chromatic Diffusion* (ChD) stage are used for extracting color information that will be used in a final *Diffusion* process for final texture smoothing.

The second type of cells in the visual cortex includes *Simple Cells* and *Complex Cells* involved in the task of edge detection to favour object segmentation. In the BioSPCIS architecture, this task is done at the *Boundary Detection* (BoD) stage. Different cells in the striate cortex area are sensitive to different orientations and in the BoD stage every input channel is decomposed into several direction-specific channels of information. The number of processing channels can be adjusted for maximum performance although it makes the segmentation process slower. Moreover, in the BoD stage feedforward and feedback interactions take place in a competitive-cooperative loop of layers of neurons. This loop is used to enhance boundaries along the edge dominant orientation and maximize the probability of defining convex segmented regions through an edge cooperation dynamics. These computations are based on those performed by cortical cells at the entry of the V2 zone. The output signal from the BoD stage is a *boundary web* containing a map of contours, that is used in the ChD stage to control the color diffusion process.

Computation of receptive fields and neural interactions between channels of information within the BoD and the ChD stages are far more time-consuming than computations in the CoP stage, just as in the brain there is an increasing specificity of the requirements necessary to activate cells present in higher visual areas, due largely to the hierarchical processing of visual information. Every stage of the architecture is computed in three different scales ((s):small, (m):medium; (l):large) and the size of the cell receptive fields can be properly adjusted on every stage. A final *Scale Fusion* (ScF) stage is used to combine the output from these scales depending on the specific segmentation purpose.

2.2   Mapping Computations to the GPU

Traditionally, the frequency at which processors run has grown exponentially with Moore's Law, but physical restraints related to power consumption and the heat that can be dissipated using cooling devices have driven to newer multi-core architectures running at lower frequencies. This has led to an increased interest in translating existing algorithms and developing new ones following a parallel processing philosophy that can benefit from newer multi-core processors. Parallel processing can be done using *task-parallel, data-parallel* or a mixture of both approaches. Multi-core CPUs are well suited for the former, but not for data-parallel execution. These CPUs spend most of their transistors on logic and cache, while most of them are not devised for computations. On the other hand, commodity GPUs (Graphics Processing Units), comprised of a high number of symmetric multi core processors, have most of their transistors oriented to computations, thus are well suited for SIMD (Single Instruction Multiple Data) processing.

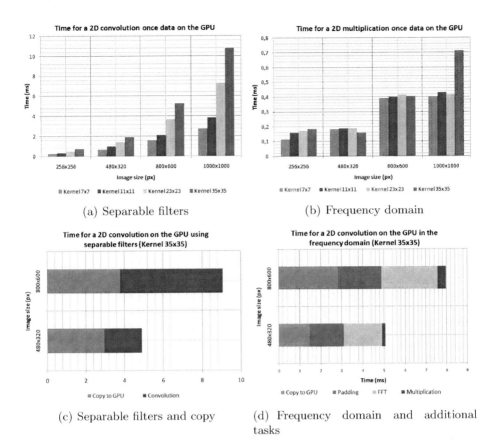

(a) Separable filters

(b) Frequency domain

(c) Separable filters and copy

(d) Frequency domain and additional tasks

Fig. 2. Time for computing a receptive field depending on its size and the size of the image

Neural processing inside the visual system can be thought as a stream processing architecture, comprised of several layers of neurons that perform different operations. Cells belonging to the same layer do exactly the same computations, but adjacent neurons receive different inputs from previous layers of neurons. Neuron response can be modeled using cell membrane electrical excitability equations [8]. A simplified equation that models neuron response is shown in (1),

$$\frac{\partial}{\partial t}V_{ij} = -A(V_{ij} - D) + (B - V_{ij})e_{ij}^{+} - (C + V_{ij})e_{ij}^{-} \tag{1}$$

where V_{ij} is the activity of a cell at position (i, j) and e_{ij}^{+} and e_{ij}^{-} are the excitatory and inhibitory inputs received by the cell. At equilibrium, $(\frac{\partial}{\partial t}V_{ij} = 0)$ the activity of a given cell can be modeled using equation (2).

$$V_{ij} = \frac{AD + Be_{ij}^{+} - Ce_{ij}^{-}}{A + e_{ij}^{+} + e_{ij}^{-}} \tag{2}$$

In the BioSPCIS architecture, neural computations are based on those performed by some cells present in the first stages of the mammalian visual system *(retinal ganglion cells, dLGN cells, simple and hipercomplex cells)* and modeled using equation (2). These signals, that are generated along the different stages of the BioSPCIS architecture and that participate in the segmentation process, are the result of computing weighted neural interactions in a neighborhood between different neuron layers or channels of information (receptive fields) and can be computed using equation (3).

$$e_{ij} = f(i,j) \otimes g(i,j) = \sum_{m=0}^{M-1} \sum_{n=0}^{N-1} f[m][n] \cdot g[i+m][j+n] \tag{3}$$

where f is the activity of a layer of neurons inside the BioSPCIS architecture, and g is the bidimensional *kernel* that modulates weighted connections among receptive field cells, M and N are *width* and *height* dimensions of the receptive field. For circular symmetry receptive fields, $g[i][j])$ in expression (3) can be expressed as $g[i][j] = g_{row}[i] \cdot g_{col}^T[j]$, so that convolution becomes separable and it can be computed using equation (4). Also, it is well known that it is possible to compute equation (3) using Fourier properties in the frequency domain. Consider $F(u,v)$ and $G(u,v)$ to be the spectral FFT *(Fast Fourier Transform)* signals of $f(i,j)$ and $g(i,j)$. In that case, the convolution operation between $f(i,j)$ and $g(i,j)$ in the space domain is equivalent to a multiplication in the frequency domain, as it is shown in equation (5).

On a *stream processing* architecture, different processing units can compute the activity of several cells in a parallel fashion. In a GPU-based implementation, *shaders* or *kernels* can be programmed to compute excitatory and inhibitory inputs and final response for a given cell inside a *fragment* or a *stream processor* using equations (3) and (4).

$$e_{ij} = \sum_{n=0}^{N-1} g_{col}[j+n] \sum_{m=0}^{M-1} f[i][j]g_{row}[i+m] \tag{4}$$

$$e_{ij} = f(i,j) \otimes g(i,j) = F(u,v) \cdot G(u,v) \tag{5}$$

Using separable filters when possible, and for filters larger enough, is faster than using equation (3) for computing the activation of the receptive field in the space domain. In figure 2(a), times for computing equation (4) using CUDA for a given layer of cells are shown, depending on the size of the image and the size of the receptive field to be computed. In figure 2(b) time for computing the same receptive field using equation (5) are shown. The GPU used is a GF GTX285. It can be observed how using frequency multiplication, the time needed to compute the activation of a receptive field do not depend too much on the size of the filter, unless $fftSize = imageSize + KerSize - 1$ has to be chosen to be the next power of two to avoid spectral aliasing, like in the case of applying a receptive field of size 35x35 px over an image of size 1000x1000 px in figure 2(b).

Worth to mention is that these graphs do not include the time needed to perform the copy of the data from the CPU to the GPU. Also, before the convolution can be made in the frequency domain through a multiplication, additional tasks for padding data and computing the FFT are needed. The figures 2(c) and 2(d) do take these times into account. Moreover, the model described in [5] includes normalizations to solve the noise-saturation dilemma in the form of equation (2). To perform these normalizations, an additional IFFT *(Inverse Fast Fourier Transform)* is needed. In the final implementation different layers of neurons are computed using a different method, in order to maximize the throughput, computing the FFT and the IFFT only when necessary, and keeping data on the GPU all along the process, and copying it back to the CPU only in a final step.

First layers of the architecture are computed using CUDA implementations based on equation (4): these include computation in the CoP stage of cells of types I, II and III , which have symmetric kernels in the form of Differences of Gaussians (DoG). Also, CDOC cells in the ChD stage are computed this way. In the BoD interaction between neurons are computed in the frequency domain using (5). Finally, the diffusion process in the ChD stage is computed using an iterative kernel that computes the activity dispersion around every neuron after a configurable number of iterations Δt. Although Δt can be configured, experimental tests showed a convergence of the diffusion in one thousand of iterations for images in the Berkeley Segmentation database, employed for tests in Section 3.1. These images are of size 481x321 px.

3 Experimental Results

The BioSPCIS architecture was developed using C++ and CUDA and has been tested on UNIX and Windows platforms. In this section we present experimental results achieved executing the architecture on a GeForce GTX 285 GPU. In section 3.1 the BioSPCIS architecture is benchmarked using images from the Berkeley Segmentation Dataset [10] and results are compared with those provided by other segmentation algorithms. Section 3.2 summarizes timing tests performed over the BioSPCIS GPU-based architecture.

3.1 Berkeley Segmentation Tests

In order to give a quantitative measurement of the quality of Boundary Detection (BoD) stage in the BioSPCIS architecture, several tests were performed using *The Berkeley Segmentation Dataset* (BSDS) [10]. This database uses human segmented images to provide ground truth boundaries. A benchmark can be used to compare BioSPCIS boundary map to the ground truth boundaries and obtain *precision* and *recall* measurements.

The F-measure obtained can be used to compare the efficiency of our architecture with respect to other algorithms published in the BSDS web page [2]. The quality of the results obtained with the BioSPCIS architecture change depending on the images to be segmented. Results of the benchmark over a total of 20 images using the same parameters are showed in Table 1. For each image, this table

Table 1. Comparison of F-measures obtained by humans, the worst and best algorithm published and the BioSPCIS arhitecture

Image	Hum	min	max	Best Algorithm	BioSPCIS
3096	0.74	0.68	0.78	(Global Probability of Boundary)	0.76
14037	0.77	0.62	0.77	(Global Probability of Boundary)	0.69
21077	0.82	0.57	0.76	(Ultrametric Contour Maps)	0.69
37073	0.78	0.67	0.84	(xren, Boosted Edge Learning)	0.84
42049	0.96	0.75	0.91	(xren, Boosted Edge Learning)	0.90
62096	0.90	0.59	0.87	(Ultrametric Contour Maps)	0.68
143090	0.84	0.59	0.74	(Global Probability of Boundary)	0.65
145086	0.85	0.63	0.85	(Boosted Edge Learning)	0.82
147091	0.87	0.46	0.79	(Brightness / Texture Gradients)	0.74
157055	0.78	0.63	0.80	(xren, Ultrametric Contour Maps)	0.68
167062	0.95	0.45	0.92	(Ultrametric Contour Maps)	0.87
175043	0.50	0.66	0.75	(Global Probability of Boundary)	0.74
182053	0.75	0.59	0.80	(Global Probability of Boundary)	0.74
182053	0.75	0.59	0.80	(Global Probability of Boundary)	0.74
210088	0.54	0.43	0.60	(Ultrametric Contour Maps)	0.56
216081	0.86	0.74	0.88	(Ultrametric Contour Maps)	0.78
227092	0.65	0.56	0.87	(Global Probability of Boundary)	0.86
253027	0.77	0.72	0.79	(Global Probability of Boundary)	0.81
285079	0.69	0.65	0.77	(Ultrametric Contour Maps)	0.69
299086	0.82	0.68	0.80	(xren, min-cover)	0.71
Average	**0.78**	**0.61**	**0.80**		**0.75**

includes the F-measures for humans, the F obtained by the worst algorithm, the F obtained by the best algorithm for that image, the name of that algorithm, and the F obtained by the BioSPCIS architecture. The average of the maximum F's obtained for the 20 images, considering always the best algorithm, is 0.80. The average F obtained using the BioSPCIS algorithm is 0.75, which is quite near the human average of 0.78, and is a very nice measure considering that the same algorithm and parameters have been used for all the images. A total of 109 parameters can be configured in the architecture, the time needed for the computation is larger for larger scales, due to the increasing size of receptive fields computed using separable filters. The diffusion process is faster for larger scales, because convergence is reached earlier.

3.2   GPU Implementation Performance

Table 2 shows the time needed for processing the BioSPCIS architecture over an image of size 481x321 px. This is the size of the images processed in section 3.1. One image is fully processed in approximately 3 seconds. All the

Table 2. Time (ms) to process an image of the Berkeley Database of size 481x321 px on the GPU-based BioSPCIS architecture

		Scale (s)	Scale (m)	Scale (l)	TOTAL
CoP	Types I,II and III	29,23	37,16	44,24	
BoD	Simple and Complex	241,42	243,87	252,48	
	Loop	445,18	438,47	449,01	
ChD	CDOC	24,45	29,60	42,75	
	Diffusion	321,56	315,23	305,87	
TOTAL		1061,84	1064,33	1094,35	3220,52

100 test images contained in the Berkeley database can be computed in approximately 5 minutes, or in 8 minutes if intermediate outputs are saved to the hard disk for further analysis.

4 Conclusion and Future Work

This paper presents a bio-inspired multiple scale neural model for image segmentation. Design and modeling of the full architecture has been made according to psycophysical studies of the mammalian visual system and using a shunting multiplicative model that can be implemented to be executed on a stream processor. Implementation of the architecture was developed for optimal execution on Graphics Processing Units (GPUs), exploiting data parallelism and memory bandwidth. Boundary detection properties of the architecture were benchmarked against hand-labelled segmentations using a well-known segmentation dataset. The BioSPCIS architecture provided nice results and was able to outperform other color segmentation algorithms in some of the tested images. Also, due to the GPU implementation, we are able to process 100 images from the dataset in only 5 minutes. Future work include exploring the possibility of using a multi-GPU approach for accelerating computations and incorporate these human-vision-based feature extraction tool to all our neural models, so that also image recognition processes will be accelerated.

Acknowledgments

This work has been partially supported by the Spanish Ministry of Science and Innovation under project TIN2010-20529.

References

1. Antón-Rodríguez, M., Díaz-Pernas, F.J., Díez-Higuera, J.F., Martínez-Zarzuela, M., González-Ortega, D., Boto-Giralda, D.: Recognition of coloured and textured images through a multi-scale neural architecture with orientational filtering and chromatic diffusion. Neurocomputing 72, 3713–3725 (2009)
2. BSDS: The berkeley segmentation dataset and benchmark, http://www.eecs.berkeley.edu/Research/Projects/CS/vision/grouping/segbench/ (last visit July 2010)
3. Carey, J.: Brain Facts: A primer on the brain and nervous system. The Society For Neuroscience (2006)
4. Díaz-Pernas, F.J., Antón-Rodríguez, M., Díez-Higuera, J.F., Martínez-Zarzuela, M., González-Ortega, D., Boto-Giralda, D.: Texture classification of the entire brodatz database through an orientational-invariant neural architecture. In: Mira, J., Ferrández, J.M., Álvarez, J.R., de la Paz, F., Toledo, F.J. (eds.) IWINAC 2009. LNCS, vol. 5602, pp. 294–303. Springer, Heidelberg (2009)
5. Díaz-Pernas, F., Antón-Rodríguez, M., Martínez-Zarzuela, M., Díez-Higuera, J.F., González-Ortega, D., Boto-Giralda, D.: Multiple scale neural architecture for enhancing regions in the colour image segmentation process. Expert Systems 28, 70–96 (in press)

6. Gobron, S., Devillard, F., Heit, B.: Retina simulation using cellular automata and gpu programming. Mach. Vision Appl. 18(6), 331–342 (2007)
7. Hérault, J., Durette, B.: Modeling visual perception for image processing. In: Sandoval, F., Prieto, A.G., Cabestany, J., Graña, M. (eds.) IWANN 2007. LNCS, vol. 4507, pp. 662–675. Springer, Heidelberg (2007)
8. Hodgkin, A.: The Conduction of the Nerve Impulse. Springfield (1964)
9. Hubel, D., Wiesel, T.: Receptive fields and functional architecture of monkey striate cortex. J. Physiology 195(1), 215–243 (1968)
10. Martin, D., Fowlkes, C., Tal, D., Malik, J.: A database of human segmented natural images and its application to evaluating segmentation algorithms and measuring ecological statistics. In: Proc. 8th Int'l Conf. Computer Vision, vol. 2, pp. 416–423 (July 2001)
11. Martínez-Zarzuela, M., Díaz-Pernas, F., Antón-Rodríguez, M., Díez-Higuera, J., González-Ortega, D., Boto-Giralda, D., López-González, F., De La Torre, I.: Multiscale neural texture classification using the gpu as a stream processing engine. Mach. Vision Appl. (January 2010),
   http://dx.doi.org/10.1007/s00138-010-0254-3
12. Martínez-Zarzuela, M., Pernas, F.D., Higuera, J.F.D., Antón-Rodríguez, M.: Fuzzy ART neural network parallel computing on the GPU. In: Sandoval, F., Prieto, A.G., Cabestany, J., Graña, M. (eds.) IWANN 2007. LNCS, vol. 4507, pp. 463–470. Springer, Heidelberg (2007)
13. Martínez-Zarzuela, M., Pernas, F.D., de Pablos, A.T., Antón-Rodríguez, M., Higuera, J.F.D., Boto-Giralda, D., Ortega, D.G.: Adaptative resonance theory fuzzy networks parallel computation using CUDA. In: Cabestany, J., Sandoval, F., Prieto, A., Corchado, J.M. (eds.) IWANN 2009. LNCS, vol. 5517, pp. 149–156. Springer, Heidelberg (2009)
14. Mingolla, E., Ross, W., Grossberg, S.: A neural network for enhancing boundaries and surfaces in synthetic aperture radar images. Neural Networks 12, 499–511 (1999)
15. Obermayer, K., Blasdel, G.G.: Geometry of orientation and ocular dominance columns in monkey striate cortex. Journal of Neuroscience 13, 4114–4129 (1993)
16. Owens, J., Luebke, D., Govindaraju, N., Harris, M., Krüger, J., Lefohn, A., Purcell, T.: A survey of general-purpose computation on graphics hardware. Computer Graphics Forum. 26(1), 80–113 (2007),
   http://www.blackwell-synergy.com/doi/pdf/10.1111/
   j.1467-8659.2007.01012x
17. Rumpf, M., Strzodka, R.: Level set segmentation in graphics hardware. In: Proceedings of the IEEE International Conference on Image Processing (ICIP 2001), vol. 3, pp. 1103–1106 (2001)
18. Schwartz, S.: Visual Perception: a clinical orientation, 3rd edn. McGraw-Hill, New York (2004)
19. Viola, I., Kanitsar, A., Gruller, M.E.: Hardware-based nonlinear filtering and segmentation using high-level shading languages. IEEE Visualization, 309–316 (2003)

Simulating a Rock-Scissors-Paper Bacterial Game with a Discrete Cellular Automaton

Pablo Gómez Esteban and Alfonso Rodríguez-Patón

Departamento de Inteligencia Artificial,
Universidad Politécnica de Madrid (UPM)
arpaton@fi.upm.es

Abstract. This paper describes some of the results obtained after the design and implementation of a discrete cellular automata simulating the generation, degradation and diffusion of particles in a two dimensional grid where different colonies of bacteria coexist and interact. This lattice-based simulator use a random walk-based algorithm to diffuse particles in a 2D discrete lattice. As first results, we analyze and show the oscillatory dynamical behavior of 3 colonies of bacteria competing in a non-transitive relationship analogous to a Rock-Scissors-Paper game (Rock bacteria beats Scissors bacteria that beats Paper bacteria; and Paper beats Rock bacteria). The interaction and communication between bacteria is done with the quorum sensing process through the generation and diffusion of three small molecules called autoinducers. These are the first results obtained from the first version of a general simulator able to model some of the complex molecular information processing and rich communication processes in synthetic bacterial ecosystems.

Keywords: bacterial computing, cellular automata, lattice-based simulation, rock-scissors-paper game, quorum sensing, particle diffusion, autoinducers.

1 Introduction

This paper describes: (1) a lattice-based simulator used to model the diffusion of particles over a two dimension grid and (2) the results obtained modeling the cyclic dominance behavior of 3 colonies of bacteria communicating through quorum sensing signals. This simulator is used in BACTOCOM project [1]. The simulator is used to model the diffusion of the so called autoinducers (small molecules generated by bacteria when they sense they are in a high density). These autoinducers are the information carriers used by bacteria to take decisions by majority (by quorum) in a communication and decision making protocol called quorum sensing [6]. Biologists know the precise biological hardware used by bacteria to generate and to sense autoinducers. And synthetic biologists try to program artificial complex communication protocols in synthetic bacterial ecosystems. We try to model and to simulate natural computations made by bacteria with artificial/digital computations.

J.M. Ferrández et al. (Eds.): IWINAC 2011, Part II, LNCS 6687, pp. 363–370, 2011.

The design and study of engineered complex bacterial behavior has been a constant issue in synthetic biology. There are already synthetic bacterial colonies interacting through quorum sensing signals and competing in a predator-prey way [3], as well as colonies of bacteria forming bright complex patterns via quorum sensing [5] or competing in a a rock-paper-scissors game [4]. Our purpose is to model, simulate and engineer sociobiological behaviors as cooperation or competition between bacteria. A simulator able to model and predict these complex behaviors would be an interesting aided design tool for microbiologists.

There exists many particle diffusion simulators, but we had some conditions to be followed. First of all, we should be working on a 2D Cellular Automata (CA) that would represent the Petri dish where bacteria will be living in. The reason behind the choice of working with a 2D CA is because our bacteria growth in a Petri dish with enough nutrients at the bottom. These bacteria growth in 2 dimensions. Third dimension (the formation of a biofilm) it's not necessary to model our bacteria environment. We also choose to work with discrete space and discrete time. All the events performed in the simulator should be executed in an asynchronous way to be more realistic.

There are some previous simulators of biofilm (colonies of bacteria in 3 dimensions) growth, like [12] which was one of the first bacterial simulators developed, or [13] and [14] which both of them are individual-based simulators, one using a discrete approach and the other one using a continuous space approach. Other simulators like [11] uses particle concentration to simulate the diffusion process or [9] which introduces an hybrid simulator combining two approaches: cellular automata and reaction-diffusion equations. Our simulator has been inspired from other previous simulators like BacMIST [10] and JCASim [7], [8]. BacMIST uses the idea of random walks to simulate the diffusion process and introduces the concept of "diffusivity" which allow the particles to be spread in the environment with more or less ease, depending on the nature of the particle or the viscosity of the media. JCASim introduces a discrete simulation of diffusion with the so called Block-Cellular-Automata (Block-CA).This diffusion algorithm will be explain in detail in the next section.

2   Description of the Simulator

In this section, the algorithms (emission, diffusion and degradation of autoinducers) inside the simulator will be explained.

The simulator is based on a 2D discrete cellular automata that represents the environment (a Petri dish) where the bacteria are placed. Each of the sites on the grid, can be filled with at most one bacteria and one autoinducer particle. In each time-step, among all the bacteria in the grid, one will be selected to perform an event in a probabilistic way (this is what it's called an iteration). This event could be reproduction, conjugation (transmission of DNA circular strands between a donor bacteria and a receiver bacteria) or autoinducer emission. Each event has its own conditions: reproduction can take place only if there is at least one empty site in the neighborhood, and conjugation takes place only if there is a receiver and a donor in the vicinity. The general description of the simulation workflow is as follows.

Quorum Sensing Simulator Workflow

Initialization: Select the size of the grid and define initial values of the parameters (sigmoidal function rate, quorum sensing threshold, conjugation rate, diffusion rate autoinducer, decay rate autoinducer).

- **Step 0**: Input. Situate an initial colony of 3 bacteria on the grid (in this initial version of the simulator prototype the initial colony can only formed by 3 different bacteria situated in the centre of the grid in a well-mixed manner or spatially separated at a short or large distance). These 3 bacteria emit 3 different autoinducers and form a Rock-Paper-Scissor bacterial game. Rock autoinducers repress light emission in Scissors bacteria, Scissors autoinducers repress light emission in Paper bacteria and Paper repress Rock light emission.

- **Step 1**: FOR ALL bacteria in the grid:
  1.1 Select an event randomly (reproduction or conjugation) and
  1.2. Calculate the number of Autoinducer (AI) particles in the vicinity of every bacteria and
  1.3 Depending on the value calculated in 1.2 and on the QS threshold parameter, decide the QS AI emission (emit or not an AI auto-inducer) and activate (or not) all other QS-dependent behaviors of the bacteria (in the prototype of the simulator only the light emission).

- **Step 2**. Increase time: $t := t + 1$.

- **Step 3**. FOR ALL AI particles: Apply decay function to decide which particles are degraded.

- **Step 4**: Diffuse all the AI particles

- **Step 5**. GO TO Step 1.

This QS simulator works in a synchronous way. All the bacteria in Step 1 must be selected before increasing time in step 2. All the AI particles must be degraded in Step 3 before applying diffusion in Step 4.

Autoinducers Emission Process: The process by which a bacterium x activates its autoinducers emission follows a sigmoid probabilistic function $f(x,k) = 1/(1+e^{-(K-m)/s})$, where K is the number of autoinducers of the same specie in the neighborhood, m is a user customizable parameter (called Sigmoid function rate in the interface) and s is a fixed parameter which determines the function shape. By default, s has a value of 0.6 estimated by empirical test.

Decay Process and degradation of AI particles: All the particles in the environment follow an exponential decay probabilistic function $f(t) = 1 - e^{-K*t}$, being K the decay rate and t the variable time. As t increases the probability that a particle disappears tends to 1.

Diffusion Process Algorithm: We follow the approach called Block-Cellular-Automaton (Block-CA), which is an approach to simulate random walks in discrete cellular automata. This diffusion algorithm starts making a tessellation of the grid using square blocks with 9 cells. Then, the content of chosen pairs of cells inside those blocks are exchanged (or not) with a probability of $\frac{1}{2}$ multiplied by the diffusivity rate of the media. This process is repeated 8 times with new pairs of adjacent cells selected in a new orientation (rotated clockwise).

3   Results

To test the correct behavior of the system and to see the effect of the diffusion of the autoinducer particles we run some simulations. We are interested in the dynamic behavior of interacting bacteria populations so we decided to simulate and replicate a Rock - Scissors - Paper (RSP) model. An example of this bacterial game is also described in the Southampton's work presented in the 2009 iGem contest [17]. The RSP model is a non-hierarchical competitive system performed by three different populations which usually compete for a resource. The competition is established by some simple rules: Rock crushes Scissors, Scissors cut Paper and Paper wraps Rock. In the Southampton's RSP model [17] they designed three types of an engineered bacteria which operates with different autoinducers. At the beginning of the experiment all the three bacteria are lighting and sending their own autoinducers. Using its own quorum sensing circuit, each population of bacteria will stop lighting and sending autoinducers it they detect a high concentration of their antagonist autoinducer. In other words, Scissors bacteria light emission will be stopped if they detect, via quorum sensing, a high concentration of Rock autoinducers, and so on. We wanted to see whether our simulator could or not replicate this cyclic dominance behavior. In the first simulation we made, all the bacteria were well mixed in the environment. The quorum sensing threshold that established if a bacteria becomes inactive was fixed to 0.3. That means, that there must be at least 3 particles of the corresponding population of autoinducers in its neighborhood to become activated. All the autoinducers in the simulation have the same decay rate (this establish an autoinducer half-life of 1386 iterations).

The next two figures show the number of active bacteria (bacteria emitting light) in each of the three populations (Fig. 1 and Fig. 2).

The difference between Fig. 1 and Fig. 2 is the diffusivity rate (propensity to diffuse). As one can observe in the figures, the oscillatory behavior is modified by the diffusivity rate. As the three populations are well mixed in the environment, they receive autoinducers from every population at the same time.

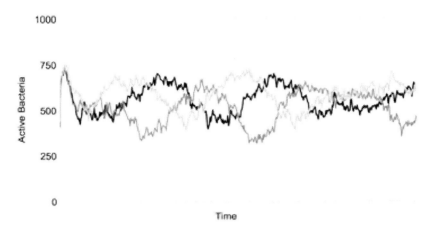

Fig. 1. Oscillations in the number of active bacteria following a Rock - Scissors - Paper cyclic dominance behavior. The diffusivity rate is 0.1 and the three bacteria are in a well mixed environment. A small diffusivity rate means that the movement of the autoinducer particles all along the grid is slow.

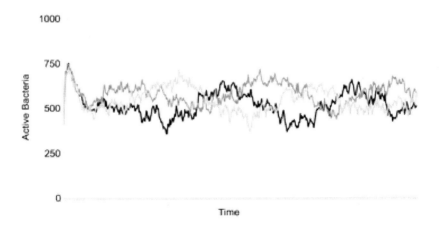

Fig. 2. Oscillations in the number of active bacteria following a Rock - Scissors - Paper cyclic dominance behavior. The diffusivity rate is 1 (autoinducer particles movement is faster). The system maintains the cyclic oscillatory behavior but with less amplitude in the oscillations.

In Fig. 3 and Fig. 4 the same simulation is performed but now the three bacteria colonies are not well mixed but spatially isolated one from the others. The RSP cyclic dominance is lost due to the distance and a low diffusivity rate of 0.1.

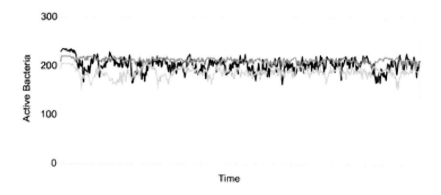

Fig. 3. The bacterial ecosystem is spatially separated and shows a soft oscillatory behavior but not Rock-Scissors-Paper cyclic oscillations. The bacteria are isolated and the diffusivity rate is 1.

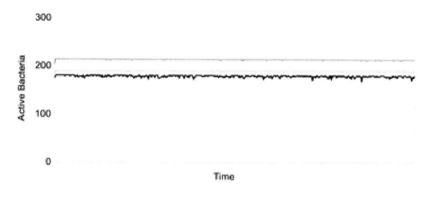

Fig. 4. The bacterial ecosystem is spatially separated and shows almost no oscillations in the number of active bacteria. The bacteria are isolated and the diffusivity rate is 0.1.

4   Conclusions

This paper describes the results obtained after the design and implementation of a new cellular automata simulating particle diffusion on bacterial ecosystems. To achieve the main goals established, a random walk-based algorithm has been used to diffuse particles in a virtual 2 dimension environment. A particle decay and a bacteria degradation process has been implemented following an exponential function. Some others functionalities has been implemented, like bacteria autoinducers emission or behavior activation via quorum sensing which allows the bacteria to perform many different events as for example, light emission.

The next step we plan to develop with this simulator is to combine it with other simulator modules that our research group has been working on. With that improvement, we will be able to simulate and study more complex bacterial ecosystems and its sociobiological interactions [15] .

Acknowledgements

Research was partially supported by project BACTOCOM (Project reference: 248919) funded by grant from the European Commission under Seventh Framework Programme (FET Proactive Program), by the Ministerio de Ciencia e Innovación (MICINN), Spain, under project TIN2009 − 14421, and by the Comunidad de Madrid and UPM (grant to the LIA research group).

References

1. BACTOCOM project, www.bactocom.eu
2. Llosa, M., de la Cruz, F.: Bacterial conjugation: a potential tool for genomic engineering. Research in Microbiology 156(1), 1–6 (2005)
3. Frederick, Song, H., Ozaki, J., Collins, C.H., Barnet, M., Arnold, F.H., Quake, S.R., You, L.: A synthetic Escherichia coli predator-prey ecosystem. Molecular Systems Biology (April 2008)
4. Kerr, B., Riley, M.A., Marcus: Local dispersal promotes biodiversity in a real-life game of rock-paper-scissors. Nature 418, 171–174 (2002)
5. Basu, S., Gerchman, Y., Collins, C.H., Arnold, F.H., Weiss, R.: A synthetic multicellular system for programmed pattern formation. Nature 434, 1130–1134 (2005)
6. Waters, C.M., Bassler, B.L.: Quorum sensing: cell-to-cell communication in bacteria. Annual review of cell and developmental biology 21(1), 319–346 (2005)
7. Freiwald, U., Weimar, J.R.: The Java based cellular automata simulation system, JCASim (2002)
8. Weimar, J.R.: Simulating reaction-diffusion cellular automata with JCASim. Discrete Modelling and Discrete Algorithms in Continuum Mechanics (2001)
9. Bandman, O.: A hybrid approach to reaction-diffusion processes simulation. In: Malyshkin, V.E. (ed.) PaCT 2001. LNCS, vol. 2127, pp. 1–6. Springer, Heidelberg (2001)
10. Chang, I., Gilbert, E.S., Eliashberg, N., Keasling, J.D.: A three-dimensional, stochastic simulation of biofilm growth and transport-related factors that affect structure. Microbiology 149, 2859–2871 (2003)
11. Kim, T.-H.H., Jung, S.H.H., Cho, K.-H.H.: Investigations into the design principles in the chemotactic behavior of Escherichia coli. Biosystems 91(1), 171–182 (2008)
12. Picioreanu, C., van Loosdrecht, M.C.M., Heijnen, J.J.: A new combined differential-discrete cellular automaton approach for biofilm modeling: Application for growth in gel beads. Biotechnol. Bioeng. 57(6), 718–731 (1998)
13. Kreft, J.-U., Booth, G., Wimpenny, J.W.T.: BacSim, a simulator for individual-based modelling of bacterial colony growth. Microbiology 144(12), 3275–3287 (1998)
14. Ginovart, M., López, D., Valls, J.: INDISIM, an individual-based discrete simulation model to study bacterial cultures. Journal of Theoretical Biology 214(2), 305–319 (2002)

15. Czárán, T.L., Hoekstra, R.F., Pagie, L.: Chemical warfare between microbes promotes biodiversity. Proceedings of the National Academy of Sciences of the United States of America 99(2), 786–790 (2002)
16. Melke, P., Sahlin, P., Levchenko, A., Jönsson, H.: A cell-based model for quorum sensing in heterogeneous bacterial colonies. PLoS Computational Biology 6(6), e1000819 (2010)
17. Southampton University Team. iGem (2009),
    http://2009.igem.org/Team:Southampton

Mobile Robot Localization through Identifying Spatial Relations from Detected Corners

Sergio Almansa-Valverde[1], José Carlos Castillo[1],
Antonio Fernández-Caballero[1,2], José Manuel Cuadra Troncoso[3],
and Javier Acevedo-Rodríguez[4]

[1] Instituto de Investigación en Informática de Albacete (I3A), n&aIS Group,
Campus Universitario s/n, 02071-Albacete, Spain
[2] Departamento de Sistemas Informáticos, Universidad de Castilla-La Mancha,
Campus Universitario s/n, 02071-Albacete, Spain
[3] Departamento de Inteligencia Artificial, E.T.S.I. Informática,
Universidad Nacional de Educación a Distancia, 28040-Madrid, Spain
[4] Universidad de Alcalá,
Departamento de Teoría de la Señal y Comunicaciones,
Alcalá de Henares, Spain

Abstract. In this paper, the Harris corner detection algorithm is applied to images captured by a time-of-flight (ToF) camera. In this case, the ToF camera mounted on a mobile robot is exploited as a gray-scale camera for localization purposes. Indeed, the gray-scale image represents distances for the purpose of finding good features to be tracked. These features, which actually are points in the space, form the basis of the spatial relations used in the localization algorithm. The approach to the localization problem is based on the computation of the spatial relations existing among the corners detected. The current spatial relations are matched with the relations gotten during previous navigation.

1 Introduction

A mobile robot must possess the capacity of self-localization while navigating in an environment [1]. An appearance-based approach for place recognition involves matching scenes based on selected features observed within the current local map or sensor view. The combination of a location and descriptor vector is termed a key point [3]. Place recognition then becomes a matter of identifying places by associating key points, or deciding that a place has not previously been seen. In a paper [16] the performance of a variety of corner (point) detecting algorithms for feature tracking applications is assessed. The overall observation of the results suggest that the Harris corner detector [9] is very suitable for tracking features in long sequences. Also, the overall empirical results revealed that the Kanade-Lucas-Tomasi (KLT) [17] and Harris detectors provided the best quality corners (qualitatively and quantitatively). In a recent review [12], detectors and local descriptors of local features for computer vision are described in a comprehensive way.

J.M. Ferrández et al. (Eds.): IWINAC 2011, Part II, LNCS 6687, pp. 371–380, 2011.
© Springer-Verlag Berlin Heidelberg 2011

Time-of-flight (ToF) range cameras give depth information per pixel which make them ideal for background foreground segmentation, as in general the depth defines the subject from background in a much more basic way than the light intensity does [8], [14], [6]. Intensity images are on the other hand affected by colors, lighting, reflections and shadows in almost every normal scenario [13]. Thanks to the larger vertical field of view of ToF cameras, difficult obstacles (like tables) are better detected by a ToF camera than by a 2D laser scanner [18]. Most of the applications extract planar regions using both intensity and depth images. In a paper [15] different methods are explored to improve pose estimation. The normal of the extracted planes is also used [10] to detect badly conditioned plane detection, as horizontal planes in a staircase. Also a corner filtering scheme combining both the intensity and depth image of a ToF camera has been proposed [7].

Our approach uses the Harris corner detection algorithm, similarly to [5], [4], mainly as its computational cost is lower than other approaches like SIFT [11]. In this case, the ToF camera is exploited as a gray-scale camera for localization. The ToF camera allows to create a gray-scale image representing distances for the purpose of finding good features to be tracked. These features, which actually are points in the space, form the basis of the spatial relations used in the localization algorithm. The approach to the localization problem is based on the computation of the spatial relations existing among the aforementioned points in the space. The current spatial relations are matched with the ones created in previous iterations of the algorithm during autonomous navigation. The robot position is estimated in accordance to the localization information provided through the identified spatial relations.

2   Mobile Robot Localization Algorithm

To estimate the robot position in the environment, only the information provided by the ToF camera is used. In this case, the images provided by the ToF camera are considered as traditional images. That is, the depth information provided is considered as gray-level values of a traditional image. From these image pixel values, a series of characteristic points (or corners) are extracted to create spatial relations to be placed in the map that the robot has been previously given. When new spatial relations are created from detected corners, they are compared to the previously created ones and identified to calculate their location. A flow diagram of the localization algorithm is presented in Fig. 1.

2.1   Detection of Corners

To exploit the visual characteristics of the ToF camera, a gray-scale image is created from the distance information, I_d. The ToF camera provides as output a matrix where each position represents the three-dimensional coordinates (x, y, z) (in meters) of a system. Here the camera is the origin of coordinates, x varies along the horizontal axis, y varies along the vertical axis and z is the distance

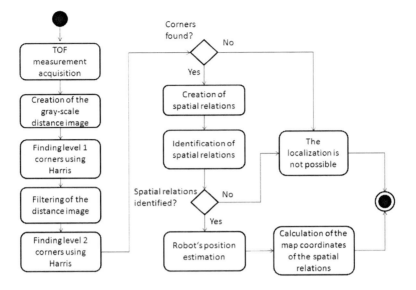

Fig. 1. Flow diagram of the localization algorithm

from the plane defined by the x and y axes. In this case, the gray level of each image pixel depends on the z value (depth) of the coordinates provided by the ToF camera for this point. Therefore, the image represents the scene using darker gray levels for near objects, and lighter levels for distant objects. Creating an image representation of the distance measurements enables the use of corner detection algorithm. Concretely, the Harris algorithm has been used for this purpose.

Applying Harris algorithm to the distance image results in a list of corners. Sometimes the difference between distances (and therefore, between gray levels) in the image is too slight, which makes it difficult to find enough significant corners. A hierarchy of two levels of corners is implemented to tackle this problem. The first level of corners is composed by those found in the initial distance image, I_d. Usually not many corners are found due to the low contrast of these images, although the gotten corners are quite resistant to noise. Some filters are performed on image I_d to get the second level of corners. A first filter equalizes the image histogram to enhance the contrast. After that, as noise is also enhanced with the equalization, a smoothing Gaussian is applied to the equalized result, obtaining I_f. In order to implement Gaussian smoothing, a transformation matrix is created from the values of the Gaussian distribution, as follows:

$$G(x,y) = \frac{1}{2\pi\sigma^2} e^{-\frac{x^2+y^2}{2\sigma^2}} \tag{1}$$

where x and y are the coordinates of the value in the Gaussian distribution and σ is the standard deviation. Matrix $G(x,y)$ is applied to every pixel of the image, setting the new value of each pixel to a weighted average of its neighboring pixels.

Using Harris algorithm on I_f returns more corners than on I_d, but they will be less resistant to noise. Hence, spatial relations based on corners belonging to level 1 are more trusted. So, they have higher priority than those based on corners belonging to level 2. But the last ones are indispensable for a correct localization as level 1 rarely contains enough corners to achieve a good localization. Fig. 2a and Fig. 2b show the corners found at levels 1 and 2, respectively. Each corner possesses the coordinates with respect to the image, $(x_{ToF}, y_{ToF}, z_{ToF})$ from the camera.

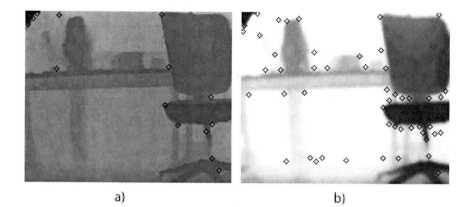

a)   b)

Fig. 2. Hierarchy of corners. (a) Level 1 corners on non-filtered image I_d. (b) Level 2 corners on filtered image I_f.

2.2   Creation of Spatial Relations

Once the corners are extracted, spatial relations connecting every pair of points from the same level are established. Some information is associated to these spatial relations in order to define and to identify them. The spatial relations information contains the following attributes: priority, according to the level of the corners belonging to the spatial relation, and distance in meters between the two points, N, calculated as follows:

$$N = \sqrt{(x_{ToF_1} - x_{ToF_2})^2 + (y_{ToF_1} - y_{ToF_2})^2 + (z_{ToF_1} - z_{ToF_2})^2} \qquad (2)$$

where the first corner has $(x_{ToF_1}, y_{ToF_1}, z_{ToF_1})$ as camera coordinates, and the second one has coordinates $(x_{ToF_2}, y_{ToF_2}, z_{ToF_2})$. The third and fourth attributes are the slopes between x and y coordinates and between x and z coordinates of the vector connecting the two corners in the space. The slope between x and y is calculated as follows:

$$S_y = \frac{x_{ToF_1} - x_{ToF_2}}{y_{ToF_1} - y_{ToF_2}} \qquad (3)$$

And the slope S_z between x and z is calculated in a similar manner. The inclination of the vector in the space is represented with these two slope values (the slope between y and z, S_x, can be calculated from the other two slopes). There is enough information to identify spatial relations comparing the last three attributes.

2.3   Identification of Spatial Relations

The next step in the localization algorithm is the identification of new spatial relations to calculate their location in the robot's map (that is, if some points have known positions in the map, new points can also be placed if there are spatial relations between them), to finally estimate the robot position. For this purpose, the attributes of distance and slope are compared to identify the spatial relations.

When identifying a spatial relation it is not necessary to find another identical one, but similarity tolerance values τ_{si} and τ_{sl} in the distance between the points and in the slopes, respectively, have been included. These tolerance values are mandatory because the ToF camera usually provides noisy distance information; so, different measurements of the same scene would result on slightly different distance information, thus affecting corners and spatial relations.

Being R_1 a spatial relation obtained from two points of the current observation and R_2 a spatial relation from two previously observed points, the matching between R_1 and R_2 can be formulated as follows:

$$R_1 = R_2 \Leftrightarrow \begin{cases} N_1 - N_2 < \tau_{si} \\ S_{y1} - S_{y2} < \tau_{sl} \\ S_{z1} - S_{z2} < \tau_{sl} \end{cases} \tag{4}$$

being N_1, S_{y1}, S_{z1} the attributes previously explained of R_1, and N_2, S_{y2}, S_{z2} the attributes of R_2.

The number of spatial relations grows with the size of the environment, raising the probability of creating very similar spatial relations. This might cause erroneous identification of spatial relations and, in consequence, lead to an erroneous robot localization. To minimize identification errors, the number of identifications is restricted to a maximum of 10, although the number of spatial relations is not limited. The robot position is estimated multiple times using them one by one. Next, Chebyshev's inequality [2] is applied to the estimated robot positions to discard outliers caused by erroneous identifications. Finally, the median value (calculated from the rest of estimations) constitutes the final robot pose estimation. Chebyshev's inequality assures that in a data sample almost all the values are close to its mean value. It defines intervals depending on the parameter k to fix the deviation tolerance, and formulates that:

$$P(Q - \mu > k\sigma) \leq \frac{1}{k^2} \tag{5}$$

being Q a random variable of mean μ and typical deviation σ. The right side of the equation represents the highest possible percentage of values in the sample that will not belong to the interval. For example, for $k = 1.6$ the Chebyshev's inequality guarantees that at least 60% of the values will belong to the interval, which means that these values are close enough to the rest of the sample values.

After the spatial relations -created at time instant t- are identified by the comparison with the previously created ones, two important pieces of information are available. Firstly, there are the ToF coordinates of the corners belonging to the new spatial relation, which represent a connection between the position of the robot and the corners. And secondly, the map coordinates of the corners belonging to the previously created spatial relation equivalent to the new one, are present. As the spatial relations are considered equivalent, the new spatial relation has the same map coordinates than the older one.

3   Data and Results

In this section, the results and configuration of the experiments are presented. The test environment consists of a laboratory containing several obstacles (mainly composed of tables and several objects under them) as shown in Fig. 3. Also notice the presence of chairs in the scenario. The ToF camera used in the research and experiments is a Mesa Imaging camera model SR4000.

Fig. 3. Laboratory test scenario partial view

There are two factors which have complicated the localization performance. First, the presence of the aforementioned obstacles in the scenario, and second, the height interval that determines which objects will be included. Fig. 4 presents the ToF observation made at a random iteration. It is presented with the approximate height interval used in the experiment and the corners found by Harris algorithm. In the figure, it is shown that several objects (including the wall) are found within the same height interval.

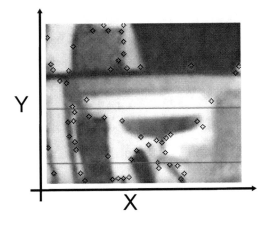

Fig. 4. Different obstacles detected with the selected height interval

To understand the difference between the representation of a group of obstacles in the same direction and the representation of only one obstacle (the most intuitive case would be a wall), Fig. 5 shows a comparison of the representation of multiple objects (see Fig.5a), and the representation of a wall (as shown in Fig. 5.b). The figure also presents the ToF images and the relative position of the robot.

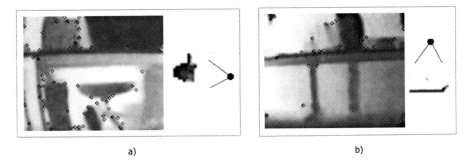

Fig. 5. Comparison of the representation of different kinds of obstacles. (a) Left: ToF image of several obstacles at different heights. Right: representation in the map (coordinates $(-2, 1)$) and position of the robot. (b) Left: ToF image of a single obstacle (a wall). Right: representation in the map (coordinates $(1, -3)$) and position of the robot.

ToF images are presented in Fig. 4 and Fig. 5, all of them with the respective corners found by the Harris algorithm. As explained before, spatial relations are created from each pair of corners at each level. Fig. 6 shows the spatial relations created for only one corner in a random ToF image captured during this test. Of course, spatial relations from the rest of corners and from the other level corners

a)   b)

Fig. 6. Example of spatial relations. (a) ToF image with second level corners. (b) Spatial relations corresponding to a random corner.

are also created. Thus, the localization is based on the identification of these spatial relations, having a group of at most 10 identifications per iteration.

The information of the localization for one of the iterations of the current test is presented next. Initially the robot coordinates are $(0, 0, 0)$. In this case, the robot has been moved to coordinates $(0, -0.4)$ and has an orientation of -1.047 radians (or 330 degrees). Firstly, the comparison between new and previous spatial relations' attributes is carried out to achieve the identifications. Table 1 presents the results of the identification of the spatial relations for the current iteration. As aforesaid, S_y is defined as the slope between x and y, S_z as the slope between x and z, and N as the distance between the points (measured in meters). Each row of the table presents, on the one hand the attributes of the spatial relations observed in the current iteration and, on the other one the attributes of the spatial relations previously detected and stored, and compared to the first ones.

Table 1. Identification of spatial relations

#	Observed spatial relation			Stored spatial relation			Level
	S_{y1}	S_{z1}	N_1	S_{y2}	S_{z2}	N_2	
1	0.148803	0.045169	0.638492	0.148148	0.046868	0.639393	1
2	0.010794	0.026259	0.977325	0.014683	0.025786	0.975971	1
3	0.341993	0.471470	0.499058	0.338043	0.471988	0.495946	1
4	1.024206	0.832313	0.740300	1.022245	0.835643	0.737851	1
5	0.257143	0.257885	0.620442	0.257143	0.259601	0.618227	1
6	0.562114	0.330744	0.558678	0.562114	0.331833	0.557799	1
7	0.257143	0.274846	0.602269	0.257143	0.271577	0.605315	1
8	0.470173	0.456345	0.532023	0.469658	0.457272	0.531060	1
9	0.017528	0.003643	0.550909	0.018519	0.008466	0.546147	1
10	0.100000	0.103997	0.370625	0.100000	0.099922	0.371421	1

All the identifications in the table fulfill equation (4) as it can be checked by comparing the attributes of the spatial relations in the table. There are two significant pieces of information in the table. First, all of the identified spatial relations belong to the first level, and second, the maximum number of spatial relations has been identified. This information indicates, a priori, that the localization would be reliable. For example, a localization starting from a shorter group of identifications, with spatial relation from the second level would be less reliable.

4   Conclusions

In this paper, a localization algorithm for mobile robots has been introduced. The Harris corner detection algorithm is applied to images captured by a ToF camera mounted on the mobile robot. In this case, the ToF camera is exploited as a gray-scale camera. The gray-scale image represents distances for the purpose of finding good features to be tracked. These features form the basis of the spatial relations used in the localization algorithm. The approach to the localization problem is based on the computation of the spatial relations existing among the corners detected. The current spatial relations are matched with the relations gotten during previous navigation.

The paper has explained the three steps present in the algorithm, namely, "detection of corners", "creation of spatial relations", and "identification of spatial relations". An experiment in a complex laboratory has shown the goodness of the proposal. Our future work consists now in taking a next step towards simultaneous localization and mapping (SLAM).

Acknowledgements

This work was partially supported by the Spanish Ministerio de Ciencia e Innovación under project TIN2010-20845-C03, and by the Spanish Junta de Comunidades de Castilla-La Mancha under projects PII2I09-0069-0994 and PEII09-0054-9581.

References

1. Begum, M., Mann, G.K.I., Gosine, R.G.: Integrated fuzzy logic and genetic algorithmic approach for simultaneous localization and mapping of mobile robots. Applied Soft Computing 8, 150–165 (2008)
2. Bennett, G.: Probability inequalities for the sum of independent random variables. Journal of the American Statistical Association 57(297), 33–45 (1962)
3. Bosse, M., Zlot, R.: Keypoint design and evaluation for place recognition in 2D lidar maps. Robotics and Autonomous Systems 57, 1211–1224 (2009)
4. Böhm, J.: Orientation of image sequences in a point-based environment model. In: Sixth International Conference on 3-D Digital Imaging and Modeling, pp. 233–240 (2007)

5. Davison, A.J., Murray, D.W.: Simultaneous localization and map-building using active vision. IEEE Transactions on Pattern Analysis and Machine Intelligence 24(7), 865–880 (2002)
6. Fernández-Caballero, A., López, M.T., Mira, J., Delgado, A.E., López-Valles, J.M., Fernández, M.A.: Modelling the stereovision-correspondence-analysis task by lateral inhibition in accumulative computation problem-solving method. Expert Systems with Applications 33(4), 955–967 (2007)
7. Gemeiner, P., Jojic, P., Vincze, M.: Selecting good corners for structure and motion recovery using a time-of-flight camera. In: Proceedings of the IEEE/RSJ International Conference on Intelligent Robots and Systems, pp. 5711–5716 (2009)
8. Guethmundsson, S.A., Pardas, M., Casas, J.R., Sveinsson, J.R., Aanaes, H., Larsen, R.: Improved 3D reconstruction in smart-room environments using ToF imaging. Computer Vision and Image Understanding 114, 1376–1384 (2010)
9. Harris, C., Stephens, M.: A combined corner and edge detector. In: The Fourth Alvey Vision Conference, pp. 147–151 (1988)
10. Hedge, G., Ye, C.: Extraction of planar features from Swissranger SR-3000 range images by a clustering method using normalized cuts. In: Proceedings of the IEEE/RSJ International Conference on Intelligent Robots and Systems, pp. 4034–4039 (2009)
11. Klippenstein, J., Zhang, H.: Quantitative evaluation of feature extractors for visual SLAM. In: Proceedings of the Fourth Canadian Conference on Computer and Robot Vision, pp. 157–164 (2007)
12. Li, J., Allinson, N.M.: A comprehensive review of current local features for computer vision. Neurocomputing 71, 1771–1787 (2008)
13. López, M.T., Fernández-Caballero, A., Mira, J., Delgado, A.E., Fernández, M.A.: Algorithmic lateral inhibition method in dynamic and selective visual attention task: Application to moving objects detection and labelling. Expert Systems with Applications 31(3), 570–594 (2006)
14. López-Valles, J.M., Fernández, M.A., Fernández-Caballero, A.: Stereovision depth analysis by two-dimensional motion charge memories. Pattern Recognition Letters 28(1), 20–30 (2007)
15. May, S., Droeschel, D., Holz, D., Fuchs, S., Malis, E., Nüchter, A., Hertzberg, J.: Three-dimensional mapping with time-of-flight cameras. Journal of Field Robotics 26(11-12), 934–965 (2009)
16. Tissainayagam, P., Suter, D.: Assessing the performance of corner detectors for point feature tracking applications. Image and Vision Computing 22, 663–679 (2004)
17. Tomasi, C., Kanade, T.: Detection and tracking of point features. Carnegie Mellon University Technical Report CMU-CS-91-132 (1991)
18. Weingarten, J.W., Gruener, G., Siegwart, R.: A state-of-the-art 3D sensor for robot navigation. In: Proceedings of the IEEE/RSJ International Conference on Intelligent Robots and Systems, vol. 3, pp. 2155–2160 (2004)

Improving the Accuracy of a Two-Stage Algorithm in Evolutionary Product Unit Neural Networks for Classification by Means of Feature Selection

Antonio J. Tallón-Ballesteros[1], César Hervás-Martínez[2],
José C. Riquelme[1], and Roberto Ruiz[3]

[1] Department of Languages and Computer Systems,
University of Seville, Spain
`atallon@us.es`
[2] Department of Computer Science and Numerical Analysis,
University of Córdoba, Spain
[3] Area of Computer Science,
Pablo de Olavide University, Seville, Spain

Abstract. This paper introduces a methodology that improves the accuracy of a two-stage algorithm in evolutionary product unit neural networks for classification tasks by means of feature selection. A couple of filters have been taken into consideration to try out the proposal. The experimentation has been carried out on seven data sets from the UCI repository that report test mean accuracy error rates about twenty percent or above with reference classifiers such as C4.5 or 1-NN. The study includes an overall empirical comparison between the models obtained with and without feature selection. Also several classifiers have been tested in order to illustrate the performance of the different filters considered. The results have been contrasted with nonparametric statistical tests and show that our proposal significantly improves the test accuracy of the previous models for the considered data sets. Moreover, the current proposal is much more efficient than a previous methodology developed by us; lastly, the reduction percentage in the number of inputs is above a fifty five, on average.

1 Introduction

The classification problem has been dealt by several machine learning techniques. Algorithms which construct classifiers from sample data, such as neural networks, radial basis functions, and decision trees, have attracted growing attention for their wide applicability. The explosion of available information complicates this problem. Moreover, redundancy or noise may be present on data. Neural networks models play a crucial role in pattern recognition [2]. For many practical problems, the possible inputs to an Artificial Neural Network (ANN) can be huge. There may be some redundancy among different inputs. Pre-processing

J.M. Ferrández et al. (Eds.): IWINAC 2011, Part II, LNCS 6687, pp. 381–390, 2011.

is often needed to reduce the number of inputs to an ANN. The application of feature selection (FS) approaches has become a real prerequisite for model building due to the multi-dimensional nature of many modeling task in some fields. Theoretically, having more features should give us more discriminating power. However, this can cause several problems: an increased computational complexity and cost; too many redundant or irrelevant features; and degradation in the classification error estimation.

Our objective is to improve the accuracy and to reduce the complexity (measured by means of the number of inputs) in classification tasks of the models of Evolutionary ANNs (EANNs) with product units (PUs) that have been employed to date by us. The computational cost is very high if Evolutionary Algorithms (EAs) with different parameter settings are employed for the training of the above-mentioned networks. However, in this paper we use a specialization of an EA called TSEA (two-stage evolutionary algorithm) [11]. First of all, FS is applied to the data sets in order to eliminate noisy and irrelevant variables. In this way, the complexity could be reduced and the accuracy could be increased. The reduction in the number of inputs could decrement the number of nodes in the hidden-layer and, hence, also simplify the associated model.

This paper is organized as follows: Sect. 2 describes some concepts about evolutionary product unit neural networks (PUNNs), the TSEA and FS; Sect. 3 presents the description of our proposal; Sect. 4 details the experimentation process; then Sect. 5 shows and analyzes the results obtained; finally, Sect. 6 states the concluding remarks.

2   Methodology

2.1   Evolutionary Product Unit Neural Networks

The single-hidden-layer feed-forward network architecture is the most popular one. Multiplicative neural networks contain nodes that multiply their inputs instead of adding them. This class of neural networks comprises such types as sigma-pi networks and PU networks. The latter type was introduced by R. Durbin and D. Rumelhart [4]. The methodology employed here consists of the use of an EA as a tool for learning the architecture and weights of a PUNN model [8]. We have used models of PUNN [11] with a three-layer $k : m : j$ architecture, with k nodes in the input layer, m ones and a bias one in the hidden layer and j nodes in the output layer. The transfer function of each node in the hidden and output layers is the identity function. Next, we are going to describe briefly the TSEA applied. A full explanation of it can be read in Sect. 3 of [11]. TSEA is used to design the structure and learn the weights of PUNNs in two sequential phases. In the first stage, TSEA evolves two populations for a small number of generations in order to explore the search space and to refine a bit random individuals. The best half individuals of each one are merged in a new population that follows the full evolutionary cycle in the second stage to perform exploitation. The main parameters of the TSEA are the maximum number of generations (gen) and the maximum number of nodes in the hidden layer (neu).

The minimum number of nodes is an unit lower than neu. The remaining parameters will be described further on. At the end of the TSEA, it returns the best PUNN model with a number of nodes between neu and $neu + 1$ in the hidden layer. We have considered a standard soft-max activation function, associated with the g network model, given by:

$$g_j(\mathbf{x}) = \frac{exp\ f_j(\mathbf{x})}{\sum_{j=1}^{J} exp\ f_j(\mathbf{x})} \quad j = 1, ..., J \tag{1}$$

where J is the number of classes in the problem, $f_j(\mathbf{x})$ is the output of node j for pattern $\mathbf{x}$ and $g_j(\mathbf{x})$ is the probability that this pattern belongs to class j. Given a training set $D = (x_i, yi)\ i = 1, ..., N$, a function of cross-entropy error is used to evaluate a network g with the instances of a problem, which is reflected in the following expression:

$$l(g) = -\frac{1}{N} \sum_{i=1}^{N} \sum_{j=1}^{J} (y_i^j ln(g_j(\mathbf{x}_i))) \tag{2}$$

and substituting g_j defined in (2),

$$l(g) = -\frac{1}{N} \sum_{i=1}^{N} \left(-\sum_{j=1}^{J} y_i^j f_j(\mathbf{x_i}) + ln(\sum_{j=1}^{J} exp\ f_j(\mathbf{x}_i)) \right) \tag{3}$$

where y_i^j is the target value for class j with pattern $\mathbf{x_i}$ ($y_i^j = 1$ if $\mathbf{x_i} \in$ class j and $y_i^j = 0$ otherwise), $f_j(\mathbf{x_i})$ is the output value of the neural network for the output neuron j with pattern $\mathbf{x_i}$. Observe that soft-max transformation produces probabilities that sum to one and therefore the outputs can be interpreted as the conditional probability of class membership. On the other hand, the probability for one of the classes does not need to be estimated because of the normalization condition. Usually, one activation function is set to zero; in this work $f_J(\mathbf{x}_i) = 0$ and we reduce the number of parameters to estimate. Thus, the number of nodes in the output layer is equal to the number of classes minus one in the problem. Since the EA objective is to minimize the chosen error function, a fitness function is used in the form $A(g) = (1 + l(g))^{-1}$.

The TSEA loops are repeated until the maximum number of generations, in each case, is reached or until the best individual or the population mean fitness does not improve during gen-without-improving generations (20 in this paper).

Parametric and structural mutations have been used and follow the expressions and details given in [11]. Table 1 summarizes the main TSEA parameters.

2.2   Feature Selection

FS is the problem of choosing a small subset of features that ideally is necessary and sufficient to describe the target concept. There are various ways in which FS algorithms can be grouped according to the attribute evaluation measure:

Table 1. TSEA general parameters

Parameter	Value
Population size(N)	1000
gen − without − improving	20
Interval for the exponents w_{ji}/coefficients β_j^l	$[-5, 5]$
Initial value of α_1	0.5
Initial value of α_2	1
Normalization of the input data	$[1, 2]$
Number of nodes in node addition and node deletion operators	$[1, 2]$

depending on the type (filter or wrapper technique) or on the way that features are evaluated (individual or subset evaluation). The filter model [7] relies on general characteristics of the data (such as distance, consistency, and correlation) to evaluate and select feature subsets without involving any mining algorithm. The wrapper model requires a predetermined mining algorithm and uses its performance as evaluation criterion.

In feature subset selection, it is a fact that two types of features are generally perceived as being unnecessary: features that are irrelevant to the target concept, and features that are redundant given other features. In a previous work, we proposed BIRS (Best Incremental Ranked Subset) [10] method to obtain relevant features and to remove redundancy. Now, we combine BIRS with TSEA. The features selected are considered as input variables to the classifier. BIRS belongs to a hybrid category where the selection process is divided into two stages: in the first one, features are evaluated individually, providing a ranking based on a criterion; in stage two, a feature subset evaluator is applied to a certain number of features in the previous ranking following a search strategy. BIRS can use any evaluator in the two phases. In the cited work, BIRS uses as a subset evaluator CFS (Correlation-based Feature Selection) [5] and CNS (consistency based measure) [7] -that are established on correlation and consistency concepts- at the second phase, and SOAP (Selection Of Attributes by Projection) [9] measure and the own subset evaluator at the first phase as a ranking evaluator.

Therefore, in the experiments, spBI_CFS indicates that SOAP is employed as an individual measure in the first part of BIRS, and CFS is employed as a subset evaluator in the second part. In the same way, cnBI_CNS denotes that CNS evaluator will be used in both parts of the BIRS algorithm.

3   Proposal Description

Our attention is focused on evolutionary PUNNs for classification problems. The current paper presents TSEAFS methodology, a mixture of two FS methods with TSEA. First of all, some feature selectors are applied independently to the training set of all datasets in order to obtain a list of attributes, for each of them, considered for training and test phases. In this way, two reduced sets (reduced training and test sets) are generated, where only most relevant features are included. It is important to point out that the FS is performed only with training data; the reduced test set has the same features as the reduced training set. These reduced sets are taken as input to TSEA. TSEAFS operates with two filters as independent feature selectors. As a result of the FS stage, a list of relevant features is obtained with each of the FS methods for each data set. There are two different configurations in TSEA, named 1* and 2*. The TSEAFS features are the following: a) PUNN have been employed, with a number of neurons in the input layer equal to the number of variables in the problem after FS; a hidden layer with a number of nodes that depends on the data set to be classified and the number of selected features; and the number of nodes in the output layer equal to the number of classes minus one because a softmax-type probabilistic approach has been used; b) two experiments have been performed for each problem with two different values for α_2, that is associated with the residual of the updating expression of the output-layer coefficients, controls the diversity of the individuals and has a great impact over the performance [11]; c) two different configurations (1 * ♯ and 2 * ♯) are applied to subsets obtained with each of the selectors, for each dataset. The parameters of each configuration are $neu♯$, $gen♯$ and α_2. The first two ones take specific values depending on the dataset and the last one depends on the configuration number (1 * ♯,...). Table 2 shows the main aspects of TSEA/TSEAFS configurations.

Table 2. Description of the TSEA/TSEAFS configurations

Methodology	Config.	Num. of Neurons in each pop.	Size of each pop.	Num. Gener. in each pop.	α_2
TSEA	1*	neu and $neu + 1$	1000	$0.1 * gen$	1
TSEA	2*	neu and $neu + 1$	1000	$0.1 * gen$	1.5
TSEAFS	1 * ♯	$neu♯$ and $neu♯ + 1$	1000	$0.1 * gen♯$	1
TSEAFS	2 * ♯	$neu♯$ and $neu♯ + 1$	1000	$0.1 * gen♯$	1.5

4   Experimentation

Table 3 describes the data sets employed. All of them are publicly available at the UCI repository [1]. The following seven have been used: *Breast* Cancer, Statlog (*Heart*), *Hepatitis*, Molecular Biology (*Promoter* Gene Sequences), *Waveform* database generator (version 2), Wine Quality (*Winequality* − *red*) and *Yeast*. The size of the data sets ranges from over one hundred to five thousand. The number of features depends on the problem and varies between eight and fifty eight, while the number of classes is between two and ten. Since we are using neural networks, all nominal variables have been converted to binary ones; due to this, sometimes the number of inputs is greater than the number of features. Regards the number of inputs (*In.*) it ranges between eight and one hundred fourteen. Also, the missing values have been replaced in the case of nominal variables by the mode or, when concerning continuous variables, by the mean, taking into account the full data set. These data sets have in common that present error rates in test accuracy about 20% or above with reference classifiers such as

Table 3. Summary of the data sets used and parameter values for TSEA and TSEAFS methodologies

Data set	Size	Train	Test	Feat.	In.	Cl.	Neu; Gen	Neu♯; Gen♯
Breast	286	215	71	9	15	2	9; 500	9; 300
Heart	270	202	68	13	13	2	6; 500	4; 20
Hepatitis	155	117	38	19	19	2	3; 300	3; 300
Promoter	106	80	26	58	114	2	11; 500	6; 300
Waveform	5000	3750	1250	40	40	3	3; 500	3; 500
Winequality − red	1599	1196	403	11	11	6	6; 300	4; 300
Yeast	1484	1112	372	8	8	10	11; 1000	11; 1000

C4.5 or 1-NN.The experimental design uses the cross validation technique called stratified hold-out that consists of splitting the data into two sets: training and test set, maintaining the class distribution of the samples in each set approximately equal as in the original data set. Their sizes are approximately $3N/4$ and $N/4$, where N is the number of patterns in the problem.

Regards to TSEA methodology, the concrete values of neu and gen parameters depend on the data set and are shown in the eighth column of Table 3. With respect to the number of generations, we have defined three kinds of values: small (300), medium (500) and large (1000). We have given, in some cases, values of our choice to the two parameters depending on the complexity of the data set (number of classes, inputs, instances,...). Other times the values are based on a previous work [11]. In TSEAFS, again there are two parameters, neu♯ and gen♯, whose value is defined for each data set. The last column of the Table 3 presents the values of them along with the ones of TSEA to have a general view of the differences. In TSEAFS the number of neurons is upper bounded by TSEA value. It is important to note that aforementioned values of the parameters concern to the base configuration $(1 * /1 * ♯)$. The gen♯ parameter takes values similar to gen with the exception of Heart in whose case is very small (20) since the search converges quickly. The values of the remaining configurations are presented further on.

Table 4 depicts the methods used in the experimentation. There are two ones with and one without feature selection that belong respectively to TSEAFS (the current proposal) and TSEA methodologies. The feature selectors are filters. Last column defines an abbreviated name for each of them that is employed in next sections.

Table 4. List of methods employed in experimentation with and without feature selection

Feature Selector	Ranking Method	Subset Evaluation	Methodology	Abb.Name
−	None	None	TSEA	FS0
spBI_CFS	spBI	CFS	TSEAFS	FS1
cnBI_CNS	cnBI	CNS	TSEAFS	FS2

5   Results

This section details the results obtained, measured in Correct Classification Ratio (CCR) in the test set or in the test subset depending on that FS has been considered or not. First of all, we present the results obtained with TSEA and TSEAFS. After that, a statistical analysis compares them to determine whether there are significant differences between applying or not FS. Later, the number of inputs is analysed.

5.1   Results Applying TSEA and TSEAFS

The results obtained by applying TSEA methodology [11] are presented, along with those obtained with TSEAFS. Table 5 shows the mean and standard deviation (SD) of the test accuracies for each data set for a total of 30 runs. The best results without and with FS appear in boldface for each data set. From the analysis of the data, it can be concluded, from a purely descriptive point of view, that the TSEAFS methodology obtains best results for all data sets. In most of cases, the SD reduction with TSEAFS is clear and it expresses more homogeneous results compared to TSEA.

Table 5. Results obtained in seven data sets applying TSEA and TSEAFS

Data set	Method	Topology	$Mean \pm SD$	
			$Config\ 1*/1*\sharp$	$Config\ 2*/2*\sharp$
$Breast$	$FS0$	$15 : [9, 10] : 1$	$\mathbf{65.96 \pm 2.89}$	62.76 ± 3.08
	$FS1$	$4 : [9, 10] : 1$	$\mathbf{69.85 \pm 1.50}$	68.21 ± 1.08
	$FS2$	$2 : [9, 10] : 1$	69.01 ± 0.00	69.01 ± 0.00
$Heart$	$FS0$	$13 : [6, 7] : 1$	76.62 ± 2.33	$\mathbf{77.45 \pm 3.09}$
	$FS1$	$7 : [4, 5] : 1$	77.45 ± 2.16	77.69 ± 2.28
	$FS2$	$9 : [4, 5] : 1$	$\mathbf{78.57 \pm 1.99}$	77.79 ± 1.60
$Hepatitis$	$FS0$	$19 : [3, 4] : 1$	82.10 ± 4.44	$\mathbf{87.01 \pm 3.78}$
	$FS1$	$10 : [3, 4] : 1$	$\mathbf{90.78 \pm 1.79}$	89.29 ± 1.53
	$FS2$	$5 : [3, 4] : 1$	86.14 ± 1.81	87.45 ± 1.49
$Promoter$	$FS0$	$114 : [11, 12] : 1$	65.76 ± 8.99	$\mathbf{68.20 \pm 9.52}$
	$FS1$	$7 : [6, 7] : 1$	83.84 ± 3.83	$\mathbf{85.64 \pm 4.03}$
	$FS2$	$7 : [6, 7] : 1$	80.00 ± 2.74	76.30 ± 4.10
$Waveform$	$FS0$	$40 : [3, 4] : 2$	$\mathbf{84.46 \pm 0.92}$	82.01 ± 1.48
	$FS1$	$14 : [3, 4] : 2$	86.35 ± 0.85	$\mathbf{86.89 \pm 0.71}$
	$FS2$	$15 : [3, 4] : 2$	86.02 ± 2.16	85.67 ± 0.96
$Winequality - red$	$FS0$	$11 : [6, 7] : 5$	60.95 ± 1.58	$\mathbf{61.11 \pm 1.02}$
	$FS1$	$5 : [4, 5] : 5$	$\mathbf{61.63 \pm 1.09}$	61.25 ± 1.62
	$FS2$	$8 : [4, 5] : 5$	61.47 ± 0.95	60.87 ± 1.29
$Yeast$	$FS0$	$8 : [11, 12] : 9$	60.05 ± 1.21	$\mathbf{60.16 \pm 1.10}$
	$FS1$	$5 : [11, 12] : 9$	59.25 ± 1.44	60.06 ± 1.09
	$FS2$	$7 : [11, 12] : 9$	$\mathbf{60.78 \pm 1.29}$	59.43 ± 1.29

5.2   Statistical-Analysis

We follow the recommendations pointed out by J. Demšar [3] to perform non-parametric statistical tests. To determine the statistical significance of the differences in rank observed for each method with all data sets, a non-parametric test might be used. There are two methods, Friedman and Iman-Davenport tests. If the null-hypothesis is rejected, we can proceed with a post-hoc test. Bonferroni-Dunn has been performed.

The average ranks of all methods, without (FS0) and with FS (FS1-2) are respectively 2.86, 1.43 and 1.71. According to Iman-Davenport test results, since the statistic $F_F = 8.0$ is higher than the critical value at $(F(2, 12) = 3.89)$ the null-hypothesis is rejected. Therefore, we apply a post-hoc Bonferroni-Dunn test that compares a number of methods with a control method, by determining whether the average ranks differ by at least the CD. In our case, we make a comparison of the methods that employ FS (FS1-2) versus the control method (FS0) that does not use FS. CDs obtained by Bonferroni-Dunn test are 1.20 (at $\alpha = 0.05$) and 1.05 (at $\alpha = 0.10$). The ranking difference with FS0 are 1.43 for FS1 and 1.15 for FS2. Thus, there are significant differences between TSEA applying each of the FS methods and without FS. The statistical tests points out that PUNN performance improves significantly pre-processing the dataset with any of the FS methods employed in this paper. However, FS1 is better regarding to statistical significance level.

Analysis of the number of inputs. As previously mentioned, two FS methods implemented as filters have been applied to each dataset. Table 6 summarizes the average number of inputs of the test bed (see column labelled FS0) and those that have been obtained with the different feature selectors (see columns labelled FS1-2) along with the reduction percentage in the inputs of each selector compared to the original datasets. The maximum reduction percentage appears in boldface. The concrete value of the inputs for each case can be found in the third column of Table 5. The reduction percentage of the number of inputs is defined as:

$$Reduction\_Inputs(\%) = \left(1 - \frac{Inputs(FSi)}{Inputs(FS0)} \right) 100 \quad i = 1, 2 \qquad (4)$$

where i is the FS methods index $Inputs(j)$ represents the number of inputs of a given dataset with method j. In all cases, FS methods successfully decreased the data dimensionality by selecting, in mean, less than the quarter of the original features. Certainly, the number of selected features fluctuates between a quarter

Table 6. Number of inputs and reduction percentage for the test bed with and without feature selection

Data set	Inputs			Reduction (%)	
	FS0	FS1	FS2	FS1	FS2
Average	31.43	7.43	7.57	**59.68**	55.32

and a third of the original features. FS1 method achieves a reduction percentage, on average, of 59.68% (from 31.43 to 7.43 features in average), which is the highest overall average value obtained.

5.3   Results Obtained with a Variety of Classifiers

Now, a comparison, applying the best filter (FS1), is performed between TSEA and other machine learning algorithms. These methods are C4.5, k-nearest neighbours (k-NN), -where k is 1-, SVM [12] and LMT [6]. Since, these methods are implemented in Weka tool [13], we have conducted the experiments and used the same cross-validation, thus the same instances in each of the partitions, that in the first experiment. Regarding the parameters, the algorithms have been run with the Weka default values. We have reported in Table 7 the results with FS1 for each dataset and algorithm. Due to we have used filters for the FS, the same reduced features set is applied to all classifiers. For TSEA is reported the best mean value of the two configurations, and for the remaining algorithms the mean. In each row, the best result appears in boldface. From an analysis of the results, we can assert the following. The TSEA method obtains the best result for four out of seven data sets; and SVM two times. Furthermore, the TSEA reports the highest mean accuracy (76.08%) followed by the SVM method (75.59%).

Table 7. Results obtained in seven data sets for several classifiers with FS1

Data set	C4.5	1-NN	SVM	LMT	TSEA
Breast	69.01	**70.42**	66.20	69.01	69.85
Heart	73.53	73.53	76.47	76.47	**77.69**
Hepatitis	84.21	89.47	86.84	89.47	**90.78**
Promoters	73.08	57.69	84.62	84.62	**85.64**
Waveform	74.40	75.36	86.88	**87.04**	86.89
Winequality − red	50.87	48.88	59.80	48.64	**61.63**
Yeast	53.49	48.92	54.03	**60.22**	60.06
Average	68.37	66.33	73.55	73.64	**76.08**

6   Conclusions

This paper presented a methodology that combines FS with Evolutionary Artificial PUNN in classification problems. Specifically, a mixture of our previous TSEA methodology and FS, called TSEAFS, has been introduced. FS is performed by means of filters. The models obtained with the proposal has the advantages that are more accurate and less complex, taking into consideration the number of inputs and/or the number of nodes in the hidden-layer. Also, the current proposal is much more efficient than the previous one. An empirical study on seven UCI classification problems, that present test accuracy error rates about a twenty percent or above with C4.5 or 1-NN classifiers, has been performed to compare TSEAFS and TSEA methodologies, both of them based on

Evolutionary Artificial PUNN. Also other state-of-the-art classifiers have been tested in order to get an overall outlook.

Acknowledgments

This work has been partially subsidized by TIN2007-68084-C02-02 and TIN2008-06681-C06-03 projects of the Spanish Inter-Ministerial Commission of Science and Technology (MICYT), FEDER funds and the P08-TIC-3745 project of the "Junta de Andalucía" (Spain).

References

1. Asuncion, A., Newman, D.J.: UCI Machine Learning Repository (2007)
2. Bishop, C.M.: Pattern Recognition and Machine Learning. Springer, Heidelberg (2006)
3. Demsar, J.: Statistical comparisons of classifiers over multiple data sets. J. Mach. Learn. Res. 7, 1–30 (2006)
4. Durbin, R., Rumelhart, D.: Products units: A computationally powerful and biologically plausible extension to backpropagation networks. Neural Computation 1(1), 133–142 (1989)
5. Hall, M.A.: Correlation-based feature selection for discrete and numeric class machine learning. In: Proceedings of the Seventeenth International Conference on Machine Learning (ICML 2000), pp. 359–366. Morgan Kaufmann, San Francisco (2000)
6. Landwehr, N., Hall, M., Frank, E.: Logistic model trees. Machine Learning 59(1-2), 161–205 (2005)
7. Liu, H., Setiono, R.: A probabilistic approach to feature selection - a filter solution. In: Proceedings of the Thirteenth International Conference on Machine Learning (ICML 1996), pp. 319–327. Morgan Kaufmann, San Francisco (1996)
8. Martínez-Estudillo F.J., Hervás-Martínez, C., Gutiérrez, P.A., Martínez-Estudillo, A. C.: Evolutionary product-unit neural networks classifiers. Neurocomputing 72(1-2), 548–561 (2008)
9. Ruiz, R., Riquelme, J.C., Aguilar-Ruiz, J.S.: Projection-based measure for efficient feature selection. J. Intell. Fuzzy Syst. 12(3-4), 175–183 (2002)
10. Ruiz, R., Riquelme, J.C., Aguilar-Ruiz, J.S.: Incremental wrapper-based gene selection from microarray expression data for cancer classification. Pattern Recognition 39(12), 2383–2392 (2006)
11. Tallón-Ballesteros, A.J., Hervás-Martínez, C.: A two-stage algorithm in evolutionary product unit neural networks for classification. Expert Systems with Applications 38(1), 743–754 (2011)
12. Vapnik, V.N.: The nature of Statistical Learning Theory. Springer, Heidelberg (1995)
13. Witten, I.H., Frank, E.: Data Mining: Practical Machine Learning Tools and Techniques. Data Management Systems, 2nd edn., Morgan Kaufmann, San Francisco (2005)

Knowledge Living on the Web (KLW)

Miguel A. Fernandez[1,4,5], Juan Miguel Ruiz[2], Olvido Arraez Jarque[5],
and Margarita Carrion Varela[3]

[1] Instituto de Investigación en Informática de Albacete (I3A), n&aIS Group,
Campus Universitario s/n, 02071-Albacete, Spain
Miguel.FGraciani@uclm.es
[2] Clinica Garaulet, 02002-Albacete, Spain
[3] Instituto Municipal de Deportes (IMD), Sport Medicine Center,
c/ Avda. de España s/n, 02002-Albacete, Spain
[4] Departamento de Sistemas Informáticos, Universidad de Castilla-La Mancha,
Campus Universitario s/n, 02071-Albacete, Spain
[5] Facultad de Relaciones Laborales y Recursos Humanos de Albacete,
Universidad de Castilla-La Mancha,
Campus Universitario s/n, 02071-Albacete, Spain

Abstract. The amount of information currently available on the Internet is absolutely huge. The absence of semantic organization in the web resources hinders the access and use of the information stored and in particular the access to the computational systems in an autonomous manner.

Ontologies make use of formal structures mainly based in logics to define and allow reuse of the knowledge stored on the Internet. This article presents a different line of work which does not provide an alternative to already existing ontologies but is intended to complement one another by coexisting on the Internet.

The foundations of our proposal lie in the idea that the nature of knowledge, which is evolutionary, emergent, self-generative, is closer to the characteristics associated to life rather than to those pertaining to logics or mathematics.

We attempt to use the Internet to generate a virtual world where knowledge can grow and evolve according to artificial life rules utilizing users as replicators, as well as the use they make of such knowledge on the Internet.

We have developed a representation scheme called KDL (Knowledge Description Language), a DKS server application (Domain Knowledge Server) and a KEE user interface (Knowledge Explorer and Editor) which allow Internet users register, search, use and refer the knowledge housed on the KLW Web.

1   Introduction

Within the field of Artificial Intelligence among others, we have aimed at creating artificial systems capable of storing, inferring and generating knowledge. The achievements made and the goals fulfilled are highly valuable and widely known.

J.M. Ferrández et al. (Eds.): IWINAC 2011, Part II, LNCS 6687, pp. 391–399, 2011.

Artificial Intelligence has developed knowledge engineering, connectionist and evolutive theories among others. However, for some time the need for a global information system which could house all the existing knowledge did not seem to be essential. The place where all this knowledge resided were human minds and written formats. Later digital and multimedia formats for independent devices or connected to small computer networks were also developed. The exchange of information was performed mainly among people. Sharing this information was not an easy task, not because of the nature of knowledge in itself but owing to access and transmission mechanisms.

The appearance of the Internet brings about a sudden change in this situation. On the one hand the information is available all over the planet and in an immediate way. On the other hand the amount of information available is continuously growing at an exponential rate. The Internet provides by far the widest repository of information ever in history. Any element can appear in our screen almost instantaneously just with a single click.

However, not everything is so perfect. We have free access to a brutal amount of data which keeps growing with no order or organization at all. But there is no tool available to help us find what we are looking for on the Internet: there is neither any type of classification system as those used by librarians for centuries or the subject indexes of encyclopedias, nor even a simple alphabetical index.

Just the wonderful concept of hypertext (or hypermedia) serves us as a mechanism to surf in the immense ocean of information, searching for the small longed-for island, not knowing its coordinates.

Fortunately browsers are developed as a tool to search for information in the huge repository which makes up the Web. However, although their contribution is sufficient and invaluable, they are syntactic searches and therefore we lack semantic searching mechanisms. Moreover, results can hardly be contrasted. It is true that we can usually solve our problem but we are unable to see the remaining information available on the Internet which the browser did not show.

From the creation of the Internet even its developers [1][2], saw the need to give semantic load to the information stored. This does not mean that the information contained does not store semantics. The point is that the information is stored in natural language and therefore its semantics is accessible for the human mind but not for the information systems (computers) which contain and process such information.

2   The Renewal of Ontologies

Facing this lack of semantics in the mechanisms the Internet uses to store the information and aiming at an efficient use of the information stored on the Web by using automatic computational systems, ontologies appear as the most adequate mechanism which enables to organize the information we want to use and share [6].

However, the application of ontologies to the nature of the knowledge contained on the Internet, is not essential [7]. The large amount and the globality of information, the existence of multiple agents and domains, the extremely rapid evolution of the network mechanisms, among others, are some of the problems which make the use of ontologies -as we know them-more complicated. The restrictions imposed by the logics on which ontologies are based, are helpful for knowledge control and inference but represent an important drawback concerning the possibilities of knowledge representation. Although this is not an easy task, it is essential that the ontologies by which we aim at establishing the semantic web [5] can embrace a great number of leisure functions, information and business, among others [4]. For all these reasons, researchers must search for new elements to address the problem of semantics on the web.

3   Ontologies and the Internet

The need to introduce semantics on the Internet has led to define mechanisms which allow the development of ontologies within the Web [8] [9].

Using the TCP-IP and HTTP basic protocols and the basic XML format, the RDF is created as a mechanism to identify and locate entities on the Web. Subsequently the RDF allowed the creation of the OWL, among others, which is currently one of the most popular ontologies on the Internet.

This type of ontology repeats the structure of the ontologies previously developed with other systems in devices not connected to the Internet. Technology allows accessibility from any point in the network, also facilitating the declaration and the use of the elements in the ontology.

The features, benefits and problems associated to ontologies in other formats different from those on the Internet can also be found in OWL based ontologies [12]. However, nowadays the ontologies which previously resided on isolated machines, are coexisting in the network, thus emerging the possibility (and the need) to fuse them together. This new situation also brings about new problems in the task of attempting to create structures for the knowledge in the network using formal ontologies.

4   Previous Considerations

It should be clearly stated that from our point of view, ontologies fulfill their objective and in fact, the formalism they provide is essential for the coherent application of the algorithms and theorems linked to the processes of structurization and inference associated.

As shown next, our proposal presents a line of work of different nature which is not incompatible with the existence and use of ontologies on the Internet.

The problem we face is how to obtain automatically - using the internet - a global knowledge base which can be used by the computational systems using such base and processing its information: a global information system which can be accessed by all information systems sharing information and inference and computational processes.

5   KLW: Our Proposal

This work raises a completely different paradigm as a way to achieve the goals previously stated. We propose the configuration of the Internet as an original breeding ground where knowledge can develop, taking into account the characteristics of what is understood as life.

The foundations of our proposal lie in the idea that the nature of knowledge, which is evolutionary, emergent, self-generative, is closer to the characteristics associated to life rather than to those pertaining to logics or mathematics. In fact, knowledge really grows and develops in living mechanisms, for life has allowed knowledge to develop with all its splendor.

Some theories, as memetics, Richard Dawkins [3], make an interpretation of knowledge by attributing ideas (called memes) characteristics inherent to living beings. Concepts such as "the sea", "my house" or "superman" are memes, which are born, evolve, replicate, die; using the human mind as ground as well as the well known written or audiovisual media.

We propose the use of a different format ground - the Internet network- to define a virtual world inhabited by a new entity of artificial life, the NETNE, the equivalent to memes in the memetics theory or to concepts in natural language.

It should be specified that here the word network here comprises not only the physical and logical network but it also includes users, their knowledge, the information stored and the use of the said network.

According to Dawkins, the sea, my house or superman are memes housed in the structures of natural intelligence. They can exist because the human mind creates them, replicates them or forgets them. Therefore they emerge from biological life, a different world, the world of memes or concepts, with "independent life", but with bilateral relationship from which both parts mutually benefit taking into account the characteristics of what is considered as life.

In our proposal, the sea, my house or superman are netnes which exist using the nature of the Internet network. We propose the incorporation of the resources required to the existing mechanisms on the Internet in order to provide a logical format ground with characteristics associated to artificial life entities. An artificial life system is configured in which netnes are the individuals living there, they spread, replicate and evolve, in accordance with the actions that the environment -the network- performs on them. Thus netnes will have "a life of their own" in a world which will maintain a close symbiotic relationship with the network world, emphasizing that here network also includes users, their definitions, their searches as well as the mechanisms they use.

Our proposal is not an ontology to provide a new knowledge structure in the network. Our objective is to use the network to generate knowledge. We do not intend to translate the structures that enable us to acquire knowledge to the Internet network, we propose to generate structures on the Internet which are capable of generating knowledge. This knowledge generated on the Internet contains the information that users define by means of their reference, concretion, abstraction and use. Obviously we intend to take advantage of the said knowledge which can be autonomously used by computational systems.

6   Technical Description

Once our objectives have been defined, let us present the technical description of our proposal.

First, the initial requirements established for the KLW information system must be stated.

Obviously the information system should be associated to the nature of the Internet network. It should be conceived as a global information system.

Another important requirement is both automatic and manual access capacity. For this purpose an access interface is defined in the DKS via web services. The information obtained through these web services can be processed by the automatic computational systems or can be transformed into multimedia information to be displayed on the browser window as the users´ interface.

Independence and clear separation between own domain and observer´s domain . According to our predecessors [11] we know that artificial intelligence ends where own domain ends. For this reason we stress the importance of defining the mechanisms associated to the system´s own domain and also of separating them from the observer´s knowledge. This will be reflected on what is known in the system as "interface support mechanisms" where the information of the observer domain is managed in natural language and to which the KLW mechanisms gain access to perform only syntactic tasks.

Multilanguage and multiculture. We have established that the system must be independent from any language or culture. This can be achieved by isolating the mechanisms defined in the LKW from the information it receives in natural language. The language is isolated in the "Interface Support Mechanisms".

Artificial life mechanisms. The system must define mechanisms associated to artificial life which can facilitate the evolution of the system and of its contents.

Data property and security. The servers containing information (DKS) must implement security mechanisms which can limit the access to the information depending on the identity of the user.

Marketing possibility.DKS offer the possibility to market the knowledge they contain. We do not intend that all knowledge should be charged but we believe that the prospect of obtaining returns from the work associated to knowledge definition may encourage its growth in the Network.

Possibility to monitor and trace the evolution of the elements which make up the system.

Once the initial requirements of the so-called LKW information system have been stated, let us now describe the system design.

In the KLW, the basic knowledge entity is a concept -what we have denominated netne-. Globally the system is made up of a representation scheme of the Knowledge Description Language (KDL), an application of the Domain Knowledge Server (DKS) which serves the said information by means of web services and houses the mechanisms of artificial life of the system and a user interface, the Knowledge Explorer and Editor (KEE), which allow users to browse, register and refer knowledge within the LKW.

Although a detailed definition of each element would require a more extended paper we will next present a brief description of the elements.

Concerning the KDL representation scheme, a representation structure which is close to Minsky frameworks has been defined [10]. It is based on concept instantiation. An XML scheme has also been defined which establishes the structure of XML documents containing the description of a concept (netne).

The KDL is a bit simpler than the RDF. The KDL is defined as the base to generate knowledge structures, and when developing its design we decided to free it from the structure subject- characteristic-value associated to the RDF. Although this structure, like many others, can be defined using the KDL.

With respect to the KDL, it should also be highlighted that it allows to associate an instance of the generic concept "interface support" to a concept, thus allowing the introduction of information in natural language, with no LKW semantics, independent from the rest of the concept structure and very useful for users.

Regarding the domain knowledge server (DKS), it is an application programmed in PHP running on a Linux server. This application contains the data corresponding to the associated domain, it maintains the information in KDL structure and controls the concept and references evolution as well as the information concerning the application use. The DKS application defines and implements the mechanisms associated to the behavior of artificial life. This application is also responsible for security, data property and commercialization. Some DKS are devoted to generic knowledge, others to a specific knowledge domain and others to a company´s particular information. Some may even be devoted to an individual´s personal information.DKS are housed on Internet servers and contain and distribute knowledge associated to its domain.

With respect to the KEE user interface, it allows to visualize the description of a concept and locate its instances and references. It also allows to gain access to the concepts by using the interface support in a specific language (if the interface support is prepared,desgined cumplimentada in such language). The KEE allows semantic search as it allows locating references which are more or less close to concepts related with the user´s search. The KEE obviously enables syntactic search, which can be done in different languages although the system only searches through the interface support contents of the corresponding languages. The KEE is currently developed as a web client in PHP technology, using XSLT, JavaScript, Flash and Ajax as well.

The KLW is a distributed information system designed to be developed as a DKS community, housed in machines distributed in the network. Each server will contain, using netnes format, the knowledge associated to a specific knowledge domain, as well as its relationships with the rest of the knowledge the Web will contain. Such knowledge -from the more general to the more specific- will increase in accordance with the rules of artificial life, by means of the use and references that users may develop.

Both, users -via the KEE- and computational systems - via web services- may have access, define and use such knowledge.

It is true that the nature of the KDL and of the artificial life mechanisms implemented in the DKS do not present the formalities of ontologies as we know them, but our proposal attempts to define less formal information systems, leaving up to the network´s behavior the responsibility to reach consensus for use which allow computational systems globally work in an efficient and safe manner.

In any case, the formal rigor and mechanisms of ontologies as we know them can be used and defined as LKW elements.

These ontologies defined using LKW concepts will always be necessary for those processes which require formal rigor to ensure the behavior of the inference or reasoning.

7   Discussion and Future Work

We have already presented the representation scheme, the mechanisms associated to artificial life and the network resources required concerning both the server and the user interface in order to configure an information system on the network which can be accessed by users and computational systems to register, reference, share, search and use the knowledge therein stored.

The nature of the system we propose is little formal and presents little restrictions allowing consensus behaviours on the network generate the structure of the knowledge it contains.

We are aware that this lack of formal rigor may limit the possibilities of controlled inference and leaves a door open to the appearance of concept duplicity as well as to other effects which may arise in a natural form. But that´s the way it is. Life succeeds in reaching an evolutionarily stable state which is what we attempt to obtain with the system proposed [13].

There is a wide range of operations -reference, search, computational- which may be accomplished without imposing strict restrictions to the nature of the information stored. When necessary, the possibility to define more restrictive ontologies by refering the concepts defined on the LKW.

The LKW appears as an answer to the needs detected in the shared use and referencing of the data in applications (RETRAPROD, SIGA) developed for engineering companies companies (Retailgas, Ideando S.L.) in order to maintain geographically distant ***active*** information. The first prototypes are currently running **** trabajando*** on the said applications.

The ideas and structures forged in those initial projects have been restructured through a more serene design outside the frenetic business world giving rise to the creation of the LKW which is currently in a stage of redefinition and re programming.

The KDL has already been defined, as well as the associated XML schemes, although they are still in the beta version, waiting for some development readjustments.

The DKS is in an advanced stage of development and some prototypes are already operative but property rights and marketing requirements are not complete yet. Artificial life mechanisms are still under construction although there is a first version of the web services which allow registration and access to the information.

There is also an operative prototype of the KEE which is valid enough to test the interface operation but which is still little attractive for the user. At present we are working to obtain a new more attractive graphic version which will incorporate new possibilities for use.

We are also making up an initial body of concepts on LKW. More specifically we are working on basic knowledge domains, medical knowledge, pharmaceutical knowledge, engineering and civil applications terms, person-related knowledge (medical records, professional and training characteristics, leisure and social networks information).

Therefore, this work does not present a finished application, not the possibility of providing robust assessment of results. For this reason no benchmarking has been performed. We simply attempted to define the path to follow, where we have been moving for some time with the desire to attract new travel mates.

Acknowledgements

This work was partially supported by Spanish Ministerio de Ciencia e Innovación TIN2010-20845-C03-01 and TIN2010-20845-C03-02 grants, and by Junta de Comunidades de Castilla-La Mancha PII2I09-0069-0994 and PEII09-0054-9581 grants.

We would like to acknowledge the valuable participation of the companies Retailgas S.A. and I.D. Ando Experiencias S.L. (Ideando).

References

1. Beners-Lee, T., Hendler, J., Lassila, O.: The Semantic Web: A new form of Web content that is meaningful to computers will unleash a revolution of new possibilities. Scientific American (May 2001)
2. Berners-Lee, T., Miller, E.: The Semantic Web lifts off. ERCIM News 51, 9–11 (2002)
3. Dawkins, R.: The Selfish Gene. Edt. Desmond Morris (1976)
4. Decker, S.: The social semantic desktop: Next generation collaboration infrastructure. Information Services & Use 26(2), 139–144 (2006)
5. Greaves, M.: Semantic Web 2.0. IEEE Intelligent Systems 22(2), 94–96 (2007)
6. Gruber, T.R.: Toward principles for the design of ontologies used for knowledge sharing. In: Formal Ontology in Conceptual Analysis and Knowledge Representation. Kluwer Academic Publishers, The Netherlands (1993)
7. van Harmelen, F.: The complexity of the web ontology. IEEE Intelligent Systems, 71–72 (March-April 2002)
8. Hendler, J.: Ontologies on the semantic web. IEEE Intelligent Systems, 73–74 (March-April 2002)
9. Lu, S., Dong, M., Fotouhi, F.: The semantic web: opportunities and challenges for next-generation web applications. Information Research 7(4) (April 2002)

10. Minsky, M.: The Society of Mind, March 15. Simon and Schuster, New York (1988); ISBN 0-671-65713-5
11. Mira, J.: Aspectos básicos de la Inteligencia Artificial. Edt. Sanz y Torres (1995)
12. OWL Web Ontology Language Guide. In: Smith, M.K., Welty, C., McGuinness, D.L. (eds.) W3C Recommendation, vol. 51, February 10, pp. 9–11 (2004), http://www.w3.org/TR/2004/REC-owl-guide-20040210/ (latest version), http://www.w3.org/TR/owl-guide/
13. Varela, F.J., Maturana, H.R., Uribe, R.: Autopoiesis: the organization of living systems, its characterization and a model. Biosystems 5, 187–196 (1974)

Local Context Discrimination in Signature Neural Networks

Roberto Latorre, Francisco B. Rodríguez, and Pablo Varona

Grupo de Neurocomputación Biológica, Dpto. de Ingeniería Informática,
Escuela Politécnica Superior, Universidad Autónoma de Madrid,
28049 Madrid, Spain
roberto.latorre@uam.es, f.rodriguez@uam.es, pablo.varona@uam.es

Abstract. Bio-inspiration in traditional artificial neural networks (ANN) relies on knowledge about the nervous system that was available more than 60 years ago. Recent findings from neuroscience research provide novel elements of inspiration for ANN paradigms. We have recently proposed a Signature Neural Network that uses: (i) neural signatures to identify each unit in the network, (ii) local discrimination of input information during the processing, and (iii) a multicoding mechanism for information propagation regarding the *who* and the *what* of the information. The local discrimination implies a distinct processing as a function of the neural signature recognition and a local transient memory. In this paper we further analyze the role of this local context memory to efficiently solve jigsaw puzzles.

Keywords: Bioinspired ANNs, Neural signatures, Multicoding, Local discrimination, Local contextualization.

1 Introduction

Artificial neural networks (ANN) are inspired from their biological counterparts. However, in the context of ANN, phenomena such as local recognition, discrimination of input signals and multicoding strategies have not been analyzed in detail. These phenomena do occur in living neural networks. Most traditional ANN paradigms consider network elements as indistinguishable units, with the same transfer functions, and without mechanisms of transient memory in each cell. None of the existing ANN paradigms discriminates information as a function of the recognition of the emitter unit. While neuron uniformity facilitates the mathematical formalism of classical paradigms [1,2,3,4,5] (which has largely contributed to their success [6]), some specific problems could benefit from other bio-inspired approaches.

Recent experiments in living neural circuits known as central pattern generators (CPG) show that some individual cells have neural signatures that consist of neuron specific spike timings in their bursting activity [7,8]. Model simulations indicate that neural signatures that identify each cell can play a functional role in the activity of CPG circuits [9,10,11]. Neural signatures coexist with the

J.M. Ferrández et al. (Eds.): IWINAC 2011, Part II, LNCS 6687, pp. 400–408, 2011.

information encoded in the slow wave rhythm of the CPG. Readers of the signal emitted by the CPG can take advantage of these multiple simultaneous codes and process them one by one, or simultaneously in order to perform different tasks [10,12]. The *who* and the *what* of the signals can be used to *discriminate* the information received by a neuron by distinctly processing the input as a function of these multiple codes. These results emphasize the relevance of cell diversity and transient memory for some living neural networks and suggest that local context-based discrimination is important in systems where neural signatures are present. This kind of information processing can be a powerful strategy for neural systems to enhance their capacity and performance.

We have recently proposed a neural network paradigm that makes use of neural signatures to identify each unit of the network, and multiple simultaneous codes to discriminate the information received by a cell [13]. We have called this paradigm Signature Neural Networks (SNN) The SNN network self-organization is based on the signature recognition and on a distinct processing of input information as a function of a local transient memory in each cell that we have called the local informational context of the unit. The efficiency of the network depends on a trade-off between the advantages provided by the local information discrimination and its computational cost.

In this paper we further discuss the effect of transient memory in SNN, i.e. the local informational context, to solve jigsaw puzzles.

2   Signature Neural Networks

In this section we briefly describe the SNN paradigm. For further details see [13]. Behind this new paradigm, there are four main ideas: (i) Each neuron of the network has a *signature* that allows its unequivocal identification by the rest of the cells. (ii) The neuron outputs are signed with the neural signature. Therefore, there are multiple codes in a message (*multicoding*) regarding the origin and the content of the information. (iii) The single neuron discriminates the input signals as a function of the recognition of the emitter signature and a transient memory that keeps track of the information and its sources. This memory provides a contextualization mechanism to the single neuron processing. (iv) The network self-organization relies to a large extent on the local discrimination by each unit.

2.1   SNN Definitions

The formalism requires the definition of several terms since some of them are not common in other ANN paradigms. Some of them depend on the specific problem to be solved. This provides a framework that can be applied to different problems by only customizing the open definitions. In a first approach, to review the main results of [13], we fix them to allow us a performance comparison in equivalent conditions with a traditional stochastic algorithm to solve jigsaw puzzles. However, this does not allow to take full advantage of all the power of the SNN, so later we propose different local context scenarios.

- *Neuron* or *cell*: the processing unit of the network.
- *Neuron signature*: the neuron ID in the network. This ID is used for the local information discrimination.
- *Neuron data*: information stored in each neuron about the problem.
- *Neuron information*: the joint information of the neuron signature and the neuron data of a cell.
- *Neuron neighborhood*: cells directly connected to the neuron. This concept is used to define the output channels of each neuron. The neuron neighborhood can change during the evolution of the SNN.
- *Local informational context*: transient memory of each neuron to keep track of the information and its sources. This memory consists of a subset of neuron informations from other cells received in previous iterations. The maximum size of the context ($N_{context}$) is the maximum number of elements in this set. The neuron signature and the local informational context are the key concepts of the SNN.
- *Local discrimination*: the distinct processing of a unit as a function of the recognition of the emitter and the local informational context.
- *Message*: the output or total information transmitted through a synapse between two neurons in a single iteration. The message consists of the neuron information of a subset of cells that are part of the context of the emitter plus its own neuron information. The maximum message size is equal to $N_{context}$. The input to a neuron consists of all messages received at a given iteration.
- A receptor *starts recognizing the signature* of an emitter cell during the message processing when it detects that the neuron data of the emitter is relevant to solve the problem (emitter and receptor data are compatible). The network self-organization is based on this recognition.
- *Information propagation mode*: depending on the problem, the information propagation can be monosynaptic or multisynaptic. Monosynaptic means that each neuron can receive only one input message per iteration. The information propagation is bidirectional between cells.
- A neuron *belongs to a cluster* if it recognizes the signature of all the neurons in its neighborhood. The clusters allow to simplify the processing rules of the SNN.

2.2   SNN Algorithm

The connectivity, the neuron data and the local informational contexts of all the network units are previously initialized. Depending on the problem, connectivity and neuron data initialization can be random or heuristic. Here we considered a free context initialization where initially the context of every neuron is empty. In this way the cells have no information about the rest of the network.

The algorithm consists in the iteration of three different steps for each neuron in the network until the stop condition is fulfilled. Note that the network self-organization takes place both in step 1 and 3 by modifying the network connections.

1 Process synaptic inputs
In this phase of the algorithm each neuron applies the local information discrimination:

  − First, the cell discriminates the input messages as a function of the emitter signature to determine which of them will pass to a second discrimination stage. If no signatures are recognized (a situation likely in the first iterations), all messages pass to the second stage.
  − Second, the neuron uses the memory of the processing in previous iterations stored in its local informational context to select the set of neuron informations from the messages that will be finally processed.
  − Third, the cell processes these set of neuron informations by applying its corresponding transfer functions or processing rules (which are specific of the problem to solve). If the neuron data processed is relevant to solve the problem, the cell starts recognizing the corresponding signature and establishes a new connection with the cell identified by this signature.
  − Finally, as the last step of this phase, the local informational context of the receptor is updated using the neuron information set analyzed during the processing of the input messages.

Local discrimination can lead to changes in the network connectivity. Network reconfiguration as a function of the local discrimination implies a nonsupervised synaptic learning. Clusters represent partial solutions to the problem. Neurons belonging to a cluster have the same processing rules.

2 Propagate information
During this phase neurons build and send the output messages. For this task each neuron adds its own information to the local informational context and signs the message. If the message size reaches the $N_{context}$ value, the neuron information from the oldest cell of the context is deleted from the message. The output message of a neuron is the same for all output channels.

3 Restore neighborhood
If a neuron has not reached its maximum number of neighbors, it randomly tries to connect to another neuron in the same situation (only one connection per neuron and iteration). First, it tries to connect to neurons from its local informational context and, if this is not possible, to other cells. This allows to maximize the information propagation in the network. To establish synapses with cells not belonging to the local context allows propagating information to other regions of the network.

3   Results

We have applied the SNN to solve jigsaw puzzles, a specific case of the NP-complete multidimensional sorting problem [21], and we have compared our results with a traditional stochastic method to solve the problem in terms of number of iterations and effectiveness.

The traditional stochastic algorithm frequently used to solve jigsaw puzzles works as follows:

- Choose a piece (P_1) from the set of available pieces.
- Search randomly for one piece (P_2) that fits with P_1 through one of its borders.
- Assemble both pieces in a new single piece.
- Add this new piece to the set of available pieces, deleting P_1 and P_2.
- Back to the first step until only one piece is left.

To apply the general SNN formalism for the jigsaw puzzle problem we need to apply the following restrictions:

- The number of neurons of the network is equal to the number of pieces of the puzzle.
- The neuron signature is the neuron number.
- The neuron data of each cell is one piece of the puzzle.
- The maximum number of neighbors is four, one for each side of the piece that the neuron represents (up, down, left and right). The neighbor order is important, up-down and left-right are opposite sides. The SNN has periodic boundary conditions.
- The initial structure of the network is two-dimensional with each cell connected to its four nearest neighbors.
- When two neurons contain pieces with a complementary border (borders that match correctly), they are compatible. When a neuron recognizes the signature of another cell, the network is reconfigured to move pieces to their correct positions. Neural signatures allow to identify the source of the information to achieve the correct fitting by reconfiguring the network from the starting two-dimensional structure to a multidimensional one. At the end of the self-organization the network recovers a two-dimensional structure.
- If a neuron *belongs to a cluster*, (i) it does not process the part of its informational context related to its neighbors, and (ii) it does not add its own neuron data to its output. Neurons in a cluster are only relayers of their input information.

The efficiency of the network depends on a trade-off between the advantages provided by the local information discrimination and its computational cost. Our main goal in this paper is to provide an analysis of the effect of the type of informational context on the efficiency of the SNN to solve jigsaw puzzles. Before, we briefly discuss the efficiency in terms of the size of the local informational context for the simplest case of monosynaptic information propagation and with a local discrimination that only uses a standard local informational context. We will compare these results with the traditional stochastic method to solve jigsaw puzzles.

Figure 1 displays the number of iterations needed to solve the puzzle as a function of the size of the local informational context for two different puzzle sizes. This figure compares the efficiency of the SNN and the traditional stochastic method. As can be observed, local information discrimination has a computational cost. For small puzzles, the performance of the SNN improves as the context size increases, but it is never better than the performance of the SA.

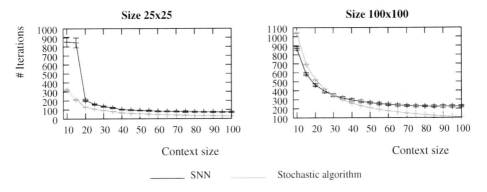

Fig. 1. Comparison between the mean number of iterations needed to solve 100 puzzles of 25x25 and 100x100 pieces with a traditional stochastic algorithm and the SNN (for details see [13]). The x-axis is the local informational context size (for the SNN) and the number of attempts to find a complementary pieces in each iteration (for the traditional method). The y-axis is the mean number of iterations needed to solve 100 different puzzles.

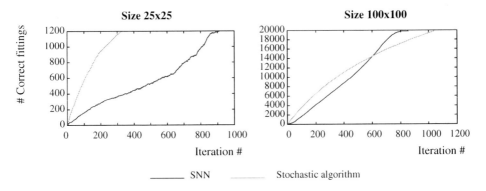

Fig. 2. Evolution of the number of correct fittings during the solution a puzzle of 25x25 and 100x100 pieces with a traditional algorithm and the SNN. In the initial iterations the effectiveness of the traditional method is better. Then the SNN improves and the effectiveness is similar for both approaches. For small puzzles the traditional method solves the problem before the SNN reaches a minimum effectiveness level. However, for large puzzles, the opposite situation occurs.

For large puzzles (right panel in Fig. 1), the performance of the SNN is better than the SA for small context sizes.

Fig. 2 shows that the SNN needs several iterations to achieve a minimum level of self-organization. For a small puzzle size (left panel), the stochastic algorithm solves the puzzle in around 300 iterations before the SNN reaches a minimum effectiveness level. However, for large puzzles this computational cost is justified as our results show that local discrimination provides a better performance (the SNN solves the 100x100 piece puzzle in about 800 iterations).

SNN capacity to solve multidimensional sorting problems can be improved by choosing the right type of local discrimination rules. In our previous description and in [13], the SNN was customized to solve the problem in equivalent conditions to the traditional stochastic algorithms. In this sense, we built the network to process one input message per iteration. However, the formalism allows a multisynaptic information propagation mode. In this scenario, neurons can process a larger amount of information in parallel. Similarly, so far the neurons in the SNN have used the local informational context as a transient memory of cells whose piece matching must be tested. Alternatively, the local context can also be used to temporary store information about cells whose matching has already been tested with a negative result and thus this information is considered not useful for the neuron.

If we build the SNN without these two features (multisynaptic propagation mode and/or the use of a negative context) and compare the results with the previous SNN scheme, the performance improves (see Fig. 3). If we further consider multiple messages per iteration, the problem can also be solved in fewer iterations. However, when we measure the performance in terms of the number of fitting tests needed to solve the problem, we see that the SNN requires a larger number of test. This is not surprising since the larger the number of message received, the larger is the number of comparisons needed for a neuron to contextualize and discriminate the information. Now, if we use a "negative context" in each neuron, this strategy not only reduces the number of iterations as expected, but also the number of iterations. Note that the large number of

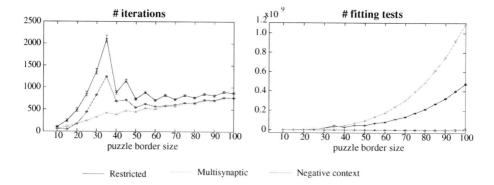

Fig. 3. Comparison between the performance of the restricted SNN (monosynaptic information propagation mode), the SNN in multisynaptic propagation mode (each neuron receives four input messages in parallel) and the SNN in monosynaptic propagation mode but with a "negative context" (half of the local informational context). The x-axis is the border size. Left panel: Performance in terms of the mean number of iterations needed to solve the puzzle. Right panel: Performance in terms of the mean number of fitting tests. In all cases the size of the local informational context is equal to 10%. All measures are calculated by solving 100 different puzzles for each border size.

iterations needed to solve 30x30 puzzle pieces is due to the low informational context for this size. This effect is reduced in the multisynaptic mode. Finally, if we consider both the multisynaptic propagation mode and the "negative context", the performance always improves significantly, even for small sized puzzles.

4 Discussion

Uniformity and simplicity of neurons has facilitated the mathematical formulation of many ANN paradigms. Local discrimination is somehow a problem to achieve a compact formalization for the SNN paradigm since this formalization depends on the specific problem that the network is trying to solve. As we have shown in this paper, a transient memory in the form of local informational context can further increase the capacity of a SNN to solve a multidimensional sorting problem.

Some concepts that underlie the strategy of SNN paradigm can be used to extend classical ANN. For example, in particular applications, we can consider having different sets of transfer functions for each unit, and make the selection of the specific function depend on the state of a local informational context. This strategy can combine synaptic and intra-unit learning paradigms and lead to achieve multifunctionality in the network.

Neuroscience research provides increasing amount of new results on the strategies of information processing in living neural systems [22]. Beyond the specific results discussed in this paper, the use of novel bio-inspired information processing strategies can contribute to a new generation of neural networks.

References

1. Fort, J.: Som's mathematics. Neural Networks 19(6-7), 812–816 (2006); Advances in Self Organising Maps - WSOM 2005
2. Anthony, M.: On the generalization error of fixed combinations of classifiers. Journal of Computer and System Sciences 73(5), 725–734 (2007)
3. Trenn, S.: Multilayer perceptrons: Approximation order and necessary number of hidden units. IEEE Trans. Neural Netw. 19(5), 836–844 (2008)
4. Auer, P., Burgsteiner, H., Maass, W.: A learning rule for very simple universal approximators consisting of a single layer of perceptrons. Neural Networks 21(5), 786–795 (2008)
5. Ilin, R., Kozma, R., Werbos, P.: Beyond feedforward models trained by backpropagation: A practical training tool for a more efficient universal approximator. IEEE Trans. Neural Netw. 19(6), 929–937 (2008)
6. White, D.A., Sofge, A. (eds.): Handbook of Intelligent Control Neural, Fuzzy, and Adaptive Approaches. Reinhold, NewYork (1992)
7. Szücs, A., Pinto, R.D., Rabinovich, M.I., Abarbanel, H.D.I., Selverston, A.I.: Synaptic modulation of the interspike interval signatures of bursting pyloric neurons. J. Neurophysiol. 89, 1363–1377 (2003)
8. Szücs, A., Abarbanel, H.D.I., Rabinovich, M.I., Selverston, A.I.: Dopamine modulation of spike dynamics in bursting neurons. Eur. J. Neurosci. 2, 763–772 (2005)

9. Latorre, R., Rodríguez, F.B., Varona, P.: Effect of individual spiking activity on rhythm generation of Central Pattern Generators. Neurocomputing 58-60, 535–540 (2004)

10. Latorre, R., Rodríguez, F.B., Varona, P.: Neural signatures: multiple coding in spiking-bursting cells. Biol. Cybern. 95, 169–183 (2006)

11. Latorre, R., Rodríguez, F.B., Varona, P.: Reaction to neural signatures through excitatory synapses in Central Pattern Generator models. Neurocomputing 70, 1797–1801 (2007)

12. Baroni, F., Torres, J.J., Varona, P.: History-Dependent Excitability as a single-cell substrate of transient memory for information discrimination. PLoS ONE 5(12), e15023 (2010)

13. Latorre, R., Rodríguez, F.B., Varona, P.: Signature Neural Networks: Definition and Application to Multidimensional Sorting Problems. IEEE Trans. Neural Netw. 22(1), 8–23 (2011)

14. Yao, F., Shao, H.: A shape and image merging technique to solve jigsaw puzzles. Pattern Recogn. Lett. 24(12), 1819–1835 (2003)

15. Freeman, H., Gardner, L.: Apictorial Jigsaw Puzzles: A Computer Solution to a Problem in Pattern Recognition. IEEE Trans. Electron. Comput. EC-13, 118–127 (1964)

16. Wolfson, H., Schonberg, E., Kalvin, A., Landam, Y.: Solving jigsaw puzzles by computer. Ann. Oper. Res. 12, 51–64 (1988)

17. Goldberg, D., Malon, C., Bern, M.: A global approach to automatic solution of jigsaw puzzles. Computational Geometry 28, 165–174 (2004)

18. Bunke, H., Kaufmann, G.: Jigsaw Puzzle Solving Using Approximate String Matching and Best-First Search. In: Chetverikov, D., Kropatsch, W.G. (eds.) CAIP 1993. LNCS, vol. 719, pp. 299–308. Springer, Heidelberg (1993)

19. Kong, W., Kimia, B.B.: On solving 2D and 3D puzzles using curve matching. In: Proc. IEEE Computer Vision and Pattern Recognition (2001)

20. Levison, M.: The Siting of Fragments. Computer Journal 7, 275–277 (1965)

21. Demaine, E.D., Demaine, M.L.: Jigsaw puzzles, edge matching, and polyomino packing: Connections and complexity. Graph. Comb. 23(1), 195–208 (2007)

22. Rabinovich, M.I., Varona, P., Selverston, A.I., Abarbanel, H.D.I.: Dynamical principles in neuroscience. Reviews of Modern Physics 78, 1213–1265 (2006)

Spider Recognition by Biometric Web Analysis

Jaime R. Ticay-Rivas[1], Marcos del Pozo-Baños[1],
William G. Eberhard[2], Jesús B. Alonso[1], and Carlos M. Travieso[1]

[1] Signals and Communications Department Institute for
Technological Development and Innovation in Communications
University of Las Palmas de Gran Canaria Campus University of Tafira, 35017,
Las Palmas de Gran Canaria, Las Palmas, Spain
{jrticay,mpozo,ctravieso}@idetic.eu
[2] Smithsonian Tropical Research Institute and Escuela de Biologia Universidad de
Costa Rica Ciudad Universitaria, Costa Rica
william.eberhard@gmail.com

Abstract. Saving earth's biodiversity for future generations is an important global task. Spiders are creatures with a fascinating behaviour, overall in the way they build their webs. This is the reason this work proposed a novel problem: the used of spider webs as a source of information for specie recognition. To do so, biometric techniques such as image processing tools, Principal Component Analysis, and Support Vector Machine have been used to build a spider web identification system. With a database built of images from spider webs of three species, the system reached a best performance of 95,44 % on a 10 K-Folds cross-validation procedure.

Keywords: Spider webs, spider classification, principal component analysis, support vector machine.

1 Introduction

Nowadays, biodiversity conservation has became a priority for researchers [1]. Knowledge about species is critical to understand and protect the biodiversity of life on Earth. Sadly, spiders have been one of most unattended groups in conservation biology [2]. These arachnids are plentiful and ecologically crucial in almost every terrestrial and semi-terrestrial habitat [3] [4] [5]. Moreover, they present a series of extraordinary qualities, such as the ability to react to environmental changes and anthropogenic impacts [5] [6].

Several works have studied the spider behaviour. Some of them analyse the use of the way spiders build their webs as a source of information for species identification[7] [8]. Artificial intelligent systems has been proven to be of use for the study of the spider nature. [9] proposed a system for spider behaviour modelling, which provides simulations of how a specific spider specie builds its web. [10] recorded how spiders build their webs in a controlled scenario for further spatio-temporal analysis.

J.M. Ferrández et al. (Eds.): IWINAC 2011, Part II, LNCS 6687, pp. 409–417, 2011.
© Springer-Verlag Berlin Heidelberg 2011

Because spider webs carry an incredibly amount of information, this work proposed the used of them as a source of information for specie identification. To our extend, this is a novel problem. Different biometric techniques have been used for this purpose, including image processing tools for prepare images, Principal Component Analysis for feature extraction and Support Vector Machines for classification.

The remainder of this paper is organized as follow. First, the database is briefly presented. Section 3 explain how images were preprocessed in order to extract the spider webs from the background.The Principal Component Analysis, and the Support Vector Machine are introduced in sections 4 and 5. Next, experiments and results are shown in detail. Finally, the conclusions derived from the results are presented.

2   Database

The database contains spider web images of three different species named Allocyclosa, Anapisona Simoni, and Micrathena Duodecimspinosa. Each class has respectively 28, 41, and 39 images, which makes a total of 108 images. Some examples can be seen in figure 1.

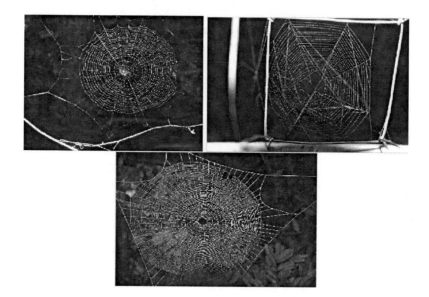

Fig. 1. Examples of spider web images from the data base corresponding to Allocyclosa, Anapisona Simoni, and Micrathena Duodecimspinosa respectively

3   Preprocessing

As can be seen in figure 1, spider web images were taken in both controlled and uncontrolled environments. Thus, the preprocessing step was vital in order to isolate the spider webs and remove possible effects of background in the system's results.

To enhance the contour of cobweb's threats an increase of colour contrast was first applied. Then, images were multiplied by two to further intensify the spider webs in relation to the background. Once the threads stood out enough, the images were binarized and cleaned up by morphological transformations. Finally, the spider webs were manually cropped following two criterion. As a results, two full set of spider web images were obtained. One containing the full web and the other showing only the central area. Examples of these sets can be seen in figures 2 and 3 respectively. Finally, all images were normalized to dimensions 20 x 20 pixels.

Fig. 2. Examples of full spider web images after preprocessing corresponding to Allocyclosa, Anapisona Simoni, and Micrathena Duodecimspinosa respectively

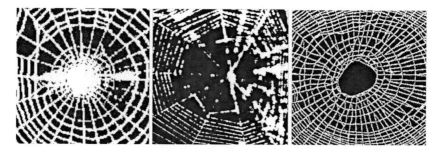

Fig. 3. Examples of centre spider web images after preprocessing corresponding to Allocyclosa, Anapisona Simoni, and Micrathena Duodecimspinosa respectively

4   Feature Extractor: Principal Component Analysis

The Principal Component Analysis (PCA) was first introduced by Karl Pearson in 1901 [11]. This tool is widely used in biometrics for feature extraction and data dimension reduction [12] [13] [14]. PCA basically finds an orthogonal linear transformation that, when applied to the original data, provides a new representation maximizing its variance.

Let X be a matrix made of data vectors x_i of length p nd zero empirical mean (columns representing samples and rows representing parameters), and let α_i be the projection vectors of length p obtained by PCA. Then, the principal component vectors (PCs)z_i are defined as:

$$z_i = \alpha X = \sum_{j=1}^{P} \alpha_{ij} \tag{1}$$

To represent the fact that PCA obtains the direction of maximum growth as represented in figure 4, it can be stated that PCs maximizes the expression $var[z] = var[\alpha^T X] = \alpha^T X X^T \alpha$. Moreover, these vectors are sorted such that the first vector has the maximum variance and the last vector has the lowest variance. Mathematically this is $var[z_i] > var[z_j]$ with $i < j$. Finally, it can be proven that the i-th projection vectorcorresponds to the i-th eigenvector of XX^T and has eigenvalue $\lambda_i = var[\alpha_i' X]$ for $i = 1, 2...p$.

Now, it is possible to decide which directions or PCs one would like to keep. Typically, those with greatest variance are associated with the most discriminative features and therefore it is enough to specify the number of PCs kept. Although this has been proven to be a dangerous supposition [15] [16], it certainly gives a good approximation of where the best PCs are, plus it is far faster than other search methods such as genetic algorithms. Thus, in this work the number of PCs used is automatically optimized by iterations, using the validation error ratc to decide which is the best solution.

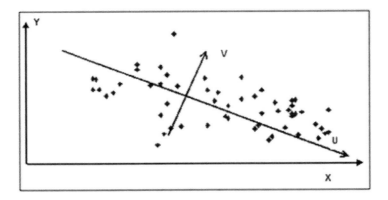

Fig. 4. U and V are the first and second directions founds by PCA respectively

5   Classification: Support Vector Machine

The Support Vector Machine is a method of structural risk minimization (SRM) derived from the statistical learning theory developed by Vapnik and Chervonenkis [17]. It is enclosed in the group of supervised learning methods of pattern recognition, and it is used for classification and regression analysis.

Based on characteristic points called Support Vectors (SVs), the SVM uses an hyperplane or a set of hyperplanes to divide the space in zones enclosing a

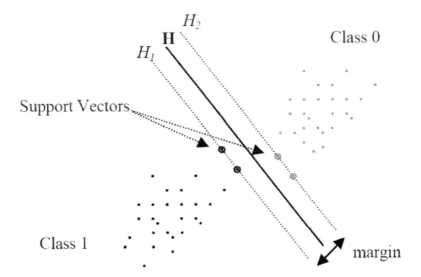

Fig. 5. Example of a separate hyperplane and its Support Vectors and margin for a linear problem

common class. Labeling these zones the system is able to identify the membership of a testing sample. The interesting aspect of the SVM is that it is able to do so even when the problem is not linearly separable. This is achieved by projecting the problem into a higher dimensional space where the classes are linearly separable. The projection is performed by an operator known as kernel, and this technique is called the kernel trick [18] [19]. The use of hyperplanes to divide the space gives rise to margins as shown in figure 5.

In this work, the Suykens' et. al. LS-SVM [20] is used along with the Radial Basis Function kernel (RBF-kernel). The regularization parameter and the bandwidth of the RBF function were automatically optimized by the validation results obtained from 10 iterations of a hold-out cross-validation process. Two samples from each class (from the training set) were used for testing and the remaining for training as we saw that the number of training samples has a big impact in the LS-SVM optimal parameters. Once the optimal parameters were found, they are used to train the LS-SVM using all available samples (training samples).

6   Experiments and Results

To sum up, the proposed system used the first M features obtained from the PCA projection of the spider webs images as inputs for a RBF-kernel LS-SVM with specific regularization and kernel parameters. These three parameters (the number of features, the regularization and the kernel parameters) were automatically optimized by iteration using validation results. To obtain more reliable results the available samples were divided into training and test sets, so that the system is trained and tested with totally different samples.

The well known K-Folds cross-validation technique were used to obtain the final results. In particular, experiments with K equal 3, 5, 7, and 10 were run. It is worth it to mention that the training and testing sets were computed for each class individually, having into account that each class has different number of samples. These experiments were performed for both datasets. This made a total of 8 experiments which results are presented in table 1.

The results show that when the full spider web is used the system became instable, not showing a clear tend when the number of training samples increase. However, with the center area of the spider web the system clearly improves when

Table 1. Results obtained for different Ks of a K-Fold cross-validation procedure with both the full spider web and the its center area

K of K-Folds Cross-Validation.	Using the full spider web.	Using only the center area of the spider web.
3	11.07% ± 2.47	10.24% ± 4.36
5	14.01% ± 12.22	6.60% ± 5.61
7	8.16% ± 8.60	5.73% ± 6.19
10	11.22% ± 8.54	4.56% ± 6.45

the number of training samples increase. Moreover, the error rate is always lower than that obtained with the full spider web.

Finally, figure 6 shows an example of how the validation error rate progresses when the number of PCs is incremented in the training process. It is interesting to note how the error rate increases from 20 to 50 features and then decreases again until 70 features are used. This shows that features presented in the way from 20 to 50 features acted as noise for the system, while features in the way from 50 to 70 provide good discriminant information even though their variance are lower.

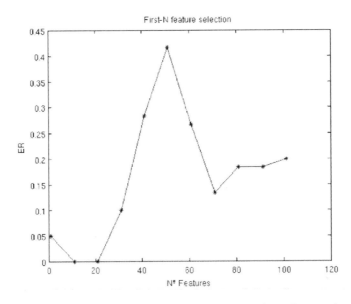

Fig. 6. An example of how the error rate progresses when the number of PCs is incremented

7   Discussion and Conclusions

This work has faced the problem of spider web recognition. A problem not considered before in the biometrics. The images were preprocessed to isolate the spider web and remove the effects of the background in the system. The resulting images were then transformed using PCA. Finally, the first M features were used as inputs for a LS-SVM classifier. Moreover, all parameters were automatically optimized by the system using the training samples.

Results of table I shows that the proposed system can achieve good performance rates. High levels of standard deviation may be caused by PCA instability, as the system does not search for the best directions, but it keeps the ones with higher variance (the first ones). Moreover, it is interesting how the results improve both in terms of pure performance and in terms of stability when only the center of the spider web is used. This suggest that discriminant information for spider identification may be contained in the center area of the spider web rather than in the periphery.

However, before make an stronger assertion about where the discriminant information is concentrate, it is necessary to perform more experiments with a wither database. On the other hand, this will allow to test the system's performance with larger training sets. Which will be interesting having into account that the results sows a clear improvement tend when the number of training samples increase. Finally, the proposed system can be easily improved using a more sophisticated feature selection such as a genetic algorithm.

Acknowledgements

This work has been supported by Spanish Government, in particular by "Agencia Española de Cooperación Internacional para el Desarrollo" under funds from D/027406/09 for 2010, and D/033858/10 for 2011.

References

1. Sytnik, K.M.: Preservation of biological diversity: Top-priority tasks of society and state. Ukrainian Journal of Physical Optics 11(suppl.1), S2–S10 (2010)
2. Carvalho, J.C., Cardoso, P., Crespo, L.C., Henriques, S., Carvalho, R., Gomes, P.: Biogeographic patterns of spiders in coastal dunes along a gradient of mediterraneity. Biodiversity and Conservation, 1–22 (2011)
3. Johnston, J.M.: The contribution of microarthropods to aboveground food webs: A review and model of belowground transfer in a coniferous forest. American Midland Naturalist 143, 226–238 (2000)
4. Peterson, A.T., Osborne, D.R., Taylor, D.H.: Tree trunk arthropod faunas as food resources for birds. Ohio Journal of Science 89(1), 23–25 (1989)
5. Cardoso, P., Arnedo, M.A., Triantis, K.A., Borges, P.A.V.: Drivers of diversity in Macaronesian spiders and the role of species extinctions. J. Biogeogr. 37, 1034–1046 (2010)
6. Finch, O.-D., Blick, T., Schuldt, A.: Macroecological patterns of spider species richness across Europe. Biodivers. Conserv. 17, 2849–2868 (2008)
7. Eberhard, W.G.: Behavioral Characters for the Higher Classification of Orb-Weaving Spiders. Evolution 36(5), 1067–1095 (1982); Society for the Study of Evolution
8. Eberhard, W.G.: Early Stages of Orb Construction by Philoponella Vicina, Leucauge Mariana, and Nephila Clavipes (Araneae, Uloboridae and Tetragnathidae), and Their Phylogenetic Implications. Journal of Arachnology 18(2), 205–234 (1990)
9. Eberhard, W.G.: Computer Simulation of Orb-Web Construction. J. American Zoologist, 229–238 (February 1,1969)
10. Suresh, P.B., Zschokke, S.: A computerised method to observe spider web building behaviour in a semi-natural light environment. In: 19th European Colloquium of Arachnology, Aarhus, Denmark (2000)
11. Jolliffe, I.T.: Principal Component Analysis, 2nd edn. Springer Series in Statistics (2002)

12. Hu, J., Si, J., Olson, B.P., He, J.: Feature detection in motor cortical spikes by principal component analysis. IEEE Transactions on Neural Systems and Rehabilitation Engineering 13(3), 256–262 (2005)
13. Zhang, Q., Leung, Y.W.: A class of learning algorithms for principal component analysis and minor component analysis. IEEE Transactions on Neural Networks 11(1), 200–204 (2000)
14. Langley, P., Bowers, E.J., Murray, A.: Principal Component Analysis as a Tool for Analyzing Beat-to-Beat Changes in ECG Features: Application to ECG-Derived Respiration. IEEE Transactions on Biomedical Engineering 57(4), 821–829 (2010)
15. Yao, H., Tian, L.: A genetic-algorithm-based selective principal component analysis (GA-SPCA) method for high-dimensional data feature extraction. IEEE Transactions on Geoscience and Remote Sensing 41(6), 1469–1478 (2003)
16. Liu, N., Wang, H.: Feature Extraction with Genetic Algorithms Based Nonlinear Principal Component Analysis for Face Recognition. In: 18th International Conference on Pattern Recognition, ICPR 2006, vol. 3, pp. 461–464 (2006)
17. Vapnik, V.: The Nature of Statistical learning Theory. Springer, New York (1995)
18. Kevman, V.: Learning and Soft Computing: Support Vector Machines, Neural Networks, and Fuzzy Logic models. The MIT Press, Cambridge (2001)
19. Schölkopfy, B., Smola, A.J.: Learning with Kernels. Support Vector Machines, Regularization, Optimization, and Beyond. The MIT Press, Cambridge (2002)
20. Suykens, J.A.K., Van Gestel, T., De Brabanter, J., De Moor, B., Vandewalle, J.: Least Squares Support Vector Machines. World Scientific, Singapore (2002); ISBN 981-238-151-1

A Prediction Model to Diabetes Using Artificial Metaplasticity

Alexis Marcano-Cedeño, Joaquín Torres, and Diego Andina

Group for Automation in Signals and Communications
Technical University of Madrid

Abstract. Diabetes is the most common disease nowadays in all populations and in all age groups. Different techniques of artificial intelligence has been applied to diabetes problem. This research proposed the artificial metaplasticity on multilayer perceptron (AMMLP) as prediction model for prediction of diabetes. The Pima Indians diabetes was used to test the proposed model AMMLP. The results obtained by AMMLP were compared with other algorithms, recently proposed by other researchers,that were applied to the same database. The best result obtained so far with the AMMLP algorithm is 89.93%.

1 Introduction

Diabetes is a major health problem in both industrial and developing countries, and its incidence is rising. It is a disease in which the body does not produce or properly use insulin, the hormone that "unlocks" the cells of the body, allowing glucose to enter and fuel them [1]. Diabetes increases the risks of developing kidney disease, blindness, nerve damage, blood vessel damage and it contributes to heart disease. The World Health Organization in 2000 indicated there were $\sim$ 170 million people with diabetes, and estimated that the number of cases of the disease worldwide will be more than doubled to 366 million by 2030 [2]. Diabetes occurs in two major forms: type 1, or insulindependent diabetes, and type 2, or non-insulin-dependent diabetes. The type 1 diabetes, is characterized by an absolute deficiency of insulin secretion. Individuals at increased risk of developing this type of diabetes can often be identified by serological evidence of an autoimmune pathologic process occurring in the pancreatic islets and by genetic markers [3, 4]. The most common form of diabetes is Type 2 [3, 4]. It is believed that some of the causes of type II diabetes are associated with: diet changes, aging, urbanization, and increasing prevalence of obesity and physical inactivity [3, 4]. This type diabetes results from insulin resistance (a condition in which the body fails to properly use insulin), combined with relative insulin deficiency. In Type II diabetes, either the body does not produce enough insulin or the cells ignore the insulin [3, 4]. Although detection of diabetes is improving, about half of the patients with Type II diabetes are undiagnosed and the delay from disease onset to diagnosis may exceed 10 years. Thus, earlier detection of Type II diabetes and treatment of hyperglycaemia and related to metabolic abnormalities is of vital importance [3, 4, 5].

J.M. Ferrández et al. (Eds.): IWINAC 2011, Part II, LNCS 6687, pp. 418–425, 2011.

There are many factors to analyze to diagnose the diabetes of a patient, and this makes the physician's job difficult. There is no doubt that evaluation of data taken from patient and decisions of experts are the most important factors in diagnosis. But, this is not easy considering the number of factors that has to evaluate [6]. To help the experts and helping possible errors that can be done because of fatigued or inexperienced expert to be minimized, classification systems provide medical data to be examined in shorter time and more detailed. Expert systems and different artificial intelligence techniques for classification systems in medical diagnosis is increasing gradually. As for other clinical diagnosis problems, classification systems have been used for diabetes diagnosis problem [7].

The main objective of the present research was to apply the artificial metaplasticity on multilayer perceptron (AMMLP) as prediction model for prediction of diabetes. The Pima Indian diabetes data set [8] was used to test the proposed model AMMLP. The results obtained by AMMLPT were compared with other algorithms, recently proposed by other researchers, that were applied to the same database.

The remainder of this paper is organized as follows. Section 2 presents a brief data mining, artificial metaplasticity. We present the data base and artificial metaplasticity. The Section 4 gives the experimental results obtained by proposed model. Section 5 compares our results to other methods from the literature. Finally,Section 6 presents the summarized conclusions.

2   Methods and Materials

2.1   Pima Indian Diabetes Data Set

The reason for using this data set is because it is very commonly used among the other classification systems that we have used to compare this study for Pima Indian diabetes diagnosis problem [8]. The data set which consists of Pima Indian diabetes disease measurements contains two classes and 768 samples. The class distribution is:

- Class 1: normal (500) (65.10%)
- Class 2: Pima Indian diabetes (268) (39.89%)

All samples have eight features. These features are:

- Feature 1: Number of times pregnant (Pregn).
- Feature 2: Plasma glucose concentration a 2 h in an oral glucose tolerance test (Gluco).
- Feature 3: Diastolic blood pressure (mm Hg) (Diasp).
- Feature 4: Triceps skin fold thickness (mm) (Thick).
- Feature 5: 2-h serum insulin (lU/ml)(Insul).
- Feature 6: Body mass index (weight in $kg/(heightinm)^2$ (Massi).
- Feature 7: Diabetes pedigree function (Predf).
- Feature 9: Age (years) (Age).
- Class: Diabetes onset within five years (0 or 1)(Yes or No)

2.2   Data Preparation

The quality of the data is the most important aspect as it influences the quality of the results from the analysis. The data should be carefully collected, integrated, characterized, and prepared for analysis. In this study, we applied the techniques of data preprocessing in order to improve the quality of the mining result and the efficiency of the mining process. In this study we analyze a data set composed of 768 data instances. A preliminary analysis of the data indicates the usage of zero for missing data. Since, it does not make sense to have the value of a variable such as plasma-glucose concentration 0 in living people; all the observations with zero entries are removed. After removing all the above said values and variables, only 763 instances remain from the data in our study.

2.3   Artificial Metaplasticity Neural Network

The concept of biological metaplasticity was defined in 1996 by Abraham [9] and now widely applied in the fields of biology, neuroscience, physiology, neurology and others [9, 10].

Recently, Ropero-Peláez [10], Andina [11] and Marcano-Cedeño [12] have introduced and modeled the biological property metaplasticity in the field of artificial neural networks (ANN), obtaining excellent results.

There are different artificial metaplasticity (AMP) models [12]. However, of all AMP models tested by the authors, the most efficient model (as a function of learning time and performance) is the approach that connects metaplasticity and Shannon's information theory, which establishes that less frequent patterns carry more information than frequent patterns [13]. This model defines artificial metaplasticity as a learning procedure that produces greater modifications in the synaptic weights with less frequent patterns than frequent patterns, as a way of extracting more information from the former than from the latter. Biological metaplasticity favors synaptic strengthening for low-level synaptic activity, while the opposite occurs for high level activity. The model is applicable to general ANNs [11, 12], although in this paper it has been implemented for a multilayer perceptron (MLP).

MLP has been used for the solution of many classification problems in pattern recognition applications [14]. The functionality of the topology of the MLP is in most cases determined by a learning algorithm, the Backpropagation (BP), based on the method of steepest descent. In the process of upgrading the connection weights, is the most commonly used algorithm by the ANN scientific community. The BP algorithm presents some limitations and problems during the MLP training [15]. Many researchers have centered their work in improving and developing algorithms for reduce those problems and increasing their advantages [15, 11, 12].

The artificial metaplasticity on multilayer perceptron algorithm (AMMLP) is included in the training algorithm by affecting the weights in each iteration step using a weight function that assumes an estimation or an hypothesis of the real distribution of training patterns. In this paper, has been used the following function to weight the weights updates in the learning phase [11, 12]:

$$f_X^*(x) = \frac{A}{\sqrt{(2\pi)^N} . e^{B \sum_{i=1}^{N} x_i^2}} \tag{1}$$

where N is the number of neurons in the MLP input layer, and parameters A and $B \in R^+$ are algorithm optimization values which depend on the specific application of the AMLP algorithm. Values for A and B have been empirically determined. Eq. (1) is a gaussian distribution, so we have assumed that X pdf is Gaussian (if it is not the case, we should use the real X pdf instead) [11, 12]. Then, $f_X^*(x)$ has high values for un-frequent x values and close to 1 for the frequent ones and can therefore be straightforwardly applied in weights updating procedure to model the biological metaplasticity during learning [12].

2.4   The AMMLP Algorithm

MLP exhibits a sigmoidal activation function with scalar output usually in the range (0,1). This property is true for all neurons of the network. All AMMLP were trained with the same training data set and tested with the same evaluation data set. The network was trained with 60% of the data (455 samples), of which 158 were diabetes and 297 were not diabetes. The testing set, composed of the remaining 40% of the data, consisted of 108 diabetes sample and 200 not diabetes sample.

100 AMMLPs with different initial weights, sampled from random values of a normal distribution (mean of 0 and a variance of 1) have been generated. In each experiment, 100 networks were trained to achieve an average result that is independent of the initial random value of the ANN values. Two different criteria were applied to stop training: a) training was stopped when the error reached 0.01 (error decreases but cannot reach to 0); b) training was performed with a fixed number of 2.000 epochs.

The AMMLP algorithm was developed in MATLAB (MATLAB version 7.6.0.324, R2008a) and on a 3.4 GHz Pentium IV computer with 2 GB of RAM.

2.5   Schema Using

The proposed model AMMLP follow the iterative phases that show Fig. 1.

1. In this step we take the data provided by the medical institute. The possible factors which affect Type-2 diabetes
2. During data preprocessing step, some inappropriate and inconsistent data are deleted.
3. The AMMLP is used to predict accuracy
4. The performance evaluation is measured using the Accuracy, Specificity, and Sensitivity.
5. Compared with existing models and algorithms

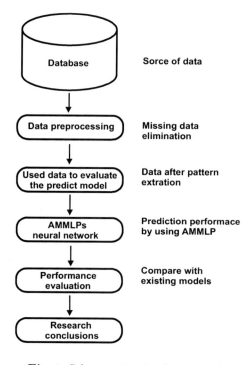

Fig. 1. Scheme using in this research

3   Results

This section present the results obtained by prediction model AMMLP.

To measure the performance of the proposed model AMMLP for the prediction of diabetic, the evaluation was performed as follows:

We determine the accuracy of prediction model using different structures of AMMLP. To evaluate this part, three criteria were applied: analysis of sensitivity, specificity and total classification accuracy prediction, confusion matrix for the best result obtained by AMMLP. These criteria are defined as:

- *Specificity*: number of correctly classified diabetes / number of total diabetes.

- *Sensitivity*: number of correctly classified not diabetes records /number of total not diabetes.

- *Total classification accuracy prediction*: number of correctly classified records/ number of total records.

- *Confusion matrix*: a confusion matrix contains information about actual and predicted classifications performed by a classifier. Performance of the classifier is commonly evaluated using the data in the matrix.

After testing different learning rate, hidden layer nodes and metaplasticity parameters, we find best networks structure and metaplasticity parameters for this case (see Table 1).

Table 1. AMMLP results obtained for different network structures and parameters of metaplasticity algorithm

Network Structure			Metaplasticity Parameters		Mean Squared Error	Clustering Accuracy (%)	
I	HL	O	A	B		Training	Testing
8	**4**	**1**	**37**	**0.2**	0.01	**82.14 %**	**89.93 %**
8	7	1	38	0.3	0.01	69.01 %	87.66 %
8	8	1	39	0.5	0.01	74.06 %	81.49 %

Table 2 and 3 show the confusion matrix and prediction accuracy obtained by AMMLP in the diabetes prediction in terms of specificity, sensitivity and total accuracy prediction.

Table 2. Confusion matrix obtained by AMMLP in the diabetes prediction

Classifier	Desired Result	Output Results	
		Diabetes	No diabetes
AMMLPs	Diabetes	82	26
	No diabetes	5	195

Table 3. Prediction accuracies obtained by AMMPL in the diabetes prediction

Classifier	Prediction Accuracies (%)		
	Specificity	Sensitivity	Total prediction accuracy
AMMLPs	75.92%	97.50%	89.93%

4   Comparison

The prediction accuracy obtained by AMMLP was compared with the best results obtained by other researchers using the same database, are summarized in Table 4.

5   Conclusion

In this paper, was applied the artificial metaplasticity on multilayer perceptron (AMMLP) as prediction model for prediction of diabetes. The proposed model AMMLP obtained an excellent accuracy in the prediction of 89.93%, the highest obtained until now. We think that the proposed model AMMLP-DT can serve as a second opinion for physicians when making their final diagnostic decisions.

Table 4. Classification accuracies of proposed Model AMMLP and other classifiers for the Pima Indians diabetes

Author (year)	Method	Accuracy (%)
Carpenter and Markuzon, (1998)	ARTMAP-IC	81.0
Deng and Kasabov, 2001	ESOM	78.4 ±1.6
Kayaer and Yildirim, 2003	GRNN	80.21
Abdel-Aal, (2005)	T-MC	77.6
Luukka and Leppälampi (2006)	PCA-Entropy	80.47
Polat and Gunes, (2007)	PCA-ANFIS	89.47
Srinivasa,Venugopal and Patnaik, (2007)	SAMGA	74.6
Ji and Carin, (2007)	POMDP	71.43
Polat, Gunes, and Aslan (2008)	LS-SVM	82.05
Kahramanli and Allahverdi (2008)	FNN	84.2
Ghazavi and Liao, (2008)	FUZZY MODELS	77.65
Termutas *et al.* (2009)	MLNN-LM	82.37
Lekkas and Mikhailov (2010)	eClass	79.37
Aibinu *et al.,* (2010)	CVNN-GDA	81.00
Dogantekin *et al.,* (2010)	LDA-ANFIS	84.61
Aibinu, Salami and Shafie (2011)	CAR	81.28
Castro, Nebot and Mugica (2011)	LR-FIR	75,39
In this study, 2011	**AMMLP**	**89.93**

References

1. Mohamed, E.I., Linderm, R., Perriello, G., Di Daniele, N., Poppl, S.J., De Lorenzo, A.: Predicting type 2 diabetes using an electronic nose-base artificial neural network analysis. Diabetes Nutrition & Metabolism 15(4), 215–221 (2002)
2. Shaw, J.E., Sicree, R.A., Zimmet, P.Z.: Global estimates of the prevalence of diabetes for 2010 and 2030. Diabetes Research and Clinical Practice 87, 4–14 (2010), doi:10.1016/j.diabres.2009.10.007
3. American diabetes asociation, http://www.diabetes.org/diabetes-basics/
4. International Diabetes Federation, http://www.idf.org
5. Temurtas, H., Yumusak, N., Temurtas, F.: A comparative study on diabetes disease diagnosis using neural networks. Expert Systems with Applications 36(4), 8610–8615 (2009), doi:10.1016/j.eswa.2008.10.032
6. Polat, K., Gunes, S., Aslan, A.: A cascade learning system for classification of diabetes disease: Generalized discriminant analysis and least square support vector machine. Expert Systems with Applications 34(1), 214–221 (2008), doi:10.1016/j.eswa.2006.09.012
7. Polat, K., Gunes, S.: An expert system approach based on principal component analysis and adaptive neuro-fuzzy inference system to diagnosis of diabetes disease. Digital Signal Processing 17(4), 702–710 (2007)
8. UCI machine learning respiratory,
   http://archive.ics.uci.edu/ml/datasets.html.
9. Abraham, W.C.: Activity-dependent regulation of synaptic plasticity(metaplasticity) in the hippocampus. In: Kato, N. (ed.) The Hippocampus: Functions and Clinical Relevance, pp. 15–26. Elsevier Science, Amsterdam (1996)
10. Kinto, E., Del-Moral-Hernandez, E., Marcano-Cedeño, A., Ropero-Peláez, J.: A preliminary neural model for movement direction recognition based on biologically plausible plasticity rules. In: Mira, J., Álvarez, J.R. (eds.) IWINAC 2007. LNCS, vol. 4528, pp. 628–636. Springer, Heidelberg (2007), doi:10.1007/978-3-540-73055-2_65

11. Andina, D., Alvarez-Vellisco, A., Jevtić, A., Fombellida, J.: Artificial metaplasticity can improve artificial neural network learning. Intelligent Automation and Soft Computing; Special Issue in Signal Processing and Soft Computing, 15(4), 681-694 (2009); ISSN: 1079-8587
12. Marcano-Cedeño, A., Quintanilla-Domínguez, J., Andina, D.: Breast cancer classification applying artificial metaplasticity algorithm. Neurocomputing, doi:10.1016/j.neucom.2010.07.019
13. Shannon, C.E.: A mathematical theory of communication. The Bell System Technical Journal 27, 379–423 (1948), doi:10.1145/584091.584093
14. Hagan, M.T., Demuth, H.B., Beale, M.: Neural network design. PWS Pub. Co., Boston (1996)
15. Leung, H., Haykin, S.: The complex backpropagation algorithm. IEEE Transactions on Signal Processing 39, 2101–2104 (1991)

Band Correction in Random Amplified Polymorphism DNA Images Using Hybrid Genetic Algorithms with Multilevel Thresholding

Carolina Gárate O., M. Angélica Pinninghoff J., and Ricardo Contreras A.

Department of Computer Science
University of Concepción, Chile
{cgarate,mpinning,rcontrer}@udec.cl

Abstract. This paper describes an approach for correcting bands in RAPD images that involves the multilevel thresholding technique and hybridized genetic algorithms. Multilevel thresholding is applied for detecting bands, and genetic algorithms are combined with Tabu Search and with Simulated Annealing, as a mechanism for correcting bands. RAPDs images are affected by various factors; among these factors, the noise and distortion that impact the quality of images, and subsequently, accuracy in interpreting the data. This work proposes hybrid methods that use genetic algorithms, for dealing with the highly combinatorial feature of this problem and, tabu search and simulated annealing, for dealing with local optimum. The results obtained by using them in this particular problem show an improvement in the fitness of individuals.

Keywords: RAPD Images, Polynomial multilevel thresholding, Hybrid genetic algorithms.

1 Introduction

Randomly Amplified Polymorphism DNA (RAPD) images are grey scale images composed by lanes and bands, which has been used in verifying genetic identity. The lanes are the vertical columns shown in Figure 1 and each one of them represents a DNA sample, except the reference lanes which are the leftmost and the rightmost lanes. The reference lanes are used to indicate the molecular weight, measured in base pairs (bp), of the DNA. The bands are the horizontal lines in each lane that represent the segments agglomeration of a DNA sample with the same bp value.

The process of producing RAPD images is affected by many physical-chemical factors [17], and generates different kind of noise, deformations and diffusion, among others, in the lanes and bands of the RAPD images. To correct these problems is important, because their effects can lead to erroneous biological conclusions.

J.M. Ferrández et al. (Eds.): IWINAC 2011, Part II, LNCS 6687, pp. 426–435, 2011.
© Springer-Verlag Berlin Heidelberg 2011

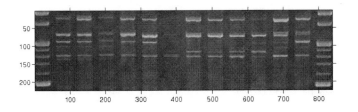

Fig. 1. A sample RAPD image with two reference lanes

The RAPD image correction is a research field not completely explored. There are software tools like ImageJ [3], Gel Compar II[1], GelQuant [5] and Quantity One [2] used for RAPD image analysis. However, the correction is done manually by an expert so the abnormalities remain undetected to the human eye. This fact is pointed out in [17] where the authors use the Radon transform to correct the whole image, and mathematical morphology with cubic spline smoothing for band detection. In [14] lanes and bands are automatically corrected by using genetic algorithms, but it is necessary to supply the number of lanes and bands to trigger the two needed processes. Besides that, in [15] genetic algorithms are combined with Tabu Search to correct the lanes of an image. On the other hand, in [18] is proposed a polynomial algorithm to support Multilevel Thresholding. This work adopts that polynomial multilevel thresholding algorithm to identify the bands in a lane, and then uses genetic algorithms (GA) to correct deformations in the bands. This problem involves a combinatorial search, and consequently it requires an heuristic approach. To optimize the GA results it is usual to hybridize it with meta-heuristics producing an Hybrid Genetic Algorithm (HGA) [1], [12], [16]. The most common meta-heuristics used are Tabu Search and Simulating Annealing, which are considered in this approach.

This article is structured as follows; the first section is made up of the present introduction; the second section defines the specific problem to be faced; the third section describes the multilevel thresholding technique. The fourth section is devoted to describe the hybrid algorithms, while the fifth section shows the results we obtained with our approach, and the final section shows the conclusions of the work.

2   Problem Definition

Our approach considers an histogram associated to a lane image, see Figure 2. The histogram is built adding the intensities per column of the lane image, allowing a better way to distinguish the bands in a lane.

Each peak in the histogram represents a band that occur in a lane. Multilevel thresholding should allow automatically detect those peaks and where they are located. Besides that, it gives information about the left and right boundaries

---

[1] Details of this software are available at
http://www.applied-maths.com/gelcompar/gelcompar.htm.

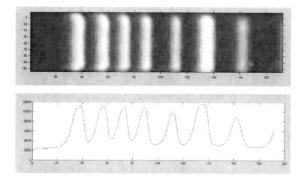

Fig. 2. Histogram associated to a Lane Image

of each band. This algorithm identifies clusters in the histogram, so we use a validity index to evaluate the formed clusters. A good evaluation should mean that the number of clusters corresponds closely to the number of bands in the lane.

The most well-known indexes are I index [13], Silhoutte index (S) [7], Dunn index (D) [7], Davies-Bouldin index (DB) [6], Calinski Harabasz index (CH) [4], SD index [10] and S_Dbw index [11].

The problem we are trying to solve can be formulated as follows:

Consider an image (matrix) $A = \{a_{ij}\}, i = 1, \ldots, m$ and $j = 1, \ldots, n$, where $a_{ij} \in Z^+$, and A is a lane of a RAPD image. Usually, a_{ij} is in the range $[0..255]$ in a grey scale image, and we use a a_{ij} to refer to an element $A(x, y)$, where x and y are the pixel coordinates.

In order to deal with band distortions, a set of templates is used. These templates are randomly created images with different distortion degrees, having curves that are in a one-to-one correspondence with bands in the original lane image. A good template is the one that reflects in a more precise degree the distortions that the lane image under consideration has.

To achieve this, we first use multilevel thresholding to automatically detect the number of bands and their location in the lane image. This information is used as input for the search of the best template with HGA. The produced templates contain curves determined by five points. A template example is shown in Figure 3.

Fig. 3. Template Example for Bands

3   Polynomial Multilevel Thresholding

To detect the number of bands within a lane automatically, we need to find out the thresholds which separate one band from each other; in all the images considered, the number of bands is higher than two, so we use an optimal multi-level thresholding (polynomial) proposed by Rueda [18]. This algorithm divides the histogram associated to a lane image (see Figure 2) in k clusters, so the thresholds are $t_1, \ldots, t_{k-1}$, each cluster represents a band, and the correspondent thresholds the left and right boundaries of these bands.

The algorithm pursues to find the ordered set $T = \{t_0, t_1, t_2, \ldots, t_{k-1}, t_k\}$ of $k + 1$ thresholds and $\forall t_i \in \{0\} \cup H$ (H denotes the histogram). To obtain T we minimize the function $\Phi(T) : H^k \times [0,1]^L \to \mathbf{R}^+$ (defined in (1)) where $L = t_k$, by using the *Minimum Error Criterion* that consists in minimizing the error between the current histogram and the approximate histogram curve of a Gaussian mixture.

$$\Phi(T) = 1 + 2 \sum_{j=1}^{k} \omega_j (log\sigma_j + log\omega_j) \tag{1}$$

In (1) $\sigma_j^2 = \sum_{i=t_{j-1}+1}^{t_j} \frac{p_i(i-\mu_j)^2}{\omega_j}$, and $\omega_j = \sum_{i=t_{j-1}+1}^{t_j} p_i$, are the variance and the cumulative probability of the cluster j, respectively [18].

After the crisp partition of the histogram we need to determine the correct number and the goodness of the clusters formed with validity indexes I, S, D, DB, CH, SD and S_Dbw.

4   Hybrid Genetic Algorithms

Hybrid genetic algorithms combine GA for a global search of solutions with meta-heuristics for a local search, to find optimal solutions in the neighborhood. It has been shown that this combination provides best solutions to certain problems.

The structure of a genetic algorithm consists of a simple iterative procedure on a population of genetically different individuals. The phenotypes are evaluated according to a predefined fitness function, the genotypes of the best individuals are copied several times and modified by genetic operators, and the newly obtained genotypes are inserted in the population in place of the old ones. This procedure is continued until a *good enough* solution is found [8].

In this work, the templates are the chromosomes, curves in a template are the genes, and a curve having a particular shape represents the value (allele) that a gene has.

Elitism was considered to keep a reduced set of the best individuals through different generations.

A good fitness means that a particular template fits better to the original image. To evaluate a template, the template and the image are put together, and a function of fitness measures the similarity.

Genetic operators: Different genetic operators were considered for this work. These genetic operators are briefly described bellow:

- Selection. Selection is accomplished by using the roulette wheel mechanism [8]. It means that individuals with a best fitness value will have a higher probability to be chosen as parents. In other words, those templates that are not a good representation of the lane image are less likely selected.
- Cross-over. Cross-over is used to exchange genetic material, allowing part of the genetic information that one individual to be combined with part of the genetic information of a different individual. It allows us to increase genetic variety, in order to search for better solutions. In other words, if we have two templates each containing $r + s$ curves, after cross-over, the generated children result in: children 1 will have the first r curves that correspond to parent 1, and the following s curves that correspond to parent 2. For children 2, the process is slightly different, in which the order the parents are considered is modified.
- Mutation. By using this genetic operator, a slight variation is introduced into the population so that a new genetic material is created. In this work, mutation is accomplished by randomly replacing, with a low probability, a particular curve in a template.

Tabu Search is one of the meta-heuristics used to hybridize the GA, is characterized for using adaptive memory to save some movements as taboo (are not allow to use this movements) and responsive exploration meaning that a bad strategic choice can yield the search to promising areas.

Tabu search (TS) is a meta-heuristic that guides a local heuristic search procedure to explore the solution space beyond local optimality. The local procedure is a search that uses an operation called *move* to define the neighborhood of any given solution. One of the main components of TS is its use of adaptive memory, which creates a more flexible search behavior. In a few words, this procedure iteratively moves from a solution x to a solution x' in the neighborhood of x, until some stopping criterion has been satisfied. In order to explore regions of the search space that would be left unexplored by the local search procedure, tabu search modifies the neighborhood structure of each solution as the search progresses [9].

A solution is a template representing a lane image, let us say the x solution; then to move from a solution x to a solution x' means that the template is modified. To modify a template we have chosen the change in the value of the shape for one or more curves in the template.

To avoid repeated movements during a certain bounded period of time, TS stores each movement in a temporal memory, which is called *tabu list*. Each element in the *tabu list* contains one band and its corresponding movement. A particular band in the list may occur more than once, but the associated movement needs to be different.

The other meta-heuristics used to hybridize the GA is Simulated Annealing (SA). This is a function optimization procedure based on random perturbations of a candidate solution and a probabilistic decision to retain the mutated solution. Simulated annealing takes inspiration from the process of shaping hot metals into stable forms through a gradual cooling process whereby the material transits from a disordered, unstable, high-energy state to an ordered, stable, low-energy state. In simulated annealing, the material is a candidate solution (equivalent to individual phenotype of an evolutionary algorithm) whose parameters are randomly initialized.

The solution undergoes a mutation and if its energy is lower than that at the previous stage, the mutated solution replaces the old one.

The temperature of the system is lowered every n evaluations, effectively reproducing the probability of retaining mutated solutions with a higher energy states. The procedure stops when the annealing temperature approaches the zero value [8].

In our problem the solution is a template, and the mutation is analogous to the tabu search movement.

In genetic algorithms the problem of local optimum is always present. By taking into account this issue, we decided to *hybridize* the procedure, i.e., to combine genetic algorithms with another strategy, in this specific work with Tabu Search and Simulated Annealing, to let potential solutions avoid those local optimum points. Both hybridization procedures consider the following strategy: the main objective of this process is to gradually improve individuals belonging to the population the genetic algorithm is working with. When the fitness measured during a certain number of iterations accomplished by the genetic algorithm doesn't vary; a reduced number of individuals is selected from the current population. One of them, at least, is the best individual of the population, while the others are randomly selected. Each one of these individuals acts as an input for triggering a tabu search procedure or a simulated annealing procedure.

Once the tabu search (simulated annealing) process is finished, the resulting individuals are re-inserted into the genetic population, and the process continues with the genetic algorithm procedure, as before. The complete process is repeated several times depending on the quality of the genetic population and the stopping process condition.

5   Experiments

To carry out the experiments, tests were separated into two parts; the first one related to Multilevel Thresholding, and the second one related to hybrid genetic algorithms.

Different techniques, functions, methods and algorithms were implemented by using the Matlab environment (version 7.10), due to its potential for working with matrices, the existence of genetic algorithms and simulated annealing toolbox, and for their capability to deal with image processing.

5.1   Multilevel Thresholding Tests

We considered six different strategies to obtain a histogram with relevant peaks:

- OI that corresponds to the original histogram of the original image;
- Same as OI but the histogram is normalized;
- SI, that corresponds to the histogram of a segmented image by using three thresholds
- SINH, same as SI but the histogram is normalized
- OIWB, that considers the image without background
- OIWBNH, same as OIWB but the histogram is normalized

Figure 4 shows the better results, that correspond to SI strategy and the SD validity index. The graphic shows 3 bars for each validity index. The first bar on each group indicates the number of exact matchings when compared to the expert solution; the second one represents the number of detected solutions that differ in one band with the expert opinion, and the last one shows the number of solution that differs in two bands with the expert opinion. In other words, bars represent standard deviations of 0, 1 and 2 respectively.

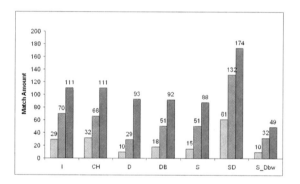

Fig. 4. Match between Expert and Multilevel Thresholding for SI Strategy

For testing, we considered 274 lanes. The SD index had the best performance, obtaining 22.26%, 48.18% and 63.50% for the standard deviations 0, 1 y 2, respectively. These values are far from representing good solutions.

Results are a consequence of images quality, and therefore of the quality of the histogram associated to the processed lane. The poor image quality obeys to the band fuzziness or to the existence of noisy images. Additionally, validity indexes seem not to be the adequate ones, because this particular problem presents well defined patterns that are very unusual when compared to the typical problems in which they are commonly used.

Although Multilevel Thresholding considering validity indexes doesn't detect the number of bands in a lane automatically, if the number of bands is known a priori, this technique successfully detects the center of the band and the left and right boundaries.

Then, we decided to ask the expert about the number of bands to work with a more accurate band position, a key issue for templates generation.

Once the position of the bands has been determined, the next step is to generate templates, by randomly creating curves and locating these curves on the bands' locations.

5.2   Genetic Algorithms Tests

For testing, we considered the same 274 lanes, as before.

To evaluate potential solutions, we used two fitness functions. The first one is the function proposed in [14] that maximizes the sum of intensities in the bands. Let us call it, *maximized fitness*.

$$\sum_{b=1}^{k} \sum_{i=1}^{m} sum(A(i, li_b : ld_b))$$ (2)

where b is the number of the band in the lane, i is the number of the row in matrix A, li_b is the left boundary for band b and ld_b is the right boundary for band b.

The second fitness function, is specifically proposed for this work, and we call it *minimized fitness*. Equation (3) shows this function.

$$sum(\Phi_{MinEr\_Original}(T) - \Phi_{MinEr\_Individual}(T))^2$$ (3)

$\Phi_{MinEr}(T)$ corresponds to equation (1). $\Phi_{MinEr\_Original}(T)$ represents the curve of the Gaussian mixture for the original lane and $\Phi_{MinEr\_Individual}(T)$ represents the curve of the Gaussian mixture for the current individual generated with the hybrid algorithm.

TS and SA heuristics are triggered after 30 generations with unchanged fitness. These processes are activated on four individuals from the current population, two of them chosen randomly, and the other two considering the individuals with the best fitness value. For testing we considered three different strategies: single genetic algorithm, genetic algorithm combined with tabu search and genetic algorithm combined with simulated annealing. For each strategy above, we

Table 1. Parameters for testing

Parameter	Value(s)
Population Size	70, 130
Num. Generations	10000
Cross-over	80%
Mutation	5%
Elitism	10%
Number of TS or SA generations	100

measured the fitness in two different ways, as defined in equations (2) and (3). In other words we grouped test cases into six test families. Table 1 shows the parameters considered.

Results have shown a similar performance on 5 of the 6 test families. For the hybridization that considers genetic algorithms and simulated annealing under the minimized fitness, results are better than in all other cases. This fact can be observed in the templates obtained where the curves follow with a higher accuracy the bands' distortion. This kind of results are illustrated in figure 5.

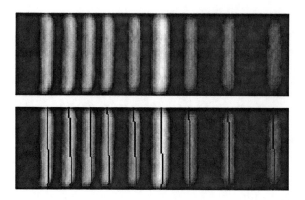

Fig. 5. Template obtained with genetic algorithm and simulated annealing

6   Conclusions

The proposed approach was tested with an important number of experiments; that allowed us to recognize the strengths and the weaknesses to be emphasized and reduced, respectively, in a future work. As a general concept, this work offers a promising combination that can be explored to solve similar problems.

The best performance is obtained by using genetic algorithms hybridized with simulated annealing. Nevertheless, results are poor, due to the fact that threshold detection need to be improved. Multilevel thresholding represents a useful technique to automatically detect bands, when the bands in an image can be clearly distinguished. For this work, most of available images didn't fulfill this condition. In spite of this, the presented approach makes a contribution that generates good results under specific constraints.

As a future work, it should be considered a particular validity index, given the specific nature of this problem, that can help to a better clusters evaluation. Besides that, a different preprocessing could help to a better peak detection, that leads to a more accurate way for determining thresholds in the histogram.

Acknowledgements. This work has been partially supported by grant DIUC 209.093.014-1.0, University of Concepción, Chile.

References

1. Ait-Aoudia, S., Mahiou, R.: Medical Image Registration by Simulated Annealing and genetic algorithms. In: Geometric Modeling and Imaging, GMAI 2007, pp. 145–148 (2007)
2. Bio Rad Laboratories. Quantity One User's Guide. Bio-Rad Laboratories (2000)
3. Burger, W.: Digital Image Processing with Java and ImageJ. Springer, Heidelberg (2006)
4. Calinski, R.B., Harabasz, J.: A Dendrite Method for Cluster Analysis. Communications in Statistics 3, 1–27 (1974)
5. Das, R., Laederach, A.: GelQuant User's Manual. Stanford University, Stanford (2004)
6. Davies, J.L., Bouldin, D.W.: A cluster separation measure. IEEE Transactions on Pattern Analysis and Machine Intelligence 1, 224–227 (1979)
7. Dunn, J.C.: Well Separated Clusters and Optimal Fuzzy Partitions. Journal of Cybernetics 4, 95–104 (1974)
8. Floreano, D., Mattiussi, C.: Bio-Inspired Artificial Intelligence. Theories, Methods, and Technologies. The MIT Press, Cambridge (2008)
9. Glover, F., Laguna, M.: Tabu Search. Springer, Heidelberg (1997)
10. Halkidi, M., Vazirgiannis, M., Batistakis, Y.: Quality scheme assessment in the clustering process. In: Zighed, D.A., Komorowski, J., Żytkow, J.M. (eds.) PKDD 2000. LNCS (LNAI), vol. 1910, pp. 265–276. Springer, Heidelberg (2000)
11. Halkidi, M., Vazirgiannis, M.: Clustering Validity Assessment: Finding the Optimal Partitioning of a Data Set. In: Proceedings of ICDM, California, USA (November 2001)
12. Li, X., Wei, X.: An Improved Genetic Algorithm-Simulated Annealing Hybrid Algorithm for the Optimization of Multiple Reservoirs. Water Resources Management 22, 1031–1049 (2008)
13. Maulik, U., Bandyopadhyay, S.: Performance Evaluation of Some Clustering Algorithms and Validity Index. IEEE Trans. Pattern Anal. Mach. Intell. 24(12), 1650–1654 (2002)
14. Pinninghoff, M.A., Contreras, R., Rueda, L.: An evolutionary approach for correcting random amplified polymorphism DNA images. In: Mira, J., Ferrández, J.M., Álvarez, J.R., de la Paz, F., Toledo, F.J. (eds.) IWINAC 2009. LNCS, vol. 5602, pp. 469–477. Springer, Heidelberg (2009)
15. Pinninghoff, M.A., Venegas, D., Contreras, R.: Genetic Algorithms and Tabu Search for Correcting Lanes in DNA Images. In: Martínez-Trinidad, J.F., Carrasco-Ochoa, J.A., Kittler, J. (eds.) MCPR 2010. LNCS, vol. 6256, pp. 144–153. Springer, Heidelberg (2010)
16. Robles, V., Peña, J.M., Pérez, M.S., Herrero, P., Cubo, O.: Extending the GA-EDA hybrid algorithm to study diversification and intensification in gAs and eDAs. In: Famili, A.F., Kok, J.N., Peña, J.M., Siebes, A., Feelders, A. (eds.) IDA 2005. LNCS, vol. 3646, pp. 339–350. Springer, Heidelberg (2005)
17. Rueda, L., Uyarte, O., Valenzuela, S., Rodriguez, J.: Processing random amplified polymorphysm DNA images using the radon transform and mathematical morphology. In: Kamel, M.S., Campilho, A. (eds.) ICIAR 2007. LNCS, vol. 4633, pp. 1071–1081. Springer, Heidelberg (2007)
18. Rueda, L.: An efficient algorithm for optimal multilevel thresholding of irregularly sampled histograms. In: da Vitoria Lobo, N., Kasparis, T., Roli, F., Kwok, J.T., Georgiopoulos, M., Anagnostopoulos, G.C., Loog, M. (eds.) S+SSPR 2008. LNCS, vol. 5342, pp. 602–611. Springer, Heidelberg (2008)

Discrimination of Epileptic Events Using EEG Rhythm Decomposition

L. Duque-Muñoz*, L.D Avendaño-Valencia**,
and G. Castellanos-Domínguez* * *

Universidad Nacional de Colombia, sede Manizales. Campus la Nubia. Km 9 Via al
Magdalena. Manizales, Caldas. Colombia
http://www.manizales.unal.edu.co/gta/signal/

Abstract. The use of time series decomposition into sub–bands of frequency to accomplish the oscillation modes in nonstationary signals is proposed. Specifically, EEG signals are decomposed into frequency sub-bands, and the most relevant of them are employed for the detection of epilepsy seizures. Since the computation of oscillation modes is carried out based on Time-Variant Autoregressive model parameters, both approaches for searching an optimal order are studied: estimation over the entire database, and over each database recording. The feature set appraises parametric power spectral density in each frequency band of the Time-Variant Autoregressive models. Developed dimension reduction approach of high dimensional spectral space that is based on principal component analysis searches for frequency bands holding the higher values of relevance, in terms of performed accuracy of detection. Attained outcomes for $k-$nn classifier over 29 epilepsy patients reach a performed accuracy as high as 95% As a result, the proposed methodology provides a higher performance when is used a optimal order for each signal. The advantage of the proposed methodology is the interpretations that may lead to the data, since each oscillation mode can be associated with one of the eeg rhythms.

1   Introduction

The electrical activity generated by the brain in normal conditions, varies depending on internal factors, such as age, sex, vigil, sleep, and external factors, like medicines and environment. For this reason, it is difficult to define a normal EEG due to the presence of multiple electrical patterns that can be found in records from healthy patients. However, there are different frequency bands, known as bands delta, theta, alpha and beta, which are activated to a greater or lesser proportion according to the state of patient [1,2]. Particularly, in the case of epilepsy, it has been shown that there is an increase of low frequency activity

---

* Leonardo Duque Muñoz is a M.eng. student lduquem@unal.edu.co
** David Avendaño Valencia is a PhD student ldavendanov@unal.edu.co
* * * Germán Castellanos Domínguez is with Electrical, Electronical and Computing Department gcastellanosd@unal.edu.co

J.M. Ferrández et al. (Eds.): IWINAC 2011, Part II, LNCS 6687, pp. 436–444, 2011.
© Springer-Verlag Berlin Heidelberg 2011

related to the delta and theta bands, and also high frequency activity, related with oscillations above 60 Hz [3].

Conventional frequency domain analysis is performed on the raw signal, which may contain irrelevant components and noise, besides of the relevant information. As the presence of these unwanted components may have a negative effect on the classification stage, reduction of unwanted spectral components on the signal might be helpful. For this reason, decomposition and selection of frequency bands on EEG signals would be of benefit on the analysis and classification of epileptic activity. Generally, the separation of frequency bands is performed using conventional filters, whose pass band corresponds to the analyzed frequency band. However, the quality of the estimate of each frequency band depends directly on the quality of the filter that is used for signal decomposition. One way to improve the estimation is to take into account changes in the dynamics of the signal using non-stationary analysis techniques such as wavelet analysis. Nevertheless, one of their problems is not know which of the bands obtained in the decomposition assist in the classification task because they present redundant information, the lack of direct interpretability, since the frequency bands obtained do not correspond to the predefined bands from the medical point of view, and how to extract the relevant information to assist in the classification task.

Another way to analyze the changing dynamics in the time of the signals is using parametric methods, such as time-varying autoregressive models (TVAR). As was previously shown in other studies [4], these models have the ability to track the dynamic changes of EEG signals while breaking the signal into frequency bands, based on the signal dynamics. The use of such decompositions in the analysis and classification of EEG signals is still an open problem, since it is unclear how to solve the problem of adjustment of each frequency bands based on TVAR model parameters, the influence of the model order, how to handle the decomposition when there are different orders for each signal of the database, and which frequency band is the most relevant in the task of epilepsy detection.

In this paper we propose a joint methodology for analysis and classification of epileptic events in EEG signals using the decomposition in frequency bands based on TVAR modeling. The methodology consists on the estimation of TVAR model for each of the signals, signal decomposition using each oscillation mode corresponding to the time-varying roots of the model and then estimation of the power spectral density of the signal in each frequency band (delta, theta, alpha and beta). These features are used to classify the different epileptic events. Obtained results show improved performance of the proposed approach compared with conventional power spectral density and discrete wavelet transform of the signal. Then it is found that the method improves the accuracy of classification and additionally allows a better understanding of the EEG signals. The results show a better performance than that obtained with wavelet analysis. The methodology has the advantage of understanding because the sub-bands obtained in the decomposition depend directly on the patient's clinical status when making the EEG.

2   Methods

2.1   Time–Varying Autorregresive Models

Let $y[t] \in \mathbb{R}$ a non–stationary signal defined for $t = 1, \ldots, N$, which can be modeled using a time-varying autoregressive model of order p, denoted as TVAR(p), described by the following linear autoregressive expression:

$$y[t] = \sum_{i=1}^{p} \theta_i[t] y[t-i] + e[t] = \boldsymbol{\theta}^{\mathsf{T}}[t] \boldsymbol{h}[t-1] + e[t]; \tag{1}$$

where $\boldsymbol{h}[t] = \{y[t-i] : i = 1, \ldots, p\}$, $\boldsymbol{h} \subset \mathbb{R}^p$ is a non–stationary process (real vector) corresponding to delayed versions of $y[t]$, $e[t] \sim \mathcal{N}(0, \sigma_e^2[t])$ is a sequence of non observed and uncorrelated innovations with zero-mean and time-varying variances $\sigma_e^2[t]$ generated the process $y[t]$, and $\boldsymbol{\theta}[t] = \{\theta_i[t]\}, \boldsymbol{\theta} \subset \mathbb{R}^p$ is the vector of parameters of the TVAR model. It can be shown that the TVAR model in Equation (1) is related to a sum of complex exponential with time-varying amplitude and exponent range:

$$\boldsymbol{y}[t] = \sigma_e^2[t] \sum_{i=1}^{p} r_i[t] \exp(-\rho_i[t]t) \tag{2}$$

$$= \sigma_e^2[t] \sum_{i=1}^{p} r_i[t] \exp(-\alpha_i[t]t) \exp(-j\omega_i[t]t) \tag{3}$$

where $\rho_i[t], i = 1, \ldots, p$ corresponds to the roots of AR polynomial for each time instant t, while α_i and ω_i correspond to the magnitude and angle of ρ_i respectively, ie. $\rho_i = \alpha_i \angle \omega_i$.

In this way the signal $y[t]$ can be decomposed into a series of p complex modes of oscillation determined by the magnitude and angles corresponding to the time–varying poles of the TVAR model. Since the signal $y[t]$ is real, then the complex poles must be located in conjugate pairs, so that complex parts are canceled. Furthermore, the amplitude modulator terms $\sigma_e^2[t]$, $r_i[t]$ and $\exp(-\alpha_i[t]t)$ can be combined as a single amplitude term $A_i^{(r)}[t]$ corresponding to real modes and $A_i^{(c)}[t]$ corresponding to the complex pairs. The basic decomposition result may now be expressed as:

$$\boldsymbol{y}[t] = \sum_{i=1}^{p_r} A_i^{(r)}[t] + \sum_{i=1}^{p_c} A_i^{(c)}[t] \cos(\omega_i[t]t) \tag{4}$$

with $p = p_r + 2p_c$. Each one of the oscillation modes associated with the TVAR model is associated with a frequency range of the bands analyzed for the EEG δ (1–4 Hz), θ (4–8 Hz), α(8–13 Hz) y β (>13 Hz).

The proposed methodology is summarized in the following steps

1. Tuning the order of the model for each signal [5].
2. Compute the oscillation modes by adjusting the TVAR model [6].
3. Compute the parametric power spectral density of each frequency band.
4. Dimension reduction by PCA.
5. Classification by nearest neighbor classifier.

The order of the model is computed using the BIC criterion, as was shown in [7]. Two different forms to use the order of the model are regarded, the first consisting of using the order in which it is possible to adequately model the bulk of the database, and the second means using the optimal order for each database record using values within the range of $7 < p < 12$. The power spectral density is computed for each isolated band as feature for the classification of EEG signals. The wavelet analysis is included as an alternative method for this task.

2.2 Wavelet Analysis

For most signal processing applications, DWT-based analysis is best described in terms of filter banks. The use of a group of filters to divide up a signal into various spectral components is termed sub-band coding. This procedure is known as multi-resolution decomposition of a signal $x[k]$. Each stage of this scheme consists of two digital filters and two down-samplers by 2. The first filter, $h[.]$ is the discrete mother wavelet, high-pass in nature, and the second, $g[.]$ is its mirror version, low-pass in nature. The down-sampled outputs of first high-pass and low-pass filters provide the detail and the approximation levels.

Selection of appropriate wavelet and the number of levels of decomposition is very important in analysis of signals using DWT. The number of levels of decomposition is chosen based on the dominant frequency components of the signal. The levels are chosen such that those parts of the signal that correlate well with the frequencies required for classification of the signal are retained in the wavelet coefficients [8]. The signal is decomposed into the details D1–D4 and one final approximation A4. These approximation and detail records are reconstructed from the Daubechies 6 (DB6) and Symlet 8 (Sym8) wavelet filter.

3 Experimental Set–Up

3.1 Data Base

The analyzed EEG signals were recorded from 29 epilepsy patients with medically intractable focal epilepsies undergoing invasive presurgical diagnostics between 1993 and 2000 at the Department of Epileptology of the University of Bonn, Germany [9]. Database consists of five sets (noted as A-E) composed of 100 single channel EEG segments. These segments were selected and extracted from continuous multichannel EEG after visual inspection to avoid artifacts. Data sets A and B consist of segments taken from scalp EEG records in five healthy people using electrode placement standard 10–20. Volunteers were woke up, relaxed with eyes open (A) and eyes closed (B), respectively. Data sets C, D and E were selected from presurgical diagnose EEG records. Signals from five patients were selected who had achieved complete control of epileptic episodes after dissection of one of the hippocampal formations, which was correctly diagnosed as the epileptogenic zone. Segments of set D were recorded in the epileptogenic zone, and segments of C in the hippocampal zone of the opposite side of the brain. While sets C and D only contain measured activity on inter–ictal intervals, set E only contains records with ictal activity. In this set all segments were selected from every record place exhibiting ictal activity. All EEG signals were recorded with an acquisition system of 128 channels, using average common reference. Data was digitized at 173, 61 Hz with 12 bits resolution. All the EEG segments from the dataset were used and they were classified into three different classes: Z and O types of EEG segments were combined to a single class, N and F types were also combined to a single class, and type S was the third class. This set is the one closest to real medical applications including three categories; normal (i.e., types Z and O) with 200 records, seizure free (i.e., types N and F) with 200 records and seizure (i.e., type S) with 100 records.

3.2 Experiments

As outlined in the methodology, the procedure is to estimate the TVAR model, using an optimal order for each sign of the database $7 < p < 12$, and a fixed order $p = 8$, calculated by BIC criterion. Then each oscillation mode is computed using the TVAR polynomial decomposition. For each oscillation mode the power spectral density is computed as characteristic for classification. An iterative process is used to find the frequency bands that maximize the accuracy of the classifier, selecting and removing some bands, using each of the bands, and then combinations of them until to combine them all. Then PCA is used in order to reduce the dimension of the feature vector. This stage comprises computation of the quantity of PCA eigenvectors which maximize the accuracy of the classifier. The method is evaluated using a $k-$nn classifier with $k = 3$ and a cross-validation procedure using 10 folds. The performance measures used are accuracy, sensitivity and specificity. Regarding wavelet decomposition, two wavelet families are analyzed Daubechies6 and Symlet8. The cross validation step is carried out in the same form.

4   Results and Discussion

Figure 1 shows the obtained decomposition in a signal corresponding to an epileptic seizure. The top plot shows the obtained decomposition using the proposed methodology, obtaining frequency bands δ, θ, α, and β. α and β band has lower amplitude, while δ wave dominates in amplitude and appears as a smoothed version of the data series whereas the second component in the θ band is lower in amplitude but still contains significant information. This decomposition is

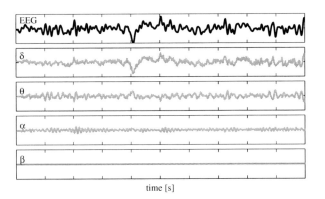

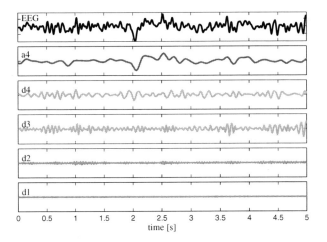

Fig. 1. Decompositions obtained by the methodology with TVAR models (top) and wavelet analysis (down)

an explicit representation of expected phenomena: the seizure EEG is dominated by slow-waves δ and θ bands, whose frequency decreases as the seizure progresses [4]. The lower plot shows the decomposition of the same signal using DWT and Symlet 8 family, showing the first four detail levels (d4–d1) and the last approximation level (a4).

As mentioned in the methodology, two ways to use the order of the model are compared, the first one consisting of using the order in which it is possible to adequately model the bulk of the database and the second means using the optimal order for each record of the database, using values within the range of $7 < p < 12$. Figure 2 shows the variation of the accuracy as the number of

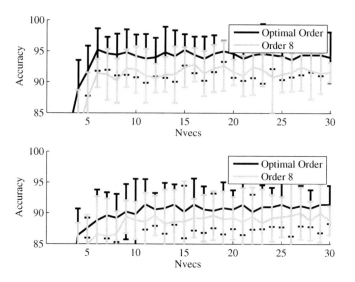

Fig. 2. Performance of the classifier as the number of vectors of PCA increased, PSD calculated on the combination of δ and θ waves (top) and PSD calculated on raw signal (down)

Table 1. Summary of best results

PSD - decomposed signal			
Method	Accuracy	Sensitivity	Specificity
BIC - record	95.20 ± 3.42	95.50 ± 4.37	95.00 ± 3.92
BIC - DB	92.80 ± 2.69	97.50 ± 2.63	94.00 ± 2.62
PSD - raw signal			
Method	Accuracy	Sensitivity	Specificity
BIC - record	91.41 ± 4.59	96.00 ± 3.49	93.37 ± 4.90
BIC - DB	89.00 ± 3.01	92.50 ± 4.74	91.00 ± 3.60
Wavelet analysis			
Method	Accuracy	Sensitivity	Specificity
db6	92.20 ± 2.57	96.50 ± 4.12	96.67 ± 3.14
Sym8	92.80 ± 2.35	96.50 ± 5.30	97.00 ± 3.31

principal components increases for each form to use the model order: black line – optimal order for each record, grey line – optimal order for the database. It can be seen that by using an optimal order for each record, the methodology presents a better performance, increasing about 2% the accuracy of the classifier, in this case, the power spectral density calculated on the combination of delta and theta waves, provides better results than the PSD calculated on the raw signal. Table 1 shows the best classification results with the proposed methodology. The best results are obtained using the optimal order for each record ($7 < p < 12$) with 14 eigenvectors of PCA and a PSD as a feature for classification, being higher than those obtained using a fixed order ($p = 8$) for the entire database, which uses 12 eigenvectors of PCA and the same feature of classification. It can be seen that for both methods, higher accuracy rates are achieved using the optimal order for each signal of the database.

The wavelet analysis also provides favorable results similar to the methodology with fixed order, for both wavelet families Daubechies and Symlet, with a higher performance with the Symlet family.

The results show that the proposed methodology has a better performance when a optimal order is selected for each record of the database unlike choosing a fixed order, doing this the accuracy increased by at least 2%. The results obtained with wavelet analysis are satisfactory because of the simple feature that was used for classification, these results can be improved by using other features more robust and appropriate for this type of analysis. Among studied wavelet families, the Symlet 8 had the best accuracy.

In [11], the extraction of PSD from time-frequency maps based on STFT and several t-f distributions are discussed for the classification of epileptic events. In particular, a accuracy of 83.8% is achieved using a knn classifier, which is remarkable lower that the performed accuracy in this paper (95.20% for 3-NN classifier).

5   Conclusions

In the present study tested a method of decomposition of non–stationary signals in frequency sub-bands, in the case of EEG signals are decomposed into sub–bands δ (1 - 4 Hz) , θ (4 - 8), α (8 - 13) and β (> 13 Hz) for the detection of epileptic signs. The method consists on obtaining the frequency responses of a TVAR model to find the desired bands, for each of these bands is calculated power spectral density, then a dimension reduction stage using PCA. The method is compared using a fixed order $p = 8$ and an optimal order for each sign of the database $7 < p < 12$, and the same test by calculating the power spectral density for the raw signal.

Decomposition of EEG signals in their rhythms was carried out, making the classification of epileptic signals. It was found that the low frequency bands (delta and theta) are discriminative when performing this task, because these rhythms are found in greater proportion in the ictal signs, when is included the analysis of the model order, the results allow better modeling signal changes to improve the performance of classification. As the wavelet analysis was included as an alternative method, it can improve, including other more robust features for classification, so that it can be used with other wavelet family or different levels of decomposition, this analysis shows similar results as the methodology that uses a fixed order for the models, it was found that the Symlet8 family provides better results than Daubechies6 family, for this type of signals.

Acknowledgments

This research is carried out under grants "Centro de Investigación e Innovación de Excelencia ARTICA", fouded by COLCIENCIAS and "Programa jovenes investigadores e innovadores 2009, convenio interadministrativo especial de coperacion No 704 de 2009 entre COLCIENCIAS y la Universidad Nacional de Colombia Sede Manizales".

References

1. Fernandez, T., Harmony, T., Rodríguez, M., Bernal, J., Silba, J., Reyes, J., Marosi, E.: EEG activation patterns during the performance of tasks involving different components of mental calculatio. Electroencephalography and Clinical Neurophysiology 94, 175–182 (1995)
2. Niemiec, A.J., Lithgow, B.J.: Alpha-band characteristics in EEG spectrum indicate reliability of frontal brain asymmetry measures in diagnosis of depression. In: 27th Annual Conference on Engineering in Medicine and Biology (2005)
3. Jung, K.-Y., Kang, J.-K., Kim, J.H., Im, C.-H., Kim, K.H., Jung, H.-K.: Spatiotemporospectral characteristics of scalp ictal EEG in mesial temporal lobe epilepsy with hippocampal sclerosis. Brain Research 1287, 206–219 (2009)
4. West, M., Prado, R., Krystal, A.D.: Evaluation and comparison of EEG traces: Latent structure in nonstationary time series. Journal of the American Statistical Association 94, 448–462 (1999)
5. Stoica, P., Selen, Y.: Model Order selection a review of information criterion rules. IEEE Signal Processing Magazine 21, 36–47 (2004)
6. Tarvainen, M.: Estimation Methods for Nonstationary Biosignals. PhD thesis, University of Kuopio (2004)
7. Avendano-Valencia, L., Godino-Llorente, J., Blanco-Velasco, M., Castellanos-Dominguez, G.: Feature Extraction From Parametric Time–Frequency Representations. Annals of Biomedical Engineering 38(8) (2010), doi:10.1007
8. Adeli, H., Zhou, Z., Dadmehr, N.: Analysis of EEG records in an epileptic patient using wavelet transform. Journal of Neuroscience Methods 123, 69–87 (2003)
9. Andrzejak, R.G., Mormann, F., Widman, G., Kreuz, T., Elger, C.E., Lehnertz, K.: Improved spatial characterization of the epileptic brain by focusing on nonlinearity. Epilepsy Research, 30–44 (2006)
10. Herman, P., Prasad, G., McGinnity, T., Coyle, D.: Comparative Analysis of Spectral Approaches to Feature Extraction for EEG-Based Motor Imagery Classification. IEEE Transactions on Neural Systems adn Rehabilitation Engineering 16, 217–326 (2008)
11. Tzallas, A.T., Tsipouras, M.G., Fotiadis, D.I.: Epileptic Seizure Detection in EEG's using Time–Frecuency Analysis. IEEE Transactions on Information Technology in Biomedicine 13, 703–709 (2009)

Methodology for Attention Deficit/Hyperactivity Disorder Detection by Means of Event-Related Potentials

Paola Castro-Cabrera[1], Jorge Gómez-García[1], Francia Restrepo[2],
Oscar Moscoso[2], and German Castellanos-Dominguez[1]

[1] Universidad Nacional de Colombia, sede Manizales
[2] Universidad Autónoma de Manizales
{pacastroc,jorgomezg,cgcastellanosd}@unal.edu.co
{franciarestrepo,oscarhma}@autonoma.edu.co
http://www.manizales.unal.edu.co/gta/signal/

Abstract. Event-related Potentials (ERPs) are voltage fluctuations in electroencephalogram that allow the examination of electrical representations of the underlying sensory and cognitive processes occurring in the brain in response to stimuli. These waveforms contain characteristic peaks and troughs, which can correspond to certain underlying processes. The determination of the functional significance of a particular ERP component involves simultaneous consideration of its eliciting conditions, polarity, latency and scalp distribution. The evaluation of these parameters, by medical specialists, leads to diagnostic of important psychiatric disorders such as attention deficit/hyperactivity disorder (ADHD). However, the measurement on these parameters is usually susceptible to the subjectivity of the medical concept. This work presents a comparison between two methodologies that consider characterization and feature extraction/selection of ERPs signals, in order to distinguish normal from ADHD patients on a feature set formed by morphological, frequency and wavelet characteristics. Moreover, tests are made on the raw signals looking for informative events that could provide an increasing on classification accuracy.

Keywords: Attention deficit/hyperactivity disorder , event-related Potentials, locally linear embedding, MANOVA, wavelet.

1 Introduction

Attention-deficit/hyperactivity disorder (ADHD) is one of the most common psychiatric disorders in school age children, which is characterized by a persistent pattern of impaired attention, impulsive behavior, and excessive motor hyperactivity [1]. These conduct manifestations directly affect academic activities, familiar dynamic and a proper performance in the social environment, which can negatively influence on the personality development of child [2]. The clinical diagnosis of this disease requires a large number of medical specialists, implying high cost that makes difficult the access to low budget population.

J.M. Ferrández et al. (Eds.): IWINAC 2011, Part II, LNCS 6687, pp. 445–453, 2011.
© Springer-Verlag Berlin Heidelberg 2011

Event-related potentials (ERPs) are defined as changes in the ongoing electroencephalogram (EEG) due to a sensory stimulus . Because these potentials are physiologically correlated with neurocognitive functions, they have been widely used for clinical-diagnostic, brain-computer interface, and especially, in investigations of perceptual and cognitive-processing deficits in children with ADHD. The most popular assessed ERP features for interpretation of cognitive processes are the areas and the peaks of the ERPs components, defined by the mean and peak to peak voltages, respectively, computed in certain windows in the time domain, which are determined by visual inspection of the average ERP waveforms [3]. The analysis of this parameters is usually performed in the time domain, whereby amplitudes and latencies of prominent peaks in the averaged potentials are measured and correlated to information processing mechanisms [4]. Although the quantification of ERP components by areas and peaks is the standard procedure in fundamental ERP research, the conventional approach has two drawbacks:

Firstly, ERPs are time-varying signals reflecting the sum of underlying neural events during stimulus processing, operating on different time scales ranging from milliseconds to seconds. Various procedures such as ERP subtraction or statistical methods have been employed to separate functionally meaningful events that partly or completely overlap in time. However, the reliable identification of these components in the ERP waveforms still remains as a problem.

Secondly, analysis in the frequency domain has revealed that EEG/ERP components in different bands (delta, theta, alpha, beta, gamma) are functionally related to information processing and behaviour. However, the Fourier transform (FT) of ERPs lacks the information about the time localization of transient neural events. Therefore, efficient algorithms for analyzing a signal in time-frequency plane are very important in extracting and relating distinct functional components.

These limitations, as well as the ones related with time–invariant methods, can be solved by using the wavelet formalism. The wavelet transform (WT) is a time–frequency representation, that has an optimal resolution both in the time and frequency domains and has been successfully applied to the study of EEG–ERP signals [5] [6]. Although ERP feature extraction from the time–frequency domain based on the discrete WT (DWT) has been growing increasingly popular, this approach can result weak to pathology detection purposes, particularly, to ADHD identification.

For this reason, this paper explores the use of morphological, frequency and wavelet features on the characterization for discrimination between normal and ADHD patients. With the aim to reduce the amount of information three different feature extraction/selection methods were applied, such as linearly local embedding (LLE), a relevance analysis using a correlation function between variables, and a sequential forward searching (SFS) using a multivariate analysis of variance (MANOVA) as cost function. Also, some tests were made on raw data to do a temporal analysis of signal in order to identify informative events capable to augment accuracy rates.

2   Experimental Framework

2.1   Database

Subjects. The experiments were carried out with 144 children belonging to educational institutions of the metropolitan area of the Manizales, from which 72 of the control group and 72 of the cases group. A written informed consent previously obtained from all parents of participants was necessary to be included into study. The subjects, with ages between 4 and 15 years old, were medically diagnosed based on clinical criteria of DSM-IV [1] and minikid criteria by a multidisciplinary specialist team consisting of a general physician, psychologist, neuropsychologist and experts in children psychiatric disorders. Both groups were tested under the same lighting and noise conditions, and were defined by the following inclusion criteria: non abnormality physical examination, normal visual and hearing ability, intellectual coefficient greater than 80 and, if necessary, pharmacologic management previously suspended. Subjects were verified to be free of some evidence of other neurological disorder.

Data Acquisition. Recordings were acquired by means of electrodes located in the midline of the head (Fz, Cz, Pz) according to 10-20 international system, at sampling rate of 640 samples per second. The examination protocol was conducted according to criteria from the Oddball paradigm in auditory and visual modalities. The first procedure involves the emission of 80 dB tone lasting 50 ms, with a frequency of 1.000 Hz for frequent stimulus and 3.000 Hz for infrequent stimulus, presented randomly every 1.5 s. In the visual modality of the test, the subject is asked to watch a monitor placed 1 m away thet shows an image with a consistent pattern (a checkerboard of 16 squares), that is the frequent stimulus. The rare stimulus is the presentation of a target in the center of the screen with the same common pattern in the background; the subject must press a button each time the unusual stimulus appears. The experiment consists of 200 stimuli, of which 80% of the stimuli are non-target and 20% remaining are target stimuli.

2.2   Methodology 1

The methodology used to detect between normal and TDAH patients by using the ERP is shown on figure 1.

An Independent Component Analysis was first performed looking for independent sources on each one of the 3 analyzed electrodes on both modalities.

The initial feature space consists of three sets of estimates and parameters that have shown optimum performance in similar studies [7]. The first group comprises 17 morphological features, which consist of parameters measured over the whole signal and are related to its shape. This set is formed by the following characteristics: latency, amplitude, latency/amplitude ratio, absolute amplitude, absolute latency/amplitude ratio, positive area, negative area, total area, absolute total area, total absolute area, average absolute signal slope, peak-peak value, peak-peak value in a time window, peak-peak slope, zero crossings, zero crossings density and slope sign alterations.

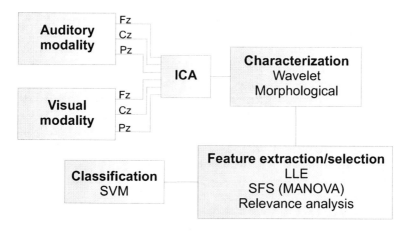

Fig. 1. Working Methodology 1

The second set of features is defined by three frequency characteristics: mode frequency, median frequency and mean frequency, which are calculated as in [7].

The third set of features uses wavelet coefficients. Firstly the database is resampled to 512 Hz, then decimating each one of the recordings to match the delta (0-4 Hz), theta (4-8 Hz), alpha (8-16 Hz), beta (16-32 Hz), and gamma (32-64 Hz) bands as in [8]. As a result 128 wavelet coefficients per electrode were obtained, and a total of 768 features (concatenating the 3 electrodes on both modalities).

Due to the big dimension of the feature space a relevance analysis was needed to avoid the curse of the dimensionality. In this apart, three different methodologies were used to choose the best subset capable of discrimate between normal and pathological patients.

The final step consisted on using a Gaussian Kernel Support Vector Machine (SVM) classifier to estimate the accuracy of the system. A leave-one-out validation technique was used to assess results.

2.3 Dimensionality Reduction Step

In pattern recognition tasks, the training and classification stages obtain a better performance whether a correct feature extraction/selection is applied. For this work, three different feature extraction/selection techniques were applied in order to choose the most relevant variables on the initial feature space of dimension p.

A first approach uses the successive evaluation of subsets of variables, using as cost function an analysis multivariate analysis of variance. The generation of searching subsets is based on the sequential forward selection technique (SFS). In MANOVA, statistical evaluation criteria used is the separability between classes.

A second approach uses a selection scheme based on relevance analysis involving correlation between features, as presented in [9]. To determine the best set of characteristics, variable selection algorithm finds the minimum subset X_r such that $P(c/X) \cong P(c/X_r)$, where $P(c/X)$ is the probability distribution function of the classes c given a dataset X and $P(c/X_r)$ is the probability distribution function of c given the feature subset X_r. In this work, the relevance measure used to find that subset, is linear correlation, which determines the linear predictability of the labels given the variable x.

The third approach uses the locally linear embedding technique (LLE), which maps the observations to a single global coordinate system of lower dimensionality, preserving the neighborhood relationships. The low-dimensional space representation of the data computed by means of LLE, provides a better local and global understanding of the observations than in the original feature space, and allows to identify the underlying structure of the input data. In this sense, it is possible to interpolate unknown samples in the embedding space using the provided dataset.

Although LLE have shown to be an appropriate technique for nonlinear dimension reduction (NLDR), specially in visualization, it has some limitations when data proceed from different manifolds or when data is divided into separated groups, which are common cases in pattern recognition. Besides, LLE does not consider class label information, which can be helpful for improving data representation on this kind of data.

For this reason, in this work is applied an extended LLE approach to deal with several manifolds, employing class labels as extra information to guide the procedure of dimensionality reduction allowing to figure out a suitable representation for each one of them.

2.4   Methodology 2

A second methodology was used working directly on the raw data looking for the most relevant section of the electrode. In this apart, each time-series data point is taken as a feature with the aim to identify temporal information on the signal, which at the same time models some manifolds of phenomena. Figure 2 shows the procedure.

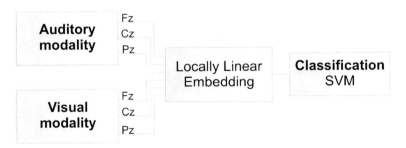

Fig. 2. Working Methodology 2

3   Results and Discussion

Figure 3 shows the graphic of the temporal evolution of the ERP signal mean, of the normal and pathological classes for the Fz, Cz, and Pz electrodes on auditory modality. The similarity of waveforms in each electrode shows the difficult that entails to find characteristic patterns for both classes, which supposes that classification tasks will be more complicated to implement.

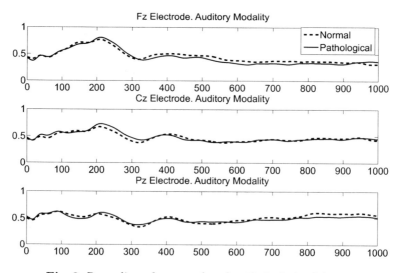

Fig. 3. Recording of a normal and pathological subjects

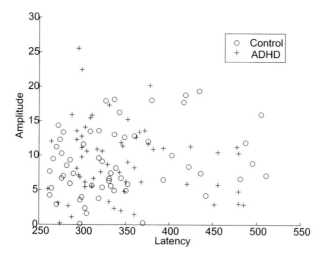

Fig. 4. Scatter plot of amplitude vs latency for Fz electrode recordings of control and ADHD subjects

In order to illustrate the lack of discriminance between the recordings of both classes, the figure 4 is presented. The two more significative characteristics in medical evaluation are plotted: amplitude and latency. Here, a significant overlap between normal and pathological classes is observed, for which reason there is no conclusive evidence of probable differentiable patterns between recordings of two subject groups.

The figure 5 shows a boxplot of the most relevant features on both classes using relevance analysis with correlation function, which was the technique that produced the best classification results. Features between-classes similarity seems to be evidence of overlapped classes, as can be also observed on the box plot.

Table 2 shows the classification results obtained by using methodology 1 with the three dimension reduction techniques. Relevance analysis was the method

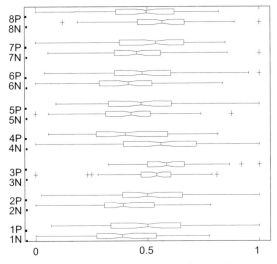

Fig. 5. Boxplot of the 8 most relevant features

Table 1. LLE on raw data

Auditory Modality		Visual Modality	
Electrode	Accuracy(%)	Electrode	Accuracy(%)
Fz	53	**Fz**	63,29
Pz	53,57	**Pz**	62,86
Cz	60,71	**Cz**	53,7

Table 2. Classification Results by using the three feature selection techniques

	Accuracy (%)
Relevance Analysis	65,14
LLE	60,10
MANOVA	59,7

that achieved the best accuracy, compared to the other techniques. The similarity of obtained results is an indicator of problems in other stages different to feature extraction/selection.

Table 1 shows the classification results obtained by using methodology 2 with the LLE on the raw data. The best performance achieved in methodology 1 was 65,14%, very similar to 63,29% with second methodology. As can be seen, working directly on the recordings does not provide any improvement compared to work on a feature set obtained from recordings.

The shown results are unreliable on detecting between healthy and ADHT patients. Two hyphotesis are provided to explain it. The first is related to the incapability of the proposed methodology, specially on the characterization step, to detect patterns that could yield valuable information to the system on correctly detecting between normal and pathological patients. That means that neither the morphological, wavelet or frequency features were capable of characterise the intrinsic dynamic of the biosignals.

A second hyphotesis upholds a no direct correlation between the ADHD and the component of event-related potentials, which will clearly explain the low classification accuracy, and the evident similarity between normal and pathological signals as shown on figure 3.

Contrary to the work presented in [10], the results have significantly decreased. The explanation could be that the database size used on this work (144 vs 46) is bigger and could provide more reliable results than in the previous work. On the other hand, another methodologies have shown higher classification performances compared to the results obtained on this work [7], [3]; however, an objective comparison is not possible due to the use of different databases and distinct recording acquisition protocols.

4   Conclusion

This paper has presented two methodologies for the detection of ADHD: The first one, employed a characterization by using morphological, frequency and wavelet features, using then, dimensionality reduction techniques in order to simplify pattern recognition tasks. On the second one, a LLE procedure was applied over the raw signals looking for informative events on the data. Both, looked for identifying normal and ADHD subjects with the best accuracy rates.

Due to the low classification results, as future work is necessary to find a better ERP characterization capable of correctly discriminating pathology from healthy subjects. Even more, an exhaustive study is needed in order to find correlation or not between ERPs and ADHD detection.

Literature have registered a great number of ERP studies but they are not directly related to ADHD automatic detection, reason for which that kind of works make a contribution to the state of the art.

Acknowledgements. Authors would like to thank Colciencias and Universidad Nacional de Colombia for the financial support on the projects *"Identificación*

Automática del Trastorno por Déficit de Atención y /o Hiperactividad sobre reg-istros de Potenciales Evocados Cognitivos" and *"Sistema de Diagnóstico Asis-tido para la Identificación de TDAH sobre Registros de Potenciales Evocados Cognitivos"*, respectively.

References

1. Association, A.P.: Diagnostic and Statistical Manual of Mental Disorders (DSM-IV)
2. Lazzaro, I., Gordon, E., Li, W., Lim, C.L., Plahn, M., Whitmont, S., Clarke, S., Barry, R.J., Dosen, A., Meares, R.: Simultaneous EEG and EDA measures in adolescent attention deficit hyperactivity disorder. International Journal of Psychophysiology 34, 123–134 (1999)
3. Bostanov, V.: Data sets Ib and IIb: Feature extraction from event-related brain potentials with the continuous wavelet transform and the t-value scalogram. IEEE Transactions on Biomedical Engineering 51(6), 1057–1061 (2004)
4. Demiralp, T., Istefanopulos, Y., Ademoglu, A., Yordanova, J., Kolev, V.: Analysis of functional components of P300 by wavelet transform. In: Proceedings of the 20th Annual International Conference of the IEEE Engineering in Medicine and Biology Society, vol. 20(4) (1998)
5. Quiroga, R.Q.: Obtaining single stimulus evoked potentials with wavelet denoise. Physica D 145, 278–292 (2000)
6. Kalatzis, I., Piliouras, N., Ventouras, E., Papageorgiou, C., Rabavilas, A., Cavouras, D.: Design and implementation of an SVM-based computer classification system for discriminating depressive patients from healthy controls using the P600 component of ERP signals. Computer Methods and Programs in Biomedicine 75, 11–22 (2004)
7. Abootalebi, V., Moradi, M.-H., Khalilzadeh, M.-A.: A new approach for EEG feature extraction in P300-based lie detection. Computer Methods and Programs in Biomedicine 57, 48–57 (2009)
8. Samar, V.J., Bopardikar, A., Rao, R., Swartz, K.: Wavelet analysis of neuroelectric waveforms: a conceptual tutorial. Brain Lang. 66(1), 7–60 (1999), http://view.ncbi.nlm.nih.gov/pubmed/10080864
9. Yu, L., Liu, H.: Efficient feature selection via analysis of relevance and redundancy. Journal of Machine Learning Research 5, 1205–1224 (2004)
10. Castro-Cabrera, P., Gómez García, J., Restrepo, F., Moscoso, O., Castellanos-Dominguez, G.: Evaluation of feature extraction techniques on event-related potentials for detection of attention-deficit/hyperactivity disorder. In: 32nd Annual International Conference of the IEEE Engineering in Medicine and Biology Society, EMBC 2010 (2010)

Methodology for Epileptic Episode Detection Using Complexity-Based Features

Jorge Andrés Gómez García[1,*], Carolina Ospina Aguirre[1],
Edilson Delgado Trejos[2], and Germán Castellanos Dominguez[1]

[1] Universidad Nacional de Colombia, sede Manizales, Colombia
jorgomezg@unal.edu.co
[2] Instituto Tecnológico Metropolitano, Medellín, Colombia

Abstract. Epilepsy is a neurological disease with a high prevalence on human beings, for which an accurate diagnosis remains as an essential step for medical treatment. Making use of pattern recognition tools is possible to design accurate automatic detection systems, capable of helping medical diagnostic. The present work presents an automatic epileptic episode methodology, based on complexity analysis where 3 classical nonlinear dynamic based features are used in conjunction with 3 regularity measures. k-nn and Support Vector Machines are used for classification. Results, superior to 98% confirm the discriminative ability of the presented methodology on epileptic detection labours.

1 Introduction

Epilepsy is the second most common serious neurological disorder in human beings after stroke [1], with a prevalence of approximately 0.6 to 0.8 % of the world population [2]. Epileptic episode present two main phenomenas: epileptiform or ictal phenomena, which are abnormal, synchronized discharges of neurons in the brain that might be accompanied by an impairment or loss of consciousness, psychic, autonomic, sensory symptoms or motor phenomena; and interictal phenomena, which are transient events in the brain that occur in-between seizures [2,3]. Due to the high incidence of the disease, an automatic diagnosis system shows itself as a valuable tool to the clinical neurophysiologist, that may facilitate a more accurate diagnosis, which, in turn, should contribute to improve the management of the patients [3].

Recently, nonlinear dynamic analysis (NDA) tools have provided a new insight, for analysing apparently irregular behaviour on biosignals that could not be explained by linear analysis techniques. Furthermore, because nonlinearity in the brain is introduced even at neuronal level; nonlinear dynamic theory has appeared as a useful resource for analysing signals generated by such nonlinear deterministic system. Typically NDA makes use of complexity features such as the Correlation Dimension (d_2), the Largest Lyapunov Exponent (λ) or the Hurst exponent (H), among others. However, those feature require the signal

---

* Corresponding author.

J.M. Ferrández et al. (Eds.): IWINAC 2011, Part II, LNCS 6687, pp. 454–462, 2011.

dynamics to be deterministic; an assumption that is not entirely valid due to stochastic components produced by effects such as noise. As a result, features based on entropy, and that do not need to assume determinism or stochasticity for its calculation, are important on pathology detection labours. In this category various entropy-based characteristics have been used successfully through numerous studies [4,5,6]. Although high detection rates have been achieved on epilectic episode detection tasks; several approaches have focused on complex methodologies using both linear and nonlinear features, while others have done on using more complex classifiers as an attempt to improve accuracy rates. This however, could lead to problems as classifier overtraining, or hiding the real NDA characterization power through other stages of the methodology.

The aim of this paper is to explore a series of complexity-based features on a methodology using k-nearest neighbour (k-nn) and Support Vector Machines (SVM) classifiers for the detection of Ictal, Interictal or Normal Phenomena on EEG signals. A non so typical regularity feature, called Gaussian Kernel Entropy (GapEn), is also used in addition to more classic NDA features such as d_2, λ, H, and other entropy-based measures as Approximate Entropy (ApEn) and Sample Entropy (SampEn). The obtained results show the good performance of GapEn on the proposed task, having also a evidence of the superiority of entropy-based features compared to classical NDA features. The proposed methodology has results which are comparable with those obtained on the state of the art, and in some cases on simpler methodological schemes.

This paper is organized as follows: Section 2 describes the methodology used on this work, including an explanation of each one of the stages. Section 3 contains the experimental framework. Section 4 presents the obtained results as the discussions. Finally, Section 5 presents the conclusions of this work.

2   Methodology

The automatic epileptic episode detection system proposed in this work is shown on Fig 1, while the most important sections are explained in the next subsections.

2.1   Characterization

Within this category a distinction should be made between two sets of characteristics: Classical complexity features (d_2, λ, H), and entropy-based or regularity features (ApEn, Sampen, GapEn).

Classic complexity Features. NDA makes use of a process, called *embedding*, that maps a given time series $s = \{x_1, x_2, ...x_n\}$, into an m-dimensional space called phase space. The reconstructed state vectors of the system are obtained by using (1), where $\mathbf{x}(.)$ is called state; and m and τ are reconstruction parameters.

$$\mathbf{x}(t) = [s(t), s(t - \tau), s(t - 2\tau), ..., s(t - (m - 1)\tau)] \tag{1}$$

The succession of states in phase space gives rise to a geometric object called attractor [7], where is now possible to extract features.

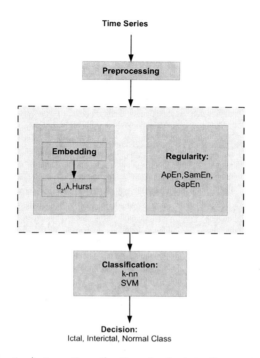

Fig. 1. Automatic epileptic episode detection system

Largest Lyapunov Exponent (λ): It is a measure of the divergence of nearby orbits in phase space, thus representing one of the basic attributes of the nonlinear dynamic systems: sensitivity to initial conditions.

Let $\mathbf{x}_1$ and $\mathbf{x}_2$ be two states in the phase space, with distance defined as $\delta_0 = \|\mathbf{x}_1 - \mathbf{x}_2\| \ll 1$; and let $\delta_{\triangle n} = \|\mathbf{x}_{1+\triangle n} - \mathbf{x}_{2+\triangle n}\|$ be the distance some time later $\triangle n$. Then λ will be determined by (2).

$$\lambda(\delta_0) = \lim_{\triangle n \to \infty} \lim_{\|\delta_0\| \to 0\|} \frac{1}{\triangle n} \log \frac{\|\delta_{\triangle n}\|}{\|\delta_0\|} \tag{2}$$

A negative λ is an indicator of a fixed point attractor. A zero value indicates stable limit cycles, while a negative λ will suggest noisy signals [7].

Correlation dimension (d_2): It quantifies with a dimension the autosimilarity of an embedded time series [7]. For a time series of length n, firstly a correlation sum is defined as in (3), where Θ is the Heaviside function, r is a tolerance measure, and $\mathbf{x}(.)$ are reconstructed state vectors as in (1).

$$C(r) = \lim_{n \to \infty} \frac{1}{n^2} \sum_{i,j=1}^{n} \Theta(r - \|\mathbf{x}(i) - \mathbf{x}(j)\|) \tag{3}$$

It expected that as $r \to 0$ then $C(r) = \phi.r^{d_2}$, where ϕ is constant reflecting the lacunarity, and d_2 is the correlation dimension.

Hurst exponent (H): It measures the degree of similarity of the system to one that presents Brownian motion [7]. The value of H varies between 0 and 1, where a value of $H > 0.5$, indicates persistent behaviour, i.e the past trends of the system remain in the future. A value $H = 0.5$ is an indicator of random walk behaviour, while $H < 0.5$ is indicator of an antipersistent system, in which past trends will reverse in the future.

Entropy-based Features. The information production rate of phase space reconstructions is defined by the Kolmogorov Entropy (H_{KS}) as in (4).

$$H_{KS} = -\lim_{\tau \to 0} \lim_{\epsilon \to 0} \lim_{n \to \infty} \frac{1}{n\tau} \sum_{k_1,\ldots,k_n} p(k_1, \ldots, k_n) \log p(k_1, \ldots, k_n) \qquad (4)$$

Because limits on which H_{KS} is defined can not be physically implemented, it is desirable to find another way of quantifying the complexity of a time series without heavy computational load, the need of large amount of information, while being robust to the presence of low amplitude noise.

Regularity measures. Several families of complexity measures have been developed to somehow overcome the problems of H_{KS}. One of first was proposed by Pincus [8] with a regularity estimator called Approximate Entropy (ApEn). It examines the time series for similar epochs, measuring the average negative logarithm of the conditional probability that two sequences that are similar for m points remain similar (within a tolerance r), at the next point [9].

ApEn is defined as in (5).

$$\text{ApEn} = \phi^m(r) - \phi^{m+1}(r) \qquad (5)$$

Where r is a tolerance measure, ϕ as in (6) and $C_i^m(r)$ as in (3).

$$\phi^m(r) = \frac{1}{n-m+1} \sum_{i=1}^{n-m+1} \log C_i^m(r) \qquad (6)$$

According to [9], ApEn is a biased regularity estimator, because of self-matching. Thus, Richman [9] proposed the Sample entropy (SampEn), defined in (7), as a way to solve that problem.

$$\text{SampEn} = -\log\left(\frac{A^m(r)}{A^{m+1}(r)}\right) \qquad (7)$$

Where A is defined as in (6), but without self-matching.

According to [10], SampEn and ApEn had problems on validity and precision due to their formulation on the non continuos Heaviside function. To solve it, the Gaussian Kernel Approximate Entropy (GapEn), that replaces the Heaviside function in the correlation sum by a Gaussian Kernel function [10] was proposed. In this case GapEn is defined as in (5), but replacing (3) by (8).

$$C_i^m(r) = \frac{\sum_{j=1,j\neq i}^{n-m+1} \exp\left(-\frac{(\|\mathbf{x}(i),\mathbf{x}(j)\|)^2}{10r^2}\right)}{n-m} \tag{8}$$

3   Experimental Setup

3.1   Database

The database was collected by the Epilepsy Clinic at the University of Bonn [11]. It consists of five sets (denoted A-E), each one composed of 100 segments of EEG channels. Sets A and B are segments taken from surface EEG recordings (scalp) in five healthy individuals using the 10-20 electrodes location schema standard. Volunteers were relaxed and in waking state with eyes open (A) and eyes closed (B), respectively. The sets C, D and E are originated from EEG recordings of presurgical diagnosis. Segments in set D were recorded from epileptogenic area, and the set C from the hippocampal formation of the opposite hemisphere of the brain. While sets C and D contained only activity measured during interictal intervals, the set E contains only ictal activity. All EEG signals were recorded with an 128 channels acquisition system, using a common averaged reference. The data were digitized at 173.61 Hz with a resolution of 12bits. The problem is then the correct discrimination of 3 classes: Normal (sets A and B), Interictal (group C and D) and Ictal (group E).

3.2   Experiments

Preprocessing on Fig 1 consists of two main stages: A normalization in amplitude, such that the dynamic range of signals remains constant; and a lowpass filter of

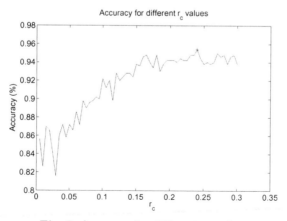

Fig. 2. Accuracy for different r_c values

40 Hz to match the brain low-frequency bands of clinical interest (delta, theta, alpha and beta waves) [12].

In order to use regularity features, is necessary to calculate the tolerance parameter r, which is typically given by $r = r_c std(.)$, where $std(.)$ is the standard deviation of the signal and r_c varies between 0 and 1, so that the estimator does not depend on the absolute amplitude of the signal [9]. As shown in Fig 2 the highest accuracy was found with $r_c = 0.24$, fixing that value as parameter.

Finally, to assess results, a 11 fold cross-validation with 70% of the samples for training and 30% for validation was used.

4   Results and Discussions

Table 1 shows the classification accuracy and standard deviation by using the k-nn and SVM classifiers, for 3 different sets of characteristics: Regularity set, composed by ApEn, Sampen, GapEn; Classic set composed by d_2, λ and H; and All features set, which is a result of combining both.

Additionally, Best combination set corresponds to those features which provided the highest accuracy rates. For the SVM tests this was achieved by excluding H from the All Features Set, while in the k-nn test it corresponded to use the All features set.

Table 1. Classification Accuracy (%) of different sets

Sets	All features	Regularity	Classic	Best Combination
SVM	97,64 ± 1,40	96,13 ± 0,86	86,68 ± 0,65	98,06 ± 0,73
k-nn	96,20 ± 0,82	94,80 ± 0,68	87,20 ± 1,68	96,20 ± 0,82

Table 2 shows the classification accuracy and standard deviation for each one of the characteristics tested individually, using k-nn and SVM classifiers.

Table 2. Classification accuracy (%) of individual features

Feature	d_2	λ	H	ApEn	SampEn	GapEn
SVM	66,72 ± 0,88	72,89 ± 1,44	57,34 ± 1,47	64,42 ± 1,74	65,86 ± 3,29	79,51 ± 1,99
k-nn	59,60 ± 0,75	66,6 ± 1,82	54 ± 1,6	55 ± 2	65,80 ± 4,37	72 ± 1,4

To analyse feature value distribution per class a boxplot of individual features is shown in Fig 3 using the SVM classifier.

The results of the Best Combination set, superior to 94% and 98% using the k-nn and SVM classifier respectively, suggest the tractability of combining both classical and regularity measures for epileptic detection tasks. Although its combination produced good system performance, individually the regularity set clearly surpassed classic set accuracy. It is also important to note that the Best Combination set on the SVM test, needed to exclude H because it diminished system performance, while that did not happened with the k-nn classifier.

As seen in table 2, H clearly presented the lowest classification rates with both classifiers. Moreover, ApEn, one of the most used regularity measures, acquainted the second worse performance. On the other hand, GapEn showed the best individual performance, even surpassing other more classic entropy-based features as ApEn and SampEn. It is also noticeable how the second best peformance was given by λ, being closely followed by SampEn.

The boxplot of Fig 3, as a tool for examining feature distribution per class, presented interesting results: GapEn clearly confirms its discriminative power, by showing non overlapped boxes between Ictal and Normal-Interictal classes, however, being overlapped the latter two. This could lead to think the possibility of obtaining good performance on a bi-class experiment, for detecting epileptic seizure or not. Boxplot also shows how λ has clearly separated boxes on the Ictal and Normal Class, having however a big overlap of the Interictal box with the other two classes. SampEn apparently shows a good performance discriminating between all three classes, having the normal class a bit overlapped with the other two. On the other hand, H again confirms its bad classification results on showing heavily overlapped and dispersed boxes.

The obtained results are comparable, or superior in some cases, to those presented on the state of the art. In [1] for example, an Extreme learning machine was used in conjunction with SampEn to discriminate between the three epileptic episodes, achieving classification rates superior to 95%. On [5] a scheme combining ApEn and Discrete wavelet entropy was used, where the authors attained a maximum accuracy of 96%.

Other works, using entropy-based features, were also compared. In [13], a scheme that used neural network models along with 3 entropies (Wavelet

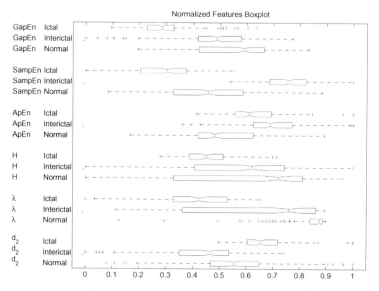

Fig. 3. Boxplot of features by class

entropy, spectral entropy and SampEn), obtained an accuracy of 99% on a bi-class problem (Seizure or not) and a 94% on a three classes problem. In [14] a 95% accuracy was obtained by using Spectral entropy, Renyi's entropy, and ApEn, with an Adaptive neuro-fuzzy inference system (ANFIS). So far, all results are superior to those obtained there.

On [15] a very similar scheme was tested, but using seven nonlinear measures: ApEn, λ, d_2, H, nonlinear prediction error, third order autocovariance, asymmetry due to time reversal, in conjunction with linear autoregressive model coefficients. The class separability was measured using Linear Discriminant Analysis (LDA) after applying Sequential Forward Selection for dimension reduction. The obtained results are slightly inferior but in some cases comparable to the ones on the latter work.

5   Conclusion

An automatic epilepsy detection system based on NDA and regularity features was presented. The obtained performance, suggests the tractability of the proposed methodology as a medical diagnosis support tool. Results demonstrated the superiority of the regularity features compared with classic NDA characteristics, specially with GapEn clearly surpassing any other feature, including other entropy-based ones. As future work, is desirable to study other entropy-based measures as well as new NDA features, focusing on the characterization

Acknowledgement

Authors would like to thank Colciencias and Universidad Nacional de Colombia for the financial support on the projects *"Identificación Automática del Trastorno por Déficit de Atención y/o Hiperactividad sobre registros de Potenciales Evocados Cognitivos"* and *"Sistema de Diagnóstico Asistido para la Identificación de TDAH sobre Registros de Potenciales Evocados Cognitivos"*, respectively.

References

1. Song, Y., Lió, P.: A new approach for epileptic seizure detection: sample entropy based feature extraction and extreme learning machine. Biomedical Science and Engineering 3, 556–567 (2010)
2. Lehnertz, K.: Nonlinear eeg analysis in epilepsy. In: Proceedings of the 23rd Annual International Conference of the IEEE Engineering in Medicine and Biology Society (2001)
3. da Silva, F.L.: The impact of eeg/meg signal processing and modeling in the diagnostic and management of epilepsy. IEEE Reviews in Biomedical Engineering (2008)
4. Improved spatial characterization of the epileptic brain by focusing on nonlinearity. Epilepsy Research 69(1), 30–44 (2006)

5. Automatic detection of epileptic seizures in eeg using discrete wavelet transform and approximate entropy. Expert Systems with Applications 36(2), Part 1 2027–2036 (2009)

6. Stam, C.J.: Nonlinear dynamical analysis of EEG and MEG: review of an emerging field. Clinical Neurophysiology 116(10), 2266–2301 (2005), http://dx.doi.org/10.1016/j.clinph.2005.06.011

7. Kantz, H., Schreiber, T.: Nonlinear time series analysis, 2nd edn. Cambridge University Press, Cambridge (2004)

8. Pincus, S.M.: Approximate entropy as a measure of system complexity. Proc. Nati. Acad. Sci. USA 88, 2297–2301 (1991)

9. Richman, J.-S., Moorman, J.-R.: Physiological time-series analysis using approximate entropy and sample entropy. Am. J. Physiol Heart Circ. Physiol. 278, H2039–H2049 (2000)

10. Xu, L., Wang, K., Wang, L.: Gaussian kernel approximate entropy algorithm for analyzing irregularity of time-series. In: Proceedings of the Fourth International Conference on Machine Learning and Cybernetics, Guangzhou, pp. 18–21 (2005)

11. Andrzejak, R.G., Mormann, F., Widman, G., Kreuz, T., Elger, C.E., Lehnertz, K.: Improved spatial characterization of the epileptic brain by focusing on nonlinearity. Epilepsy Research 69(1), 30–44 (2006)

12. Rowan, A.J., Tolunsky, E.: Primer of EEG: With A Mini-Atlas. Elsevier, Amsterdam (2004)

13. Entropies based detection of epileptic seizures with artificial neural network classifiers. Expert Systems with Applications 37(4), 3284–3291 (2010)

14. Entropies for detection of epilepsy in eeg. Computer Methods and Programs in Biomedicine 80(3), 187–194 (2005)

15. Balli, T., Palaniappan, R.: A combined linear and nonlinear approach for classification of epileptic eeg signals. In: 4th International IEEE/EMBS Conference on Neural Engineering, NER 2009, vol. 29 (2009)

Segmentation of the Carotid Artery in Ultrasound Images Using Neural Networks

Rosa-María Menchón-Lara, M-Consuelo Bastida-Jumilla,
Juan Morales-Sánchez, Rafael Verdú-Monedero,
Jorge Larrey-Ruiz, and José Luis Sancho-Gómez

Dpto. Tecnologías de la Información y las Comunicaciones
Universidad Politécnica de Cartagena
Plaza del Hospital, 1, 30202, Cartagena (Murcia), SPAIN
rmml@alu.upct.es

Abstract. Atherosclerosis is a cardiovascular disease very widespread into population. The intima-media thickness (IMT) is a reliable early indicator of this pathology. The IMT is measured by the doctor using images acquired with a B-scan ultrasound and this fact presents several problems. Image segmentation can detect the IMT throughout the artery length in an automatic way. This paper [1] presents an effective segmentation method based on the use of a neural network ensemble. The obtained results show the ability of the method to extract the IMT contour in ultrasound images.

1 Introduction

Atherosclerosis is characterized by a thickening of the arterial walls, which affects blood flow. It may progress throughout life being unnoticed or it can lead to serious cardiovascular diseases such as heart attack or stroke. Hence, it is of vital importance to diagnose and treat this disease early. One of the most reliable indicators to detect the thickening of the arterial walls is the intima-media thickness (IMT) of common carotid artery (CCA) [2].

The main advantage of the IMT analysis lies in the nature of its measurement, achieved by means of a B-mode ultrasound scan, which is a non-invasive technique that allows studying IMT in a short time on a large number of patients. However, this can lead to several problems such as the variability between observers, or the obtainment of a minimum few points along the entire length of the arterial walls. Image segmentation can detect the *IMT contour* (I5 and I7 interfaces in Fig. 1) throughout the artery length, which leads to better results and allows us to extract statistics such as the maximum, the minimum or the average IMT in a precise manner.

---

[1] This work is partially supported by the Spanish Ministerio de Ciencia e Innovación, under grant TEC2009-12675, by the Universidad Politécnica de Cartagena (Inicialización a la Actividad Investigadora, 2010), and by the Séneca Foundation (09505/FPI/08).

J.M. Ferrández et al. (Eds.): IWINAC 2011, Part II, LNCS 6687, pp. 463–471, 2011.
© Springer-Verlag Berlin Heidelberg 2011

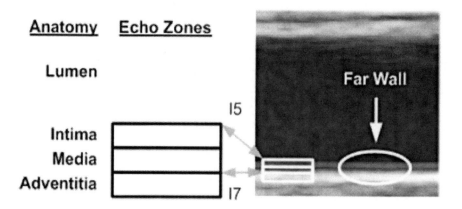

Fig. 1. Ultrasound B-scan image of the carotid artery [9]

Since Gustavsson [14] began working on the automatic measurement of IMT, various solutions have been developed [6],[12],[11]. In this paper, an efficient image segmentation technique is proposed. The problem is solved using a neural network ensemble. With the proper training, the nets are able to recognize the pixels belonging to the IMT contour. Once the networks are trained, the proposed method allows getting IMT measurements in an automatic way.

2   Image Segmentation

Segmentation is a fundamental task in image processing providing the basis for any kind of further high-level image analysis. Segmentation subdivides an image into its constituent regions or objects. In many cases segmentation accuracy determines the eventual success or failure of computerized analysis procedures, since other processing steps are based on segmented regions. For this reason, considerable care should be taken to improve the segmentation stage. Although a wide variety of segmentation techniques has been proposed, there is no one standard segmentation technique that can produce satisfactory results for all imaging applications.

Segmentation algorithms for monochrome images generally are based on one of two basic properties of image intensity values [10]: discontinuity (*edge-based segmentation techniques*) and similarity (*region segmentation techniques*). In the first category, the approach is to partition an image based on abrupt changes in intensity, such as edges in an image. The main approaches in the second category, are based on partitioning an image into regions that are similar according to a set of predefined criteria.

In medical imaging [1], automated delineation of different image components is used for analyzing anatomical structure and tissue types, spatial distribution of function and activity, and pathological regions. The applications of segmentation in medical image processing include detection of the coronary border in angiograms, multiple sclerosis lesion quantification, surgery simulations, surgical planning, measuring tumor volume and its response to therapy, functional mapping, automated classification of blood cells, studying brain development, detection of microcalcifications on mammograms, image registration, atlas-matching, heart image extraction from cardiac cineangiograms, detection of tumors, etc. As commented in the Section 1, this work focuses on measuring the intima-media thickness (IMT) of common carotid artery (CCA).

Since segmentation requires classification of pixels, it is often treated as a *pattern* recognition problem and addressed with related techniques. Especially in medical imaging [1], where variability in the data may be high, pattern recognition techniques that provide flexibility and convenient automation are of special interest.

3   Multilayer Perceptrons

MultiLayer Perceptrons (MLPs) are one of the most important types of neural nets. The architecture of the MLP is completely defined by an input layer, one or more hidden layers, and an output layer. Each layer consists of at least one neuron. The input vector is processed by the MLP in a forward direction, passing through each single layer [7]. MLPs have been applied successfully to solve some difficult and diverse problems by training them in a supervised manner with a highly popular algorithm known as the *backpropagation algorithm*. A MLP has three distinctive characteristics [4]:

1. The model of each neuron in the network includes a *nonlinear activation function*.
2. The network contains one or more layers of *hidden neurons*. These hidden neurons enable the network to learn complex tasks by extracting progressively more meaningful features from the input patterns (vectors).
3. The network exhibits a high degree of *connectivity*, determined by the synapses of the network.

It is through the combination of these characteristics together with the ability to learn from experience through training that the MLP derives it computing power.

4   Committee Machines

Complex classification problems in practice require the contribution of several neural networks for achieving an optimal solution [7]. According to the *principle of divide and conquer*, a complex computational task is solved by dividing it into a number of computationally simple tasks and then combining the solutions to those tasks. In neural network terminology, it means to allocate the learning task among a number of *experts*, which in turn split the input data into a set of subspaces. The combination of experts is said to constitute a *committee machine*.

Committee machines fall into the category of universal approximators and they may be classified into two major categories [13]:

1. *Static structures*. In this class of committee machines, the responses of several experts are combined by means of a mechanism that does not involve the input signal.
2. *Dynamic structures*. In this second class of committee machines, the input signal directly influences the mechanism that fuses the output decisions of the individual experts into an overall output decision. There are two groups of dynamic structures:

   - Mixture of experts, in which the individual responses of the experts are nonlinearly combined by means of a single gating network.
   - Hierarchical mixture of expert, in which the individual responses of the experts are nonlinearly combined by means of several gating networks arranged in a hierarchical fashion.

In the mixture of experts, the principle of divide and conquer is applied just once, whereas in the hierarchical mixture of experts it is applied several times, resulting in a corresponding number of levels of hierarchy.

In the field of image processing, the experimental results reported in the literature showed that the image classification accuracy provided by the committee machines can outperform the accuracy of the best single net. However, the above also showed that neural networks ensembles are effective only if the nets forming them make different errors. Several methods for the creation of ensembles of neural networks making different errors have been investigated. Such methods basically lie on 'varying' the parameters related to the design and to the training of neural networks [3].

5   Proposed Method

The main goal of this work is to measure the IMT in a reliable and automatic way. For this purpose, neural networks have been used to extract the IMT boundaries in ultrasound images of the carotid artery. Therefore, we propose an image segmentation task using neural networks. As mentioned above, it is logical to raise the segmentation as a pattern recognition problem. With the appropriate training, our network is able to recognize the IMT contour within an image. Given an input image, the system must classify the input pixels in two classes: on one hand, those belonging to the IMT boundaries and, on the other hand, the remaining pixels which do not belong to the region of interest. Once the segmentation is done, we will be able to measure the IMT along the arterial wall.

As is well known, the segmentation is a nontrivial task. In addition, we must take into account the ultrasound image features, which further complicate the problem. In order to improve the accuracy and robustness of our system, the results from four different networks have been combined, i.e. a committee machine has been used (see Fig. 2).

To perform the neural network training, we must assemble a consistent dataset composed of ultrasound images and the associated desired outputs (*supervised learning*), also called *target images*. The target images are binary images in which white pixels (with value '1') show the IMT boundaries (see Fig. 3).

5.1   Architecture

The artificial neural networks used in this work are standard *multilayer perceptrons* (MLPs) with a single hidden layer. The number of neurons in the input layer depends on the size of the input vectors. In a similar way, the number of neurons in the output layer depends on the size of the target vectors. However, the number of hidden neurons is a network parameter to be optimized. For this purpose there are several approaches, such as cross validation. In our case,

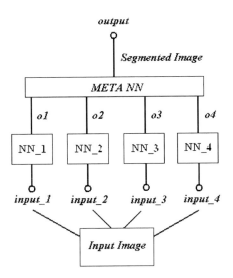

Fig. 2. Committee machine structure

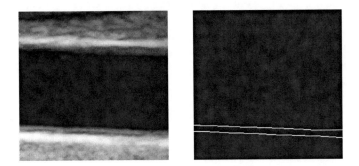

Fig. 3. Ultrasound image of the carotid artery (left) and associated target image (right)

the number of hidden neurons is fixed to ten for all the networks in the committee machine. At this point, we consider separately the networks belonging to the first stage and the *'meta'* neural network in the second stage.

Networks in the first stage: The four experts in our scheme (NN1, NN2, NN3, and NN4 in Fig. 2) take as inputs *subimages* resulting from a sampling process. A window must be moved pixel-by-pixel on the original image. For each input subimage, the network output has a single component, which is the desired (target) value associated with the central pixel in the subimage. According to this, the output layer of the networks in the first stage has a single neuron. Following the idea of creating committee machines in which the nets forming them make different errors, we train each network with a different learning set. This can be done by considering different sampling methods which will provide different training sets. Basically, what is done is to take as inputs decimated versions of the original image. Although the window size varies from one network to another, we build the input vectors with only nine pixels of the window. Thus, the number of neurons in the input layer is nine for each net. As commented above, the input vectors are constructed differently for each network. This can be seen more clearly in the Figure 4, in which the gray squares represent the selected pixels in each case. The nine pixels from a 3-by-3 window are taken in the neural network labeled as NN1. In the case of NN2 and NN3 the window size is 5-by-5, but only nine pixels are selected for the input vector. Finally, the input vectors of NN4 are built with nine pixels from a 7-by-7 window. It is important to remember that the window is shifted pixel-by-pixel on the original image. Then, the number of input vectors is equal to the number of pixels in the original image.

'Meta' neural network: This network combines the results obtained by the experts. As we can see in Fig. 2, the outputs (o1, o2, o3, and o4) are taken as inputs, i.e., four images (with the same size that the original image) arrive at the input of this net. In this case, the input vector associated with a pixel is constructed with four subimages (3-by-3), one for each input image. In this way, the input vectors have 36 components. This 'meta' net also provides a single output component, which is the desired value associated with the central pixel. According to these considerations, our 'meta' neural network has 36 neurons in the input layer and a single neuron in the output layer. Remember that the number of hidden neurons is 10 also for this net.

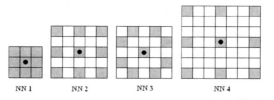

NN 1   NN 2   NN 3   NN 4

Fig. 4. Selected pixels to construct the input vector associated with the central pixel

5.2   Training

The nets in our committee machine are training under the *backpropagation* learning rule for *feedforward* artificial neural networks. Neuron activation functions of sigmoid type have been used. Standard backpropagation is a gradient descent algorithm in which the network weights are moved along the negative direction of the gradient of the performance function. This standard method is often too slow for practical problems. There are a number of variations on the basic algorithm that are based on other optimization techniques, such as conjugate gradient and Newton methods. These algorithms are faster than the standard method. In this work, the *scaled conjugate gradient algorithm* [8] has been used. It is important to note that the four experts are trained first, and then the 'meta' neural network is trained with the output results of the nets of the previous stage.

5.3   Debugging Stage

Once the segmented image is obtained at the output of our system, a debugging stage is necessary. In this image processing step, some morphological techniques are applied to the output image to remove small impurities and fill holes.

6   Results

Once the networks have been trained, the tests are performed. The proposed method has been tested with a set of images in DICOM format, which must be read and converted to grayscale format. Figure 5 shows the results of applying the whole procedure to three different images. The measurements related to these images are shown in Table 1. In Figure 6, the output images of the system can be seen. These output images are the committee machine outputs after the debugging stage. As one can note in Table 1, the IMT shows no substantial difference with the IMT values extracted by a previous application based on frequency implementation of B-spline active contours [5].

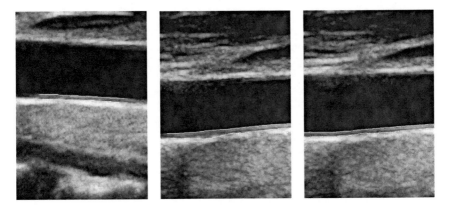

Fig. 5. Images under study together with the achieved results

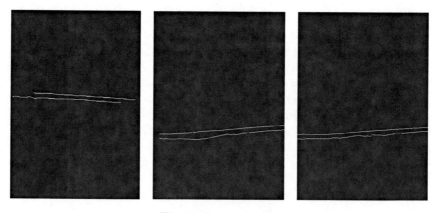

Fig. 6. Output images

Table 1. Mean values of IMT (in centimeters)

IMT (cm)	*Measured Value*	*Reference Value*
Image Fig.5(a)	0.0459	0.043
Image Fig.5(b)	0.0573	0.052
Image Fig.5(c)	0.0464	0.047

7   Conclusions

This paper proposes a segmentation method of the carotid artery based on neural networks. Segmentation is treated as a pattern recognition problem in which a network must classify the pixels to find the IMT contour. To improve the accuracy of this classification task, a committee machine is developed. The implemented system achieves the extraction of the wall interfaces. Finally, the IMT can be easily measured from the obtained results.

The suggested approach presents some advantages such as the simultaneous detection of both lumen-intima and media-adventitia interfaces. Moreover, it allows to obtain the results in a short time once the networks have been trained. The authors are working on an exhaustive validation process, including more images and patients, in order to verify these preliminary results.

Acknowledgements

The images used and the anatomical knowledge have been provided by the radiology department of "Hospital Universitario Virgen de la Arrixaca", Murcia, Spain.

References

1. Bankman, I.N.: Handbook of Medical Imaging: Processing and Analysis. Academic Press, London (2000)
2. Burke, G.L., Evans, G.W., Riley, W.A., Sharrett, A.R., Howard, G., Barnes, R.W., Rosamond, W., Crow, R.S., Rautaharju, P.M., Heiss, G.: Arterial wall thickness is associated with prevalent cardiovascular disease in middle-aged adults. Stroke 26(3), 386–391 (1995)
3. Giacinto, G., Roli, F.: Desing of effective neural network ensembles for image classification purposes. Image and Vision Computing 19, 699–707 (2001)
4. Haykin, S.: Neural Networks: A Comprehensive Foundation. Prentice-Hall, Englewood Cliffs (1999)
5. Izquierdo-Zaragoza, J.-L., Bastida-Jumilla, M.C., Verdú-Monedero, R., Morales-Sánchez, J., Berenguer-Vidal, R.: Segmentation of the carotid artery in ultrasound images using frequency-designed b-spline active contour. In: International Conference on Acoustics, Speech and Signal Processing (2011)
6. Ceccarelli, M., Luca, N.D., Morganella, A.: An active contour approach to automatic detection of the intima-media thickness. In: IEEE Int Conf. Acoustics, Speech and Signal Processing (2006)
7. Meyer-Base, A.: Pattern Recognition for Medical Imaging. Academic Press, London (2004)
8. Moller, M.F.: A scaled conjugate gradient algorithm for fast supervised learning. Neural Networks 6, 525–533 (1993)
9. Liang, Q., Wendelhag, I., Wikstrand, J., Gustavsson, T.: A multiscale dynamic programming procedure for boundary detection in ultrasonic artery images. IEEE Trans. on Medical Imaging 19(2), 127–142 (2000)
10. González, R.C., Woods, R.E.: Digital Image Processing. Prentice-Hall, Englewood Cliffs (2002)
11. Rocha, R., Campilho, A., Silva, J., Azevedo, E., Santos, R.: Segmentation of the carotid intima-media region in b-mode ultrasound images. Image and Vision Computing 28(4), 614–625 (2010)
12. Santhiyakumari, N., Madheswaran, M.: Non-invasive evaluation of carotid artery wall thickness using improved dynamic programming technique. Signal, Image and Video Processing 2, 183–193 (2008)
13. Theodoridis, S., Koutroumbas, K.: Pattern Recognition. Academic Press, London (2003)
14. Gustavsson, T., Liang, Q., Wendelhag, I., Wikstrand, J.: A dynamic programming procedure for automated ultrasonic measurement of the carotid artery. In: Proc. Computers in Cardiology, pp. 297–300 (1994)

Tools for Controlled Experiments and Calibration on Living Tissues Cultures

Daniel de Santos[1], José Manuel Cuadra[1], Félix de la Paz[1], Víctor Lorente[2],
José Ramón Álvarez-Sánchez[1], and José Manuel Ferrández[2,3]

[1] Departamento de Inteligencia Artificial, UNED, Spain
{jmcuadra,delapaz,jras}@dia.uned.es
[2] Departamento de Electrónica, Tecnología de Computadoras y Proyectos,
Universidad Politécnica de Cartagena, Spain
jm.ferrandez@upct.es
[3] Instituto de Bioingeniería, Universidad Miguel Hernández, Spain

Abstract. In recent years, numerous studies attempted to demonstrate
the feasibility of using live cell cultures as units of information processing.
In this context, it is necessary to develop both hardware and software
tools to facilitate this task. The later part is in the aim of this paper. It
presents a complete software suite to design, develop, test, perform and
record experiments on culture-based biological processes of living cells
on multi-electrode array in a reliable, easy and efficient way.

1 Introduction

RNN (Real Neural Networks) are becoming more and more interesting for the
study of learning, memory and data processing [10, 11, 22]. Therefore, we need
several devices to connect cultures of living neurons to a computer, allowing us
to send and receive impulses that we can code and interpret for a particular application, such as the movement of a robot [2, 6, 7]. In this case, computer works
like sensory receptor, taking external data and producing electrical impulses for
neuronal stimulation through a culture-computer interface and, at the same time,
taking neuronal electrical response and codifying an external effector action.

A Multi-Electrode Array (MEA)[21, 23] is a square grid of 8x8 electrodes
without the corners, i.e. 60 electrodes. It is used as a tissue-computer interface
to send electrical stimulus and to receive electrical response (see section 2.2).
Highly sensitive electrodes record not only the neuronal activity, but also environmental noise (e.g. electric current frequency) getting low signal-noise ratio
(SNR). Therefore, MEA is connected to a filter and pre-amplifier, MEA1060BC,
providing higher SNR. Living tissues need a very controlled conditions to ensure their survival (see section 2.1) among them, temperature. In consequence, a
heater, TC01, is a crucial element that maintains and monitors tissues temperature. And finally, we need a stimulator, STG1002, and a data acquisition card
(DAQ), MC_Card, to stimulate the culture with a sequence of pulses and to get
the neuronal activity of each electrode respectively. In figure 1, the distribution
of all those different devices is presented.

J.M. Ferrández et al. (Eds.): IWINAC 2011, Part II, LNCS 6687, pp. 472–481, 2011.
© Springer-Verlag Berlin Heidelberg 2011

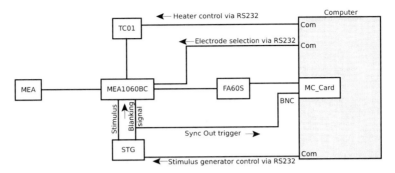

Fig. 1. Recommended structure for a system with MEA60, stimulus generator and blanking circuit

It is clear that we need to control the different instruments within the same environment [5] in order to check its actions and the consequences in the culture. Devices were designed by Multi-Channel System (MCS), and each of them is controlled with closed source software designed for MS-Windows. This situation increases the difficulty to implement new modules and their implementation in a closed loop, in which could be included, for instance, a robot [7]. Also this does not allow real-time bidirectional communication between an effector, in this case the robot, and the culture. Therefore, we seek to design a freeware application in a Linux environment that gives us control over the different devices.

2   Experimental Setup

2.1   Cell Cultures

Cell cultures are neuroblastoma with SH-SY5Y cells. A neuroblastoma is a tumor that forms in nerve tissue and typically begins in the adrenal glands. This type of tumor cells develop into primitive nerve cells, called neuroblasts, which remain in the body as remnants of the embryonic stage of development before birth.

SH-SY5Y cells thrice cloned from a SK-N-SH neuroblastoma are used[3]. SK-N-SH cells are cloned to produce SH-SY cells that produce the clone SH-SY5 cells and by cloning the latter, we obtain the SH-SY5Y cells. The cloning process is an artificial selection of cells or a group of them presenting a particular phenotype of interest.

The environment where this cells are grown consists of a mixture of DMEM, Ham's F12 and 10% fetal bovine serum. For cell growth, the environment must be at 37 °C in 5% CO_2, 95% air and humid atmosphere.

2.2   MEA

Tissues primary cells are grown directly over the MEA, that is a set of electrodes distributed in a surface to allow stimulation and record of extracellular data

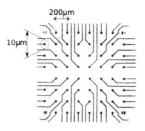

Fig. 2. Standard Multi-Electrode Array with a 8x8 electrode grid. Electrodes, composed of TiN, have a 10 μm. Gap between electrodes is 200 μm.

[4, 5, 9, 23, 25]. We use a standard MEA [18] with an array of 8x8 electrodes made of titanium nitride (TiN) and without electrodes at the corners, i.e. a total of 60 electrodes, as we can see in figure 2.

2.3   MEA1060BC

Raw data from 60 electrodes of a MEA is received through a pre-amplifier and filter with preconfigured gain and bandwidth. A blanking circuit prevents amplifier saturation that might occur during stimulation and allowing a correct reading of the electrodes later.

MEA1060BC [17] allows the simultaneous stimulation of the electrodes with two different pulses, channel A and channel B, or connecting the electrodes to read (ungrounded) or connecting the electrodes to ground (for those electrodes that we don't want to use). Multi-Channel System has developed a closed-source MS-Windows application, MEA_SELECT[1], for MEA amplifiers use and configuration.

2.4   STG

The available stimulus generator is a STG1002[15]. This stimulator allows two different pulse sequences simultaneously. We can also set a TTL signal to help us to perform synchronization between various devices or actions like, for example, stimulation-saving information.

The application offered by Multi-Channel Systems for stimulus generator configuration is called MC_STIMULUS[1] and is closed source. MC_STIMULUS is able to generate strings [14] that represent the pulse sequences required by the user, and then send them to the stimulus generator.

2.5   TC01

Heater TC01 [16] has the task of complying with the temperature restrictions being imposed by the cultures for their survival, in our case, a constant temperature of 37 °C. The TC01 is designed for use with Pt100 sensors which allow

---

[1] Available in web page: http://www.multichannelsystems.com/downloads.html.

temperature reading very accurately. The outputs of the TC01 are designed to not interfere with experiments. The heater heats actively, but the cooling is passive, so the minimum temperature is the temperature of the room. MCS provides the application TCX_CONTROL[1] for temperature control via a computer.

2.6   MC_Card

MC_Card is an A / D card that converts analog signals into digital data streams in real time, therefore it can convert the voltage obtained from the extracellular activity of neurons to manageable data by a computer. MCS provides software for MS-Windows, like MC_Rack[1], which allows the analysis, filtering, recording and displaying of the information obtained from a MC_Card. These data can be exported to a readable format with the help of MC_DataTool[1].

Due to the constraints of proprietary software to implement new modules, and the problems to its use in a closed-circuit, MeaBench was developed [24]. MeaBench is a set of semi-independent programs sharing a common library, which allows communication with data acquisition cards (DAQ), in particular, it has a module for MC_Card.

3   BRAVE Software

The proposed application is called BRAVE (acronym for "Biosignals Recorder And Visualization interfacE"). With BRAVE we have a controlled environment for performing experiments with biological cultures. Until now, we had a collection of closed-source applications each of them controlling a specific device, from now, we have an open-source application which controls everything and uses MeaBench for data acquisition. BRAVE is friendly and easy to develop new features. BRAVE is written in C++ and Qt4 and it is available for Linux operative system under the GNU General Public License (GPL).

3.1   Architecture

BRAVE is composed by four independent applications connected to a global program, as seen in figure 3. Each independent application controls a device, i.e. TC01, MEA1060BC, etc. Independent applications offer different services like

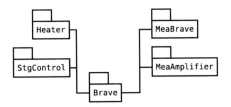

Fig. 3. Structure for the connection between standalone applications and the general one

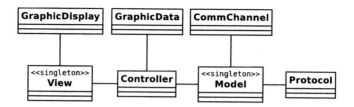

Fig. 4. Schematic class diagram of the device controllers applications

events information, warnings, data, executed operations, etc. The general application, Brave, is connected to the independent applications services and use them to show information to the researcher. A developer could extend Brave functionality making new independent applications for new devices and connecting them to Brave, or extending a set of them with new devices types.

Each controller application is designed with a Model-View Controller (MVC) design pattern, as shown in figure 4. MVC pattern has been extended to separate the communication protocol with the device communication channel used to send and receive messages. With this architecture, designing new features is easy, because we only need a new communication protocol to use a new model for a particular device.

In our case, the connection device-computer is done via standard USB. We use the library from CuteCom [19] application modified for this purpose. The scheme shown in figure 4 is just a simplified scheme and does not faithfully represent the full structure of the four controls. For example, GraphicData represents a buffer of data received by the controlled device to which the main application can request service. In the following sections we will see each independent application.

3.2   Heater

The Heater program, figure 5, fulfills the TCX-Control (section 2.5) functionality for the TC01 heater control. With Heater we can modify the current temperature and PID parameters, as well as getting information about them every second until stop reading.

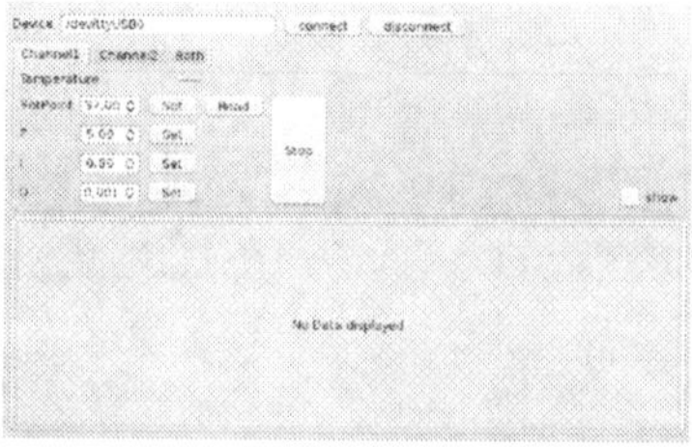

Fig. 5. GUI application Heater. Above, PID parameters and temperature settings are shown, plus the option of reading the temperature. Below, we can see the graph that is to make a temperature reading.

3.3 MeaAmplifier

MeaAmplifier is the open-source application equivalent to MEA_SELECT (section 2.3) for the MEA1060BC amplifier device. With MeaAmplifier, figure 6, you can configure the electrodes settings in one of the following ways, Channel A, Channel B, for two different stimulus sequences (A and B), or Ungrounded to read raw data or Ground for unused electrodes.

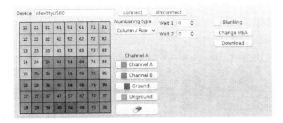

Fig. 6. MeaAmplifier GUI. The left matrix represents the MEA. Configuration options are shown on the right.

(a) StgControl Standard Tab. Show standard actions like Test connection, Reset, enable or disable debug mode, etc.

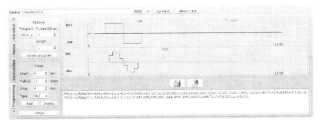

(b) StgControl Programming Tab. This tabs allows the configuration of two different pulses. The pulses wave form are displayed on the right-upper part of the window. The code for stg is presented in the lower part.

Fig. 7

3.4 StgControl

StgControl is the open-source application equivalent to MC_STIMULUS (section 2.4) for STG1002 stimulus generator. With StgControl we can manage STG1002 and configure two different pulse sequences for channel A and B with a user-friendly interface, figure 7, where you can save and open pulse sequences files. Also we can configure all parameters available in the stimulator, including expert options, but the later are under user responsibility.

3.5 MeaBrave

MeaBrave can record raw and spike data through MeaBench libraries. The data recorded can be replayed in Meabench later. Raw data type contains the voltages of the 60 MEA electrodes sampled at a frequency of 25 kHz. Spike data type contains information about action potentials, i.e. voltages higher than a threshold [1, 8, 12, 13, 20]. If you choose this option you will see real time spike data (figure 8).

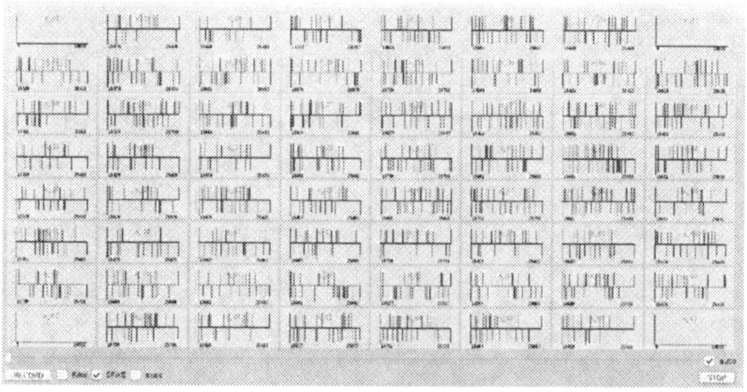

Fig. 8. MeaBench GUI shows the spikes that occurred over time at each electrode

3.6 Brave

Brave is connected to the different services we have discussed in previous sections. Actions executed while connected to a device become part of the historical record of actions in Brave. On the other side, actions tried to execute while not connected to any device are identified as an execution pattern configuration to be executed latter. This way the user can create execution patterns to build projects (figure 9). In addition to the instructions that can be created from the four applications, Brave allows two more instructions: an instruction to wait (WAIT) and a jump to an instruction (GOTO) that lets you repeat a set of instructions.

When an execution pattern is run we must make sure the devices are connected to applications. The implementation of the pattern creates a new project that stores all data in a folder with the same name. On the other hand, we can see all the data buffers of the application by selecting each one individually. In addition to the data buffers of the applications, Brave can display a special graphic that marks the time when an instruction is executed for each application, see figure 10.

Fig. 9. Brave GUI. Left part shows a execution pattern, and in the right part the history of actions executed is shown.

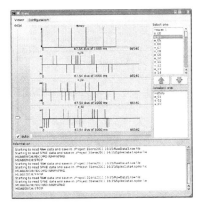

Fig. 10. Brave GUI can display only those graphics more interesting for the user. In this example we can see, top to bottom, "history" (marks when an action has occurred that may affect the result), electrodes e_01, e_02 and e_03 (following the numbering of the MEA hardware).

4   Conclusions and Future Work

We have developed a series of software tools that make it easy to test, control and design experiments, and the benchmarking or calibration of living cells cultures over a MEA. We have unified under our control software the communications with all the hardware (that come from different manufacturers) necessary for these experiments. Also, this software eases the exchange of information between different parts of the experiment. This will not only allow faster and more efficient application development, but also to gain full control over all data at each time instant. The software was developed under GNU philosophy, entirely with free software under GPL license, this will allow to incorporate any technical innovation that could occur in this area, and not only by our team, but by any scientist interested in it.

Our work is now starting to use this software for test and development of biological processes using different types of cells. Currently testing is done with cultured human neuroblastoma, and we plan to use Hippocampus, cortex dissociated and PC12.

References

1. Aertsen, A., Diesmann, M., Gewaltig, M.O.: Propagation of synchronous spiking activity in feedforward neural networks. J. Physiol. Paris 90(3-4), 243–247 (1996)
2. Bakkum, D.J., Shkolnik, A.C., Ben-Ary, G., Gamblen, P., DeMarse, T.B., Potter, S.M.: Removing Some 'A' from AI: Embodied Cultured Networks. In: Iida, F., Pfeifer, R., Steels, L., Kuniyoshi, Y. (eds.) Embodied Artificial Intelligence. LNCS (LNAI), vol. 3139, pp. 130–145. Springer, Heidelberg (2004)
3. Biedler, J.L., Roffler-Tarlov, S., Schachner, M., Freedman, L.S.: Multiple Neurotransmitter Synthesis by Human Neuroblastoma Cell Lines and Clones. American Association for Cancer Research (1978)
4. Bland, J.M., Altman, D.G.: Statistics notes: Multiple significance tests: the bonferroni method. BMJ 310(6973), 170 (1995)
5. de Santos, D., Lorente, V., de la Paz, F., Cuadra, J.M., Álvarez-Sánchez, J.R., Fernández, E., Ferrández, J.M.: A client-server architecture for remotely controlling a robot using a closed-loop system with a biological neuroprocessor. Robotics and Autonomous Systems 58(12), 1223–1230 (2010)
6. Demarse, T.B., Wagenaar, D.A., Blau, A.W., Steve, Potter, M.: The neurally controlled animat: Biological brains acting with simulated bodies. Autonomous Robots 11, 305–310 (2001)
7. Ferrández, J.M., Lorente, V., Cuadra, J.M., de la Paz, F., Álvarez-Sánchez, J.R., Fernández, E.: A hybrid robotic control system using neuroblastoma cultures. In: Graña Romay, M., Corchado, E., Garcia Sebastian, M.T. (eds.) HAIS 2010. LNCS, vol. 6076, pp. 245–253. Springer, Heidelberg (2010)
8. Frostig, R.D., Frostig, Z., Harper, R.M.: Information trains. the technique and its uses in spike train and network analysis, with examples taken from the nucleus parabrachialis medialis during sleep-waking states. Brain Research 322(1), 67–74 (1984)
9. Gross, G.W.: Simultaneous single unit recording in vitro with a photoetched laser deinsulated gold multimicroelectrode surface. IEEE Transactions on Biomedical Engineering BME-26(5), 273–279 (1979)
10. Gütig, R., Sompolinsky, H.: The tempotron: a neuron that learns spike timing-based decisions. Nature Neuroscience 9(3), 420–428 (2006)
11. Hebb, D.O.: The Organization of Behavior: A Neuropsychological Theory. Wiley, New York (1949)
12. Izhikevich, E.M.: Polychronization: Computation with spikes. Neural Computation 18(2), 245–282 (2006)
13. Izhikevich, E.M., Gally, J.A., Edelman, G.M.: Spike-timing dynamics of neuronal groups. Cereb. Cortex 14(8), 933–944 (2004)
14. Multi Channel System MCS GmbH. STG100X RS232 Communication Manual
15. Multi Channel System MCS GmbH. Stimulus Generator 1000 Series User Manual (June 2007), http://www.multichannelsystems.com/uploads/media/STG1000_Manual_01.pdf

16. Multi Channel System MCS GmbH. Temperature Controller TC01/02 User Manual (October 2007), http://www.multichannelsystems.com/uploads/media/TC01-2_Manual_04.pdf

17. Multi Channel System MCS GmbH. MEA Amplifier with Blanking Circuit for Inverse Microscopes (December 2008), http://www.multichannelsystems.com/uploads/media/MEA1060-Inv_Manual_01.pdf

18. Multi Channel System MCS GmbH. Microelectrode Array (MEA) Manual (October 2010), http://www.multichannelsystems.com/uploads/media/MEA_Manual_02.pdf

19. Neundorf, A.: CuteCom (May 2009), http://cutecom.sourceforge.net/

20. Quian Quiroga, R., Nadasdy, Z., Ben Shaul, Y.: Unsupervised spike detection and sorting with wavelets and superparamagnetic clustering. Neural Comput. 16(8), 1661–1687 (2004)

21. Taketani, M., Baudry, M.: Advances in Network Electrophysiology: Using Multi-Electrode Arrays. Springer, Heidelberg (2006)

22. Vogels, T.P., Rajan, K., Abbott, L.F.: Neural network dynamics. Annual review of neuroscience 28(1), 357–376 (2005)

23. Wagenaar, D.A., Pine, J., Potter, S.M.: An extremely rich repertoire of bursting patterns during the development of cortical cultures. BMC Neurosci. 7 (2006)

24. Wagenaar, D.A., DeMarse, T.B., Potter, S.M.: MeaBench: A toolset for multi-electrode data acquisition and on-line analysis. In: Proceedings of 2nd International IEEE EMBS Conference on Neural Engineering (March 2005), doi:10.1109/CNE.2005.1419673

25. Wagenaar, D.A., Nadasdy, Z., Potter, S.M.: Persistent dynamic attractors in activity patterns of cultured neuronal networks. Physical Review E (Statistical, Nonlinear, and Soft Matter Physics) 73(5), 051907 (2006)

Author Index